Topics in Applied Physics
Volume 116

Topics in Applied Physics

Topics in Applied Physics is a well-established series of review books, each of which presents a comprehensive survey of a selected topic within the broad area of applied physics. Edited and written by leading research scientists in the field concerned, each volume contains review contributions covering the various aspects of the topic. Together these provide an overview of the state of the art in the respective field, extending from an introduction to the subject right up to the frontiers of contemporary research.

Topics in Applied Physics is addressed to all scientists at universities and in industry who wish to obtain an overview and to keep abreast of advances in applied physics. The series also provides easy but comprehensive access to the fields for newcomers starting research.

Contributions are specially commissioned. The Managing Editors are open to any suggestions for topics coming from the community of applied physicists no matter what the field and encourage prospective editors to approach them with ideas.

Managing Editor

Dr. Claus E. Ascheron

Springer-Verlag GmbH
Tiergartenstr. 17
69121 Heidelberg
Germany
Email: claus.ascheron@springer.com

Assistant Editor

Adelheid H. Duhm

Springer-Verlag GmbH
Tiergartenstr. 17
69121 Heidelberg
Germany
Email: adelheid.duhm@springer.com

Harry Bernas (Ed.)

Materials Science with Ion Beams

With 180 Figures

Springer

Dr. Harry Bernas
Universite Paris-Sud 11
CSNSM-CNRS
91405 Orsay, France
E-mail: bernas@csnsm.in2p3.fr

Topics in Applied Physics ISSN 0303-4216

ISBN 978-3-642-26085-8 e-ISBN 978-3-540-88789-8

DOI 10.1007/978-3-540-88789-8

Production: VTEX
Cover concept: eStudio Calamar Steinen
Cover design: SPI Publisher Services

SPIN: 12221363 57/3180/VTEX
Printed on acid-free paper

9 8 7 6 5 4 3 2 1

springer.com

Foreword

Materials science is the prime example of an interdisciplinary science. It encompasses the fields of physics, chemistry, material science, electrical engineering, chemical engineering and other disciplines. Success has been outstanding. World-class accomplishments in materials have been recognized by Nobel prizes in Physics and Chemistry and given rise to entirely new technologies. Materials science advances have underpinned the technology revolution that has driven societal changes for the last fifty years.

Obviously the end is not in sight! Future technology-based problems dominate the current scene. High on the list are control and conservation of energy and environment, water purity and availability, and propagating the information revolution. All fall in the technology domain. In every case proposed solutions begin with new forms of materials, materials processing or new artificial material structures. Scientists seek new forms of photovoltaics with greater efficiency and lower cost. Water purity may be solved through surface control, which promises new desalination processes at lower energy and lower cost. Revolutionary concepts to extend the information revolution reside in controlling the "spin" of electrons or enabling quantum states as in quantum computing. Ion-beam experts make substantial contributions to all of these burgeoning sciences.

A striking feature of modern technology is the important role of the surface and near-surface regions of materials. Modern communications, complex data storage, electronic thin-film displays, biochips, digital cameras are products of innovative research employing surfaces and thin films in new and creative ways. Ion-beam technology provides a unique and exciting way of modifying the near-surface region of a solid; controlling its surface properties, adding beneficial impurities in the near-surface region, modifying the crystallinity, and providing a control and specificity that exceeds almost all other methods of surface modification. Ion-beam science and engineering have already made extraordinary impacts in current silicon technology for communications, surface hardening for structural improvements and materials modification to create solids with new properties. In addition, ion-beam science has emerged as one of the principal ways of quantifying surfaces, through a subfield known as ion-beam analysis. This background and accomplishment, and the use of these analytical tools, now comprise the underpinning for ion-beam technology to reach out in new and innovative ways to a broader array of materials, to the important nanoscience domain and to new fields of endeavor such as geology and art forensics.

The chapters in this book describe some of the forefront investigations of the creative use of ion beams in materials science. The materials list is impressive – and includes polymer surfaces used in biological application, semiconductors and semiconductor processing, magnetic structures, nanocrystals and nanoclusters that display new optical properties, and samples of geological interest – progress in quite distinct fields.

The field is still in its infancy. The basic ion–solid interaction (ion–atom interaction), is now well understood, based on close to a half century of work. However, when extended into the complex world of materials, the description is not entirely adequate for the world of solids with its complex many-body aspects. The massive array of energetic ions, defects and impurity atoms interact to form entirely new solid-state complexes, unachievable by traditional methods of solid-state chemistry. Two examples of these phenomena, highlighted in this book, concern metastable systems and complex surface patterning induced by energetic beams. Such understanding will undoubtedly lead to still newer applications.

Finally, it is appropriate to comment on the future. Progress in this field will emerge from the ingeniousness of those who understand the process and the societal needs of new materials. Many such examples are contained in the following chapters. The future will also be governed by "machine technology", advances in making more precise, smaller, lower-cost and more fashionable ion-beam facilities. All of us in this endeavor await advances in accelerator technology resulting in nanoscale beam sizes, flexible single-ion implantation, abundant neutral beams for insulator studies and cluster beams.

Ion-beam material science has established a remarkable record of accomplishment to date responding to technological needs. Each generation of technology has been improved as the ion-beam community addressed the materials limitations. The world we face of nanoscale manipulation and precise atomic control will be even more demanding and require new forms of ion-beam technology able to move the ever-expanding frontiers of materials science.

Piscataway, *L.C. Feldman*
April 2009

Preface

There are many ways to synthesize novel nanoscale materials. What is special about those involving ion-beam irradiation or implantation? Such processes are often viewed as complex, associated with inconvenient or unwanted lattice disordering or with hard-to-control compositional gradients due to collision-induced atomic motion. In addition, they, as other "directed energy" methods based on laser or electron-beam interactions, require equipment that may not be run-of-the-mill in many laboratories or industrial production lines. This book purports to show that such views are outdated.

Rather obviously, the repeated appearance of new ion-beam applications – such as occurred for over 40 years in semiconductor science and technology, and more recently in nanoscience uses of focused ion beams – provides reasons to be aware of the physics and the potential technical advantages of ion-beam interactions with materials. Increasing awareness of their importance to radiation damage and radwaste studies for nuclear-energy applications provides still further impetus.

But another, less-recognized transition pervades the basic science of ion–solid interactions and prompted this book. Although a few aspects of elementary "ion beam–matter interactions" per se (e.g., related to very high energy or highly charged ion stopping) remain to be settled, most interaction processes are now largely well understood. In the last twenty-odd years, the study of ion-beam-induced synthesis or modification of all kinds of solid-state systems has progressively initiated a cultural revolution: this area is not "ion-beam physics", it is a component of nonequilibrium thermodynamics of solids. An increasing number of scientists are intrigued by the corresponding concepts and possibilities offered.

This book is designed to encourage this trend, notably by attempting to draw the interest of the physics, solid-state chemistry and materials-science communities towards recent developments in the very diverse areas where ion-beam interactions have been used. The basic theme is: "Here is a technique that may be of use to *control* the synthesis and evolution of many solid-state systems. Because it is nonequilibrium and often involves nonlinear processes, it is not universal, but it may be combined with other physical parameters (temperature, pressure, etc.) to lead a system through novel evolution patterns. Can it contribute to your research?"

As indicated by its title, the purpose of the book is to introduce materials scientists and materials developers, as well as physicists and chemists, to novel physical and technical properties that rest on the use of ion-beam effects on materials. It is aimed at practitioners or students who are not particularly familiar with ion-beam techniques and their specific traits, and wish to obtain a reasonably accurate picture of effects on materials in order to determine if and how it can possibly augment their own potential in designing new materials or testing novel solid-state properties. For example, in several areas ion-beam interactions provide a "Third Way" [1], neither top-down nor bottom-up, in which they are used to modify materials from the nanoscale up to large scales without requiring special, often detrimental, lithographic or chemical solvent techniques. In metals and insulators, as well as in the more familiar case of semiconductor physics, new approaches that control atomic displacements, defect creation and evolution often pose interesting physics problems and lead to specific applications. Many examples are given in the following chapters to illustrate these statements.

Over the last decades, ion beams have been increasingly used to synthesize and explore properties of metallic or semiconductor alloys and compounds outside of equilibrium phase diagrams. Ion beams induce atomic replacements and mobility that often resemble – and may be modeled as – a diffusional process, albeit at temperatures well below thermally induced motion. Depending on the combination of thermodynamical and ion-beam energy deposition parameters, the outcome may be either a known or a novel phase, ordering or disordering. The ion beam may provide a further control parameter over the system. Can this be related in some way to equilibrium thermodynamics treatments? Such work has been going on in parallel with developments in thermodynamics based on the approaches of J.W. Cahn, W.W. Mullins, J.S. Langer and coworkers, and with the huge progress in Monte Carlo-type and molecular-dynamics modeling and simulations. Parallels don't meet until forced to, and the influence of modern statistical mechanics on the design and interpretation of ion-beam experiments has only been felt fairly recently. Its mounting impact is a major justification for this book, since the concepts, the language and the research challenges are increasingly common.

The first two chapters provide a brief and fairly general theoretical and experimental background, aiming to bridge the gap between ion-beam effects and quasiequilibrium thermodynamics. The remainder of the book retains the same balance, highlighting the consequences of ion-beam interactions in different materials, and employing the basic concepts and methods that are familiar to the practitioner of materials science, physics or chemistry. We show that this can succeed in diverse areas of materials science (semiconductor and surface properties, crystal and nanocluster growth and self-organized processes, optics, magnetism...), and can also apply to important aspects of geology or biology. The emphasis is on illustrative examples and reference to the primary literature for topics, methods and results that might contribute to the reader's own fields of research. We do not describe ion-beam interactions per se in detail, since excellent books that do so are referenced in the text (a most recent and complete one is [2, 3]). Several important application areas have been left out because they are well covered elsewhere. Ion beams are obviously already at the heart of such research areas as that of nuclear materials, whose manifold aspects require more specialized material (e.g., see [4] and papers in the *Journal of Nuclear Materials*). Another specific, fast-developing area that is not treated here is that of focused ion-beam physics (see, e.g., [5, 6]). Finally, the vast literature related to analysis techniques using ion beams is summarized in several very useful books [7–10].

References

1. N. Mathur, P. Littlewood, Nanotechnology – The third way. Nat. Mater. **3**, 207 (2004) viii
2. P. Sigmund, *Particle Penetration and Radiation Effects.* Springer Series in Solid State Sciences, vol. 151 (Springer, Berlin, 2007), pp. 1–437 ix
3. P. Sigmund (ed.), *Ion Beam Science: Solved and Unsolved Problems*, Mat. Fys. Medd. Dan. Vid. Selsk. 52(1), vol. 1–2 (Nov. 2006) ix
4. K. Sickafus, E. Kotomin, B. Ueberaga (eds.), *Radiation Effects in Solids.* NATO Science Series II: Mathematics, Physics & Chemistry, vol. 235 (Kluwer Academic, Dordrecht, 2006) ix
5. C.A. Volkert, A.M. Minor (eds.), *MRS Bulletin*, vol. 32 (Materials Research Soc., Pittsburgh, 2007), p. 389 and ff. ix
6. L.A. Giannuzzi, F.A. Stevie, *Introduction to Focused Ion Beams: Instrumentation, Theory, Techniques and Practice* (Springer, Berlin, 2005) ix
7. J.R. Tesmer, M. Nastasi, *Handbook of Modern Ion Beam Materials Analysis* (Materials Research Society, Pittsburgh, 1995) ix
8. L.C. Feldman, J.W. Mayer, *Fundamentals of Surface and Thin Film Analysis* (Elsevier, New York, 1986) ix
9. L.C. Feldman, J.W. Mayer, S.T. Picraux, *Materials Analysis by Ion Channeling: Submicron Crystallography* (Academic Press, New York, 1982) ix
10. W.-K. Chu, J.W. Mayer, M.A. Nicolet, *Backscattering Spectrometry* (Academic Press, New York, 1978) ix

Orsay,
April 2009

Harry Bernas

Contents

Ion Beams in the Geological Sciences

Fundamental Concepts of Ion-Beam Processing

R.S. Averback and P. Bellon

Department of Materials Science and Engineering, University of Illinois at Urbana Champaign, Urbana, IL, 61801, USA, e-mail: averback@uiuc.edu

Abstract. The basic concepts underlying the response of materials to ion-beam irradiation are outlined. These include the slowing of energetic ions, the creation of defects, sputtering, ion-beam mixing, the acceleration of kinetic processes, and phase transformations. Several examples are cited to illustrate how each of these concepts can be exploited to modify materials in ways not easily achieved, or not even possible, by more conventional processing methods. The chapter attempts to provide a physical understanding of the basic effects of ion-beam irradiation on materials, to enable readers in other areas of research to better understand the more technical chapters that follow, and to develop ideas relevant to their own disciplines. We provide references to more quantitative treatments of the topics covered here.

1 Introduction: Basic Mechanisms of Ion–Solid Interactions

The underlying principles guiding the design and processing of engineering materials have traditionally been based on the equilibrium properties of solids, gradients in the chemical potential, and atomic mobilities. Typically, a material is first excited above its ground state by such means as quenching from high temperatures, plastic deformation, vapor deposition, or even ion implantation. The material is subsequently annealed according to a predetermined time–temperature program designed to select a kinetic pathway toward a desired metastable, or even unstable, state. A specific example illustrating this concept is the processing of age-hardening alloys such as Al–Cu. In this example, the alloy is homogenized at elevated temperatures and subsequently quenched to low temperatures, where the Cu exceeds its solubility limit. Upon subsequent aging at elevated temperature, the Cu precipitates out of solution in a series of metastable phases before arriving at the equilibrium $\approx CuAl_2$, body-centered tetragonal phase. The microstructure of the alloy is thus controlled by applying specific heating programs to take advantage of different nucleation barriers of the metastable phases and their growth kinetics. Notably, the processing is irreversible and once the alloy is overaged, there is no possibility to recover the previous microstructure. Vapor deposition, mechanical processing, powder metallurgy and in-

H. Bernas (Ed.): Materials Science with Ion Beams,
Topics Appl. Physics **116**, 1–28 (2010)
DOI: 10.1007/978-3-540-88789-8_1,

deed some elements of ion implantation all involve these common principles. While ion-beam processing of materials of course shares many of these same ideas, the introduction of persistent displacement damage, or driving forces, into the processing scheme greatly enriches the materials science and provides new opportunities for synthesizing materials with unique microstructures and properties. The reasons for this will become evident later in this chapter and throughout this volume; however, a simple example will serve at present to illustrate this key point.

The example considers ion irradiation of dilute $Ni_{(1-x)}Si_x$ alloys at elevated temperatures. Assume $x < .10$ and so under equilibrium conditions the alloy forms a homogeneous single phase. The effect of irradiation is to produce vacancies and interstitials in large supersaturations, and as a result they flow to sinks, such as dislocations, grain boundaries, and surfaces, to restore equilibrium concentrations. It is well documented that interstitial atoms in Ni have a strong binding energy with Si solute, and as a consequence, the migration of an interstitial to a point defect sink drags a Si atom along with it, enriching the local concentration of Si at these sites. The system is thus driven away from equilibrium. Eventually this enrichment leads to precipitation of the Ni_3Si, γ' phase, and a two-phase alloy is formed. If the irradiation is switched off, the point-defect fluxes quickly die out, and the precipitates redissolve in the matrix by ordinary diffusion mechanisms, and equilibrium is restored.

In the remainder of this chapter, we will develop the theoretical framework to understand this example, and other materials problems involving irradiation. The discussion will center on metals, but the concepts are general and apply to most solid materials.

1.1 Electronic Excitation

As an ion impinges on a solid it begins a series of collision with both the electrons and the ion cores of target atoms. The collisions with electrons are more numerous owing to their larger number and cross section, but since their mass is small they do not much alter the trajectory of the incoming ion, nor do they usually result in atomic displacements. In most materials, therefore, these inelastic collisions can be treated simply by assuming that the electrons form a viscous background that extracts energy from the fast-moving ions and slows them down [1]. In a few special cases, however, such as swift ions (i.e., heavy ions with GeVs of energy) the high density of energy deposited along the ion track, often tens of keV per nanometer, can have significant consequences. For example, the traversal of such ions can produce a linear track of damage in various materials. These tracks can be later etched to form pores for various applications, most often in insulators [2]. In high-T_C superconductors, the damage tracks themselves can pin fluxoids, which greatly increases critical currents [3]. The energy loss of ions due to these inelastic excitations is characterized by the electronic stopping power, $S_e(E)$.

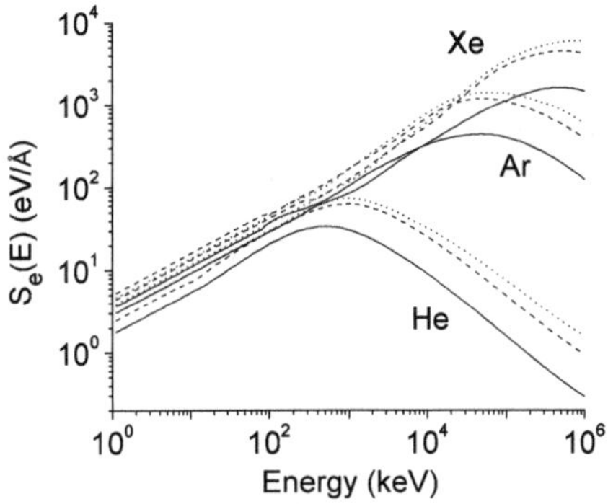

Fig. 1. Electronic stopping power as a function of energy in Si (*solid line*), Cu (*dashed line*) and Au (*dotted line*) for different ions. Data from [4]

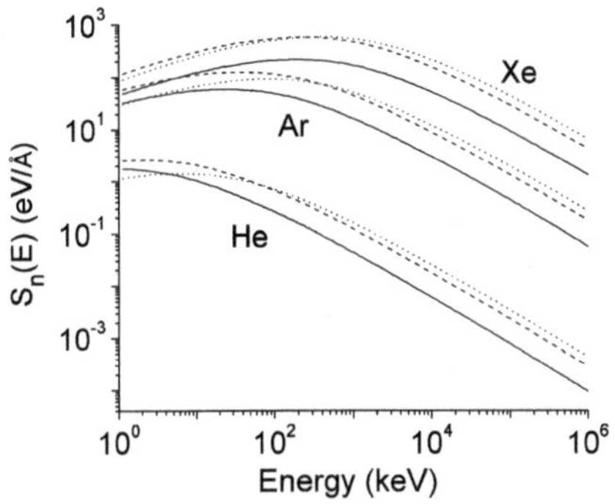

Fig. 2. Nuclear stopping power as a function of energy in Si (*solid line*), Cu (*dashed line*) and Au (*dotted line*) for different ions. Data from [4]

Figure 1 illustrates the electronic stopping as a function of energy for a few representative ion-target combinations. Notice that the maximum in $S_e(E)$ and the energy where the maximum is reached both increase with the atomic masses of the ion and target atoms.

1.2 Nuclear Collisions

Defect Production

Ions are also slowed in a solid by the elastic collisions between the projectiles and target atoms; this slowing can also be characterized by a stopping power, S_n, as shown in Fig. 2. These collisions, however, can lead to displacement damage, whereby a knocked-on atom recoils away from its initial lattice site. Typically, an atom must receive $\approx$25 eV of energy to create a stable interstitial–vacancy (Frenkel) pair. Many recoil atoms receive much higher energies, as discussed below, and these recoils can undergo a series of secondary recoils with target atoms displacing them as well, and indeed, many of these secondary recoil atoms can create yet additional displacements in tertiary recoils, and so on. In this way, a displacement cascade evolves. When the energies of recoil atoms fall below 25 eV, the atoms continue to be displaced from their lattice sites, however, the separation between the interstitial–vacancy pair is too small to avoid spontaneous recombination owing to the strong elastic interaction between the two defects. The number of Frenkel pairs produced by a projectile of energy, E_1, $\nu(E)$, can be estimated using the expression [5]:

$$\nu(E_1) = \int_{E_1}^{0} \frac{\mathrm{d}E}{S_\mathrm{n} + S_\mathrm{e}} \int_{E_\mathrm{d}}^{T_\mathrm{max}} \mathrm{d}T \frac{\mathrm{d}\sigma(E,T)}{\mathrm{d}T} \nu(T), \tag{1}$$

where $\mathrm{d}\sigma(E,T)/\mathrm{d}T$ is the differential scattering cross section for an ion of energy E to produce a recoil of energy T, T_max is the maximum energy transfer in a single collision, and $\nu(T)$ is the damage function given by [6, 7],

$$\nu(T) = \begin{cases} 0, & T < E_\mathrm{d}, \\ 1, & E_\mathrm{d} < T < 2.5E_\mathrm{d}, \\ \frac{0.8\cdot\xi(T)E_\mathrm{D}(T)}{2E_\mathrm{d}}, & 2.5E_\mathrm{d} < T, \end{cases} \tag{1.a}$$

where $\xi(T)$ is the efficiency function, E_D is the total energy of the cascade less that lost to electronic excitation, and E_d is the displacement energy, on the order of 25 eV in metals, 60 eV in ionic crystals, and 15 eV in semiconductors. For most metals, $\xi(T)$ is unity at low energies and drops smoothly to ≈1/3 at energies above 1–2 keV [5, 8]. E_D can be found from simulation.[1]

Sputtering

Atoms located in the first layer or two of the surface that receive recoil energies greater than the sublimation energy, ≈5 eV, with momentum directed away from the surface can be sputtered into the vacuum. As a consequence, the surface continually erodes during irradiation. Under prolonged irradiation at temperatures where defects are immobile, therefore, a system reaches a steady state in terms of damage and changes in alloy composition when the thickness of sputtered material is roughly equal to the depth of the implanted ions. It is easily shown from conservation of mass, moreover, that the steady-state concentration of an implanted ion, $X_\mathrm{max}/(1–X_\mathrm{max}) \approx 1/(S–1)$, where S is the number of sputtered atoms per incident ion. This expression assumes that the partial sputtering yields of different alloy components are the same, i.e., proportional to the surface composition. In general, partial sputtering yields are different, and this leads to surface compositions that differ from the bulk composition [9, 10]. For low-energy ion sputtering of an alloy target, these composition fluctuations are confined to a depth of ≈2 nm, i.e., the penetration depth of the ion, which is in strong contrast to thermal evaporation. Once sufficient material has been removed and steady state achieved, the alloy compositions are constant in time, albeit inhomogeneous just below the surface. In steady state, the alloy components must sputter at rates proportional to their bulk compositions (not their surface compositions). This

[1] See [4] for details of SRIM.

has practical significance for the growth of alloy films by sputter deposition, since it guarantees that the film composition will equal that of the sputtering target after a short transient.

Ion-Beam Mixing

One of the consequences of the displacement process is that several atoms in the vicinity of the recoil location exchange lattice sites with neighboring atoms. A simple estimate of this ballistic mixing rate can be obtained by considering the mixing in terms of a diffusion process, $\langle r^2 \rangle = n\lambda^2$, where n is the average number of random jumps each atom performs, and λ is the jump distance [11]. If we assume that s atoms each jump one atomic distance in creating a Frenkel pair, then using (1), the mean square displacement of all atoms per incident ion is simply,

$$\langle R^2 \rangle = \frac{0.8 E_\mathrm{D} s \lambda^2}{2 E_\mathrm{d}}. \tag{2}$$

After irradiation to dose ϕ, the total *damage energy per atom* is $E'_\mathrm{D} = N_0 \phi F_\mathrm{D}$, where N_0 is the atomic density, $F_\mathrm{D} = \mathrm{d}E_\mathrm{D}/\mathrm{d}x$ and x is measured normal to the surface. F_D is usually obtained by computer simulation.[2] The mean square displacement per atom, $\langle r^2 \rangle$, normalized by damage energy, can thus be written,

$$\xi = \frac{\langle r^2 \rangle}{\phi F_\mathrm{D}} = \frac{0.8 s \lambda^2}{2 N_0 E_\mathrm{d}} \tag{3}$$

Evaluation of (3) illustrates that each atom in a displacement cascade undergoes ≈0.1–1 jumps on average.

1.3 Thermal Spikes

The displacement cascade just described evolves in time over a period of a few tenths of ps. Beyond this period the atomic energies fall below 5 eV and the collisions can no longer be considered as two-body events, but rather many-bodied. An atom with ≈5 eV, notably, has a velocity on the order of the speed of sound in crystals. Indeed, as a molecular dynamics simulation of a 10-keV collision event in the ordered B_2 phase of NiAl shows in Fig. 3a, all of the atoms localized in a small volume are set into motion. Plotting the energy distribution in Fig. 3b for various instants of time illustrates that after a few tenths of a ps, the distribution becomes Maxwellian, with the maximum temperature exceeding 3000 K at this time. Depending on the energy density in this locality, the local temperature can therefore rise significantly above the melting temperature for several ps, giving rise to liquid-like diffusion and defect clustering on subsequent cooling [12]. Notice in Fig. 3a the rather well-defined solid–liquid phase boundary. At the end of the recoil event shown in

[2] See footnote 1.

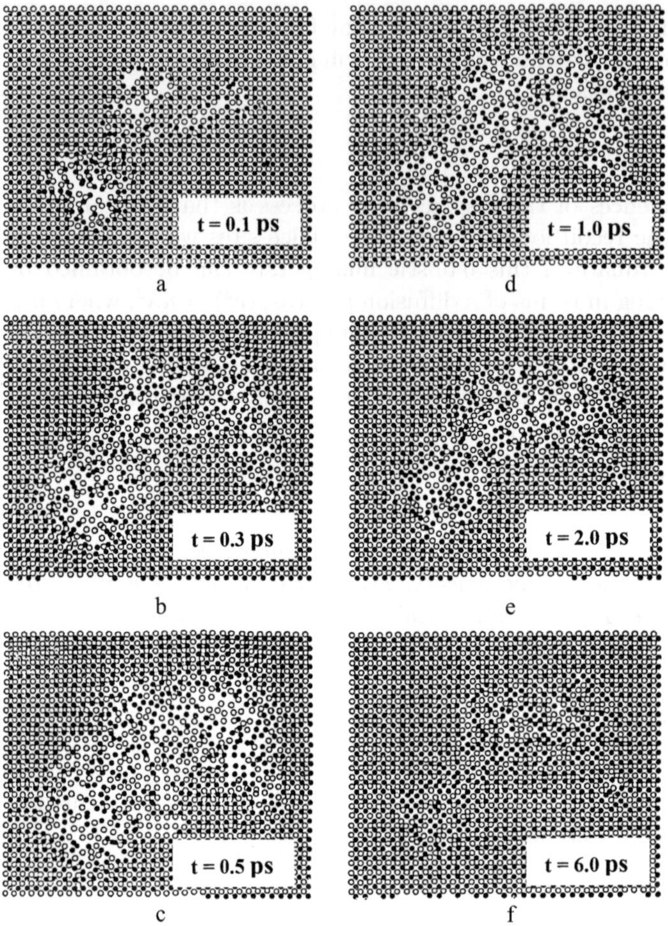

Fig. 3a. Position of atoms in a cross-sectional slice, one lattice parameter thick, during a 10-keV event in β-NiAl. After [13]

Fig. 3a–f, only 25 Frenkel pairs are created ($\xi(10\,\text{keV}) = 0.27$) while ≈2000 atoms relocate from their initial lattice sites.

A brief comment is in order concerning the analysis of thermal spikes, either theoretically or through MD simulations. Most thermal spike models do not include heat loss from the cascade that arises from thermal conduction by electron carriers. This omission assumes that equilibration times for the phonon and electron systems are long compared to the lifetimes of the thermal spikes. In most metals, however, these times are comparable, and therefore the models should be considered approximate. Attempts have been made to include the electronic system [14–16], but at present, their accuracy has not been determined.

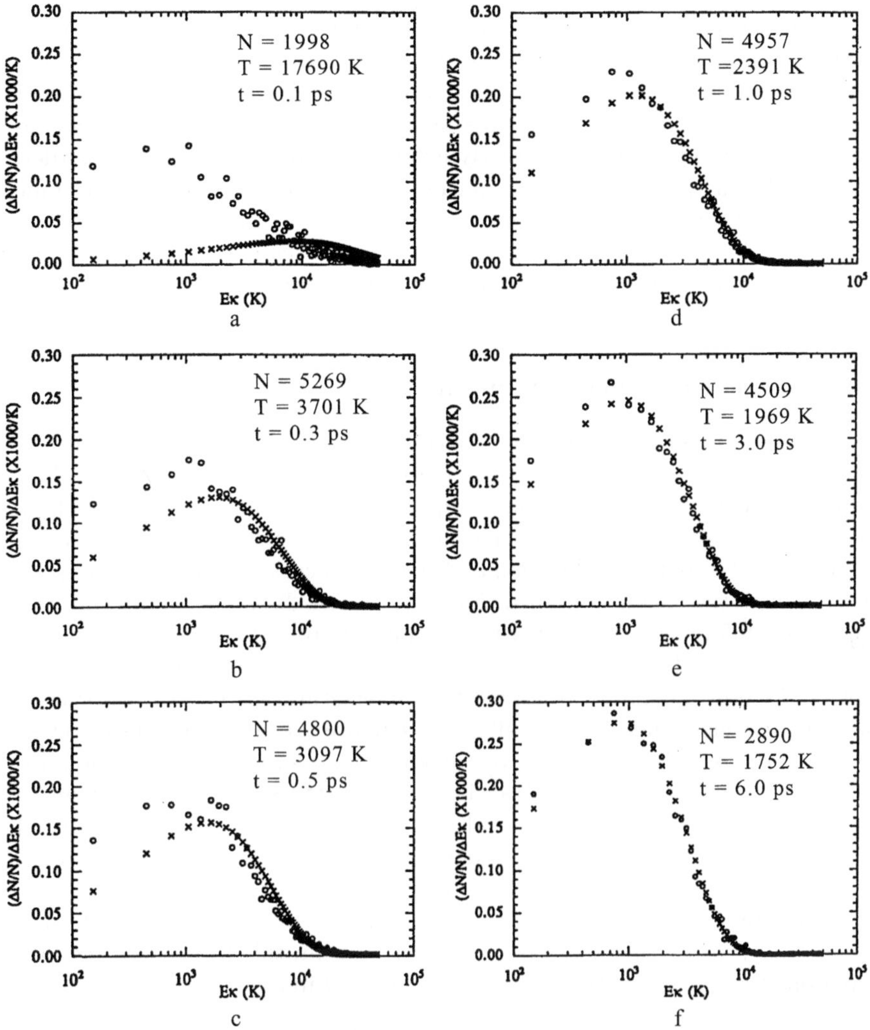

Fig. 3b. Distribution of atomic energies for different times during the evolution of a 10-keV cascade event in β-NiAl. o – actual distribution; x – Maxwellian distribution. N is the number of atoms within the cascade and T represents the average kinetic energy. After [13]

From this brief description of the implantation of an energetic ion in a solid, it is apparent how processing of materials with ions differs from that by traditional methods. Consider first the displacement process during the early phases of the cascade. The energies required to create a Frenkel pair exceed $\approx$25 eV, and therefore defect production is not sensitive to the thermochemical properties of the material. Typical point-defect concentrations at the end of cascade events are highly supersaturated, $\approx$ one per cent, which at room temperature corresponds to a chemical potential on the order of one eV for vacancies and four eV for interstitials. Atomic motion during this phase of the cascade is ballistic, i.e., atoms relocate randomly, driven by gradients in their concentrations rather than gradients in their chemical potentials. As a consequence, solubility limits are greatly extended, and ordered alloys disorder. Later in the cascade development, during the thermal spike, thermodynamic forces can become relevant. From the perspective of forming alloy phases, displacement cascades with high energy densities can be qualitatively described as rapid quenching from the melt with a quench speed on the order of $\approx 10^{14}\,\mathrm{s}^{-1}$.

1.4 Radiation-Enhanced Diffusion

At elevated temperatures the high concentrations of defects produced in the cascade can migrate throughout the material and begin to restore equilibrium. Again, this has similarities to quenching in point defects from high temperatures. For example, supersaturations of quenched in vacancies can enhance low-temperature diffusion and therefore accelerate kinetic processes, at least until the excess concentration of vacancies dissipates. There are two important differences, however. First, irradiation produces vacancies and interstitials in equal numbers, whereas quenching only produced vacancies in metals. As noted before, the interstitial supersaturation is far greater than that of vacancies, and thus the alloy can explore far larger regions of phase space. During continued irradiation, moreover, the supersaturations of point defects are continually replenished, leading to persistent net defect fluxes to sinks. These in turn can result in radiation-induced segregation (as discussed in the introduction), disordering/ordering of ordered alloys, and dimensional instabilities such as creep and void swelling.

Radiation-enhanced diffusion is typically treated within a mean-field theory approach using chemical rate equations. Within this framework the rate equations for the concentrations of vacancies and interstitials are [17]:

$$\frac{\partial c_\mathrm{i}}{\partial t} = \sigma\dot{\phi} - \frac{4\pi r}{\Omega_0}(D_\mathrm{i} + D_\mathrm{v})c_\mathrm{i}c_\mathrm{v} - \left(\sum_j K_{\mathrm{i},j}\right)D_\mathrm{i}c_\mathrm{i} + D_\mathrm{i}\nabla^2 c_\mathrm{i}, \tag{4a}$$

$$\frac{\partial c_\mathrm{i}}{\partial t} = \sigma\dot{\phi} - \frac{4\pi r}{\Omega_0}(D_\mathrm{i} + D_\mathrm{v})c_\mathrm{i}c_\mathrm{v} - \left(\sum_j K_{\mathrm{v},j}\right)D_\mathrm{v}c_\mathrm{v} + D_\mathrm{v}\nabla^2 c_\mathrm{v}, \tag{4b}$$

where σ is the cross section for producing Frenkel pairs, $\dot{\phi}$ is the ion flux, r is the capture radius for interstitial-vacancy recombination Ω_0 is the atomic volume, $D_{\mathrm{v,i}}$ are the diffusivities of vacancies and interstitials, respectively, and the $K_{j,k}$ are the strengths of the various sinks (grain boundaries, surfaces, etc.) for interstitials and vacancies. The atomic diffusion coefficient is thus obtained as,

$$D = f_\mathrm{i}c_\mathrm{i}D_\mathrm{i} + f_j c_\mathrm{v}D_\mathrm{v}, \tag{5}$$

where f_j are correlation factors, which are of order unity. During irradiation at elevated temperature, the defect concentrations typically reach their steady states long before the phase transformations take place, and spatial variations in the steady-state concentrations often remain small, so that the defect concentrations are easily obtained. Figures 4a and b show schematically the concentration of point defects and the temperature dependence of the radiation-enhanced self-diffusion coefficient. At low temperature, the migration of defects is negligible and the diffusion is controlled by ion-beam mixing. As the temperature is increased, defect diffusivities increase. The concentration of point defects in this regime is controlled by the production rate and recombination, giving rise to a temperature-dependent diffusion coefficient. At still higher temperatures, the point defects migrate to sinks, rather than recombining. While the defect diffusivities continue to increase with temperature, as before, the concentrations of defects in this sink-limited regime correspondingly decrease and the diffusion coefficient becomes independent of temperature. The reason for this behavior is simply that in the sink-limited regime, each irradiation-produced defect undergoes a fixed number of jumps to reach a sink, and this number is independent of how fast the defects perform these jumps. This behavior is very different from that arising from thermal diffusion, which is included in Fig. 4 for comparison. While (4) provides a convenient first approximation for treating alloy evolution under irradiation, many complexities arise in inhomogeneous alloys where defect and solute mobilities depend on local environments. Moreover, the assumption that the different terms in (4) act independently breaks down for inhomogeneous sink structures, such as grain boundaries and surfaces [19, 20].

In many applications irradiations are conducted at room temperature, where defects are immobile, and subsequently the implanted material is annealed to remove implantation damage, such as excess point defects and defect clusters. When these defects become mobile, they mediate diffusion, often referred to as transient-enhanced diffusion (TED) [21]. As pointed out earlier, once these defects produced at room temperature have migrated to sinks, no additional diffusion takes place, similar to rapid quenching. Since the maximum concentrations of defects that are stable in crystals do not exceed $\approx$1 at.%, TED is usually not overly significant for phase transformations in concentrated alloys. As discussed in chapter on Transient enhanced diffusion (S. Libertino & S. Coffa, U. of Catania & ST Microelectronics, Catania, Italy), however, TED can play an important role in the semiconductor industry since it leads to precipitation (and loss of electrical activity) and broadening of the depth distribution of shallow implanted dopants.

1.5 Primary Recoil Spectrum

While the preceding discussion has described irradiation effects in general terms, a material's response depends sensitively on the choice of ion mass and energy as well as the specific target. For example, irradiating a target

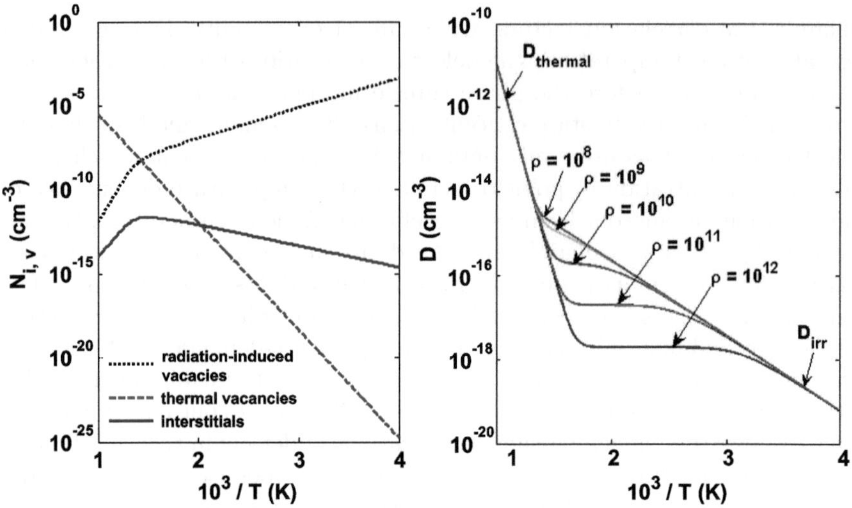

Fig. 4. (*Left*) Concentration of point defects in Cu as a function of temperature at constant flux 10^{-6} dpa s^{-1}, and a dislocation density of 10^9 cm^{-2}. (*Right*) Diffusion coefficient shown in an Arrhenius plot for the conditions in (*left*), but with different dislocation densities. After [18]

with high atomic number and low melting temperature, such as Au or PbTe, with energetic heavy ions creates cascades with high energy densities and extreme thermal spike effects. As a consequence, defect production results in high densities of point defects that often condense into immobile clusters and dislocation loops. Ion-beam mixing is extensive due to the local melting, and the surface is pocked with craters and mounds. At the other end of the spectrum, irradiation of a low-Z material such as Ti or Si with a light ion such as He results in isolated point defects, little atomic mixing, and sputtering yields much less than unity. The difference between these two irradiations arises primarily from the difference in the screening of the Coulomb interaction between the ion and target atom.

For high-energy, light ions, the screening is minimal and the interaction can be described by Rutherford scattering. The scattering cross section in this case is given by,

$$\frac{\mathrm{d}\sigma(E_1,T)}{\mathrm{d}T} = \frac{\pi m_1 Z_1^2 Z_2^2}{m_2 E_1} \cdot \frac{1}{T^2}, \tag{6}$$

where $Z_{i(j)}$ and $m_{i(j)}$ are the atomic number and mass of the projectile (target atom), respectively. The average recoil energy, therefore, is,

$$\langle T \rangle = T_{\min} \ln \frac{T_{\max}}{T_{\min}}, \tag{7}$$

where $T_{\max} = 4m_1 m_2 E_1/(m_1 + m_2)^2$ and $T_{\min} \approx E_{\mathrm{d}}$. For 1-MeV proton irradiation of Cu, for example, $\langle T \rangle \approx 200$ eV. For heavy-ion irradiations of targets with high atomic weights, screening is strong and the interaction can be approximated reasonably well by hard-sphere collisions. The scattering cross section in this case is,

$$\frac{\mathrm{d}\sigma(E_1, T)}{\mathrm{d}T} = \frac{\pi\rho_0^2}{T_{\max}}, \tag{8}$$

where ρ_0 is the hard-sphere radius, and

$$\langle T \rangle = \frac{T_{\max}}{2}. \tag{9}$$

The average recoil energy is thus seen to increase logarithmically with energy for light ions but linearly with energy for heavy ions. The primary recoil spectrum for any particular irradiation is usually obtained by computer simulation [4], which employs the so-called *universal potential* to describe the two-body interactions [22]. Physical insight into the primary recoil function can be obtained by considering the related function, $W(T)$, which is the integral fraction of damage energy (and thus also the fraction of defects, atomic mixing, or sputtered atoms) associated with cascades of all energies up to energy, T. This function thus weights the different recoils by how much damage they produce. It is plotted in Fig. 5 for various 1-MeV ions in Ni. A useful single-parameter characterization of a recoil spectrum is provided by $T_{1/2}$, defined by $W(T_{1/2}) = 0.5$, since it yields the recoil energy at which half the defects are produced in cascades of energy greater than $T_{1/2}$ and half in cascades with energies less than $T_{1/2}$. Notice that $T_{1/2}$ increases from 500 eV to over 10 keV on switching the incident ion in Ni from protons to Xe. An interesting comparison shown in Fig. 5 is the $W(T)$ functions for protons and neutrons since the projectiles have the same masses and energies. Since the former undergoes nearly purely Rutherford scattering while the later undergoes hard-sphere scattering, the $W(T)$ are very different, with $T_{1/2}$ changing from ≈400 eV to 40 keV. Lastly, we consider the spatial distribution of recoil events. Shown in Fig. 6 is a simulation of a 200 keV Cu event in Cu. The figure illustrates that the energy is distributed inhomogeneously along the path of the projectile, but with local regions of high energy density. These local regions are denoted as subcascades, suggesting that high energy events can be considered as a series of isolated, smaller events of energy E_C or less. E_C marks the energy at which cascades begin to split into subcascades. For recoils in Si, Ni, and Au, for example, $E_C \approx 5$ keV, 20 keV, and 50 keV, respectively.

2 Irradiation-Induced Stresses and Surface Effects

Changes in the state of stress in materials under irradiation derive from a number of mechanisms: accumulation of defects, the redistribution of material near surfaces, and phase transformations. We do not discuss phase

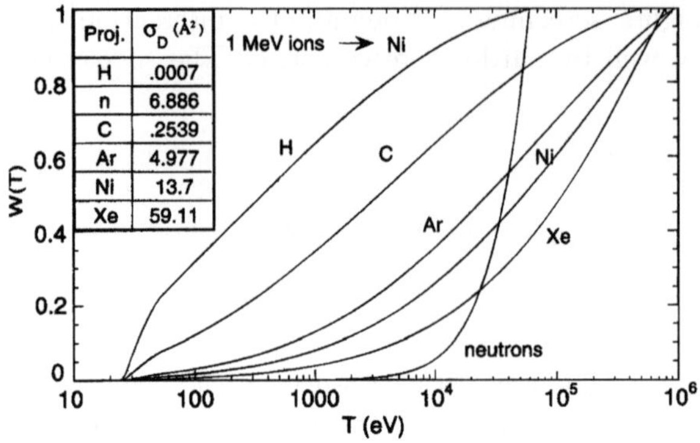

Fig. 5. The function $W(T)$ for various 1-MeV ions in Ni [12]

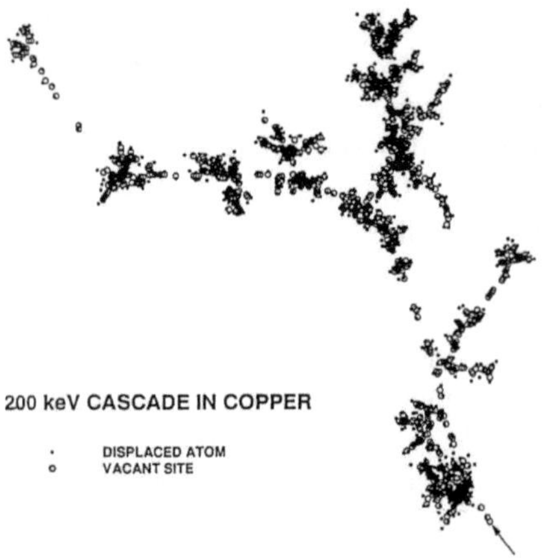

Fig. 6. Cascade evolution for a 200-keV self-ion event in Cu. (Figure courtesy of H. Heinisch)

transformations in this section, since their effect on stress will be apparent after discussing the other mechanisms.

2.1 Defect Accumulation

Irradiation of materials at low temperatures produces point defects and defect clusters. In order/disorder alloys, antisite defects can also be produced. Associated with each of these defects is an excess (relaxation) volume,[3] thus creating internal stress. For example, the relaxation volume of a Frenkel pair in metals is $\approx 1\ \Omega_0$ (atomic volume), i.e., ≈ 1–$2\ \Omega_0$ for the interstitial and $\approx -0.3\ \Omega_0$ for the vacancy [23]. As defects cluster and "collapse" into dislocation loops, the relaxation volumes for interstitials and vacancies become symmetric, $\approx +1$ or $-1\ \Omega_0$ for the two defects, respectively. The implanted ion also creates stress since it contributes one excess atom to the sample.

[3] The relaxation volume is the change in sample volume when an atom is removed or added to the interior of a crystal. It differs from the formation volume by one atomic volume.

Measurements of strain, either by lattice parameter measurements or wafer bending, in fact, provide sensitive measures of defect concentrations [24, 25].

An important example of irradiation-induced stress concerns the implantation of hydrogen atoms into Si. It has long been known that implantation of many materials with H or He ions results in high-pressure bubbles and eventually to void formation (see chapter "Voids and Nanocavities in Silicon" by Williams and Wong-Leung), blistering and exfoliation of the surface. It is now recognized that this seemingly detrimental effect can be utilized to good purpose, such as the slicing of thin plates from a thick wafer of single-crystalline material, a process called "ion cut" [26]. Typically, hydrogen is implanted at room temperature to high doses. Subsequently, the material is annealed at elevated temperatures, allowing the hydrogen to coalesce into bubbles with very narrow depth distribution. As the pressure builds, the surface layer fractures parallel to the surface, providing a thin slice of material. The implantation damage, moreover, is removed by the annealing process. By first oxidizing the wafer before slicing, the method can be used to fabricate single-crystalline Si wafers on SiO_2 for SOI devices.

2.2 Collective Behavior: Irradiation-Induced Viscous Flow

Irradiation of materials under an applied stress at elevated temperature can lead to enhanced creep rates and stress relaxation owing to the increased concentrations of point defects. While this result is not surprising in light of Sect. 1.5 on radiation-enhanced diffusion, rather unusual plastic deformation has also been observed during irradiation at temperatures where defects are immobile. Such behavior was first observed during swift ion, i.e., GeV energies, irradiation of silicate [27] and metallic [28] glasses. These studies showed anisotropic deformation in irradiated glasses; i.e., they elongate in directions perpendicular to the direction of an energetic ion beam and shrink parallel to it, conserving volume in the process. Using ions with much lower energies, Volkert noticed that epitaxial stresses in amorphous Si underwent stress relaxation during irradiation and showed that relaxation in amorphous Si followed Newtonian viscous flow [29]. Similar behavior has been observed in metallic glasses [30] and amorphous SiO_2 [31]. These behaviors are usually attributed to thermal spike effects [32, 33], although alternative mechanisms have been suggested in these glasses [34]. Before going into further detail, it is illuminative to examine the near-surface of an irradiated material during ion irradiation.

Figure 7 shows snapshots obtained from a MD simulation of 10-keV self-ion bombardment of Au at different instants of time. As the cascade event evolves, the local volume heats above the melting temperature and pressures of $\approx$1–10 GPa develop in the core. The pressure associated with the thermal expansion and the solid–liquid transformation causes mass to flow onto the surface. With time, the pressure relaxes and a small volume of liquid is left in the surface region. When the liquid cools and resolidifies, however, atomic mobility becomes negligible and many atoms are left frozen on the surface. As a consequence, there is a net flux of mass onto the surface, leaving a many vacancies below the surface, which condense into dislocation loops. Since the relaxation volume of each vacancy in a loop is ≈1 Ω_0, a permanent biaxial tensile stress is created in the surface region of the films.

While this example uses a relatively low-energy ion and the penetration is shallow, the same basic mechanism has been shown to operate during MeV irradiations as well, and in a variety of different materials [36]. It is also observed in Fig. 7f, that a mound forms on the surface around each impact, owing to the excess material. In some cases craters are also formed, surrounded by a rim. These features add roughness to an irradiated surface and generally they contribute far more roughness to a film than simple sputtering [37, 38]. Lastly, we remark that low-energy sputtering (0.5–5 keV ions) is

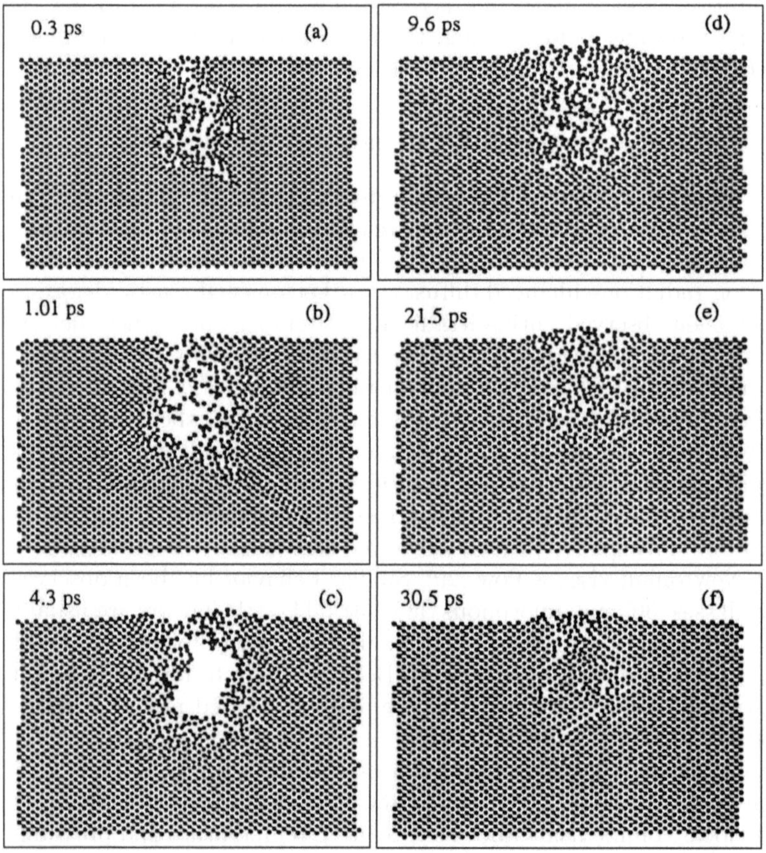

Fig. 7. Evolution of a 10-keV cascade in Au. This event is initiated by a Au impinging on the surface at 0 K. Atoms located within a cross-sectional slab 0.4 nm thick are shown. After [35]

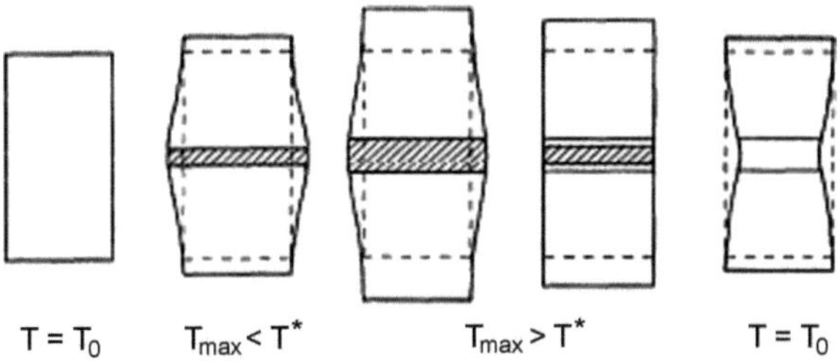

Fig. 8. Evolution of strain during the heating cycle induced by the passage of a swift ion in a glassy material. After [40]

often used in processing materials. This procedure can induce tensile stress in films due to the process just described, but it can also create compressive stress owing to end-of-range defects and the addition of the implanted atom. This idea has been used for controlling the radius of curvature of Si components employed in MEMS technology [39].

We return now to the question of anisotropic deformation in metallic glasses during swift ion irradiation. As seen in Fig. 1, the electronic stopping power for swift ions is several keV nm^{-1}, which can heat a material within a cylindrical region surrounding the track to temperatures far in excess of the glass temperature. As pointed out by Trinkaus [40], the thermal expansion creates stress within the cylinder, but owing to the elongated asymmetry, the strain is not homogeneous, but rather larger in the direction radial to the beam than parallel to it, as illustrated schematically in Fig. 8. The cylinder of material thus deforms. When the local region cools to below the glass temperature, the anisotropic deformation becomes frozen in, similar to the situation described for mound formation at surfaces. Subsequent tracks add to the deformation. This model is applicable for glasses, but not crystalline material, since for the latter, lattice sites must be conserved during crystallization of the melt. In crystals, therefore, surfaces or other sources and sinks for mass are required for the macroscopic flow described above.

3 Phase Transformations

3.1 Order-Disorder Alloys: Cu_3Au

Cu_3Au provides a model system for illustrating how ion beams can be employed to control phase stability in order–disorder alloys since the equilibrium thermochemical properties of this alloy are well established. The effect of irradiation on the order of Cu_3Au at 80 K is illustrated in Fig. 9 [41]. Here, the long-range order parameter decreases nearly exponentially with ion fluence. This behavior is understood on the basis that atomic mixing arises solely from the ballistic mixing in cascade events and that no reordering is possible during diffusion of vacancies at this low temperature. Interstitial atoms are mobile at 80 K, but their interstitialcy diffusion mechanism does not promote ordering. Since the disordering arises only from mixing within cascades, the disordering rate in this situation can be written [42],

$$\frac{\mathrm{d}S}{\mathrm{d}\phi} = -\alpha S, \tag{10}$$

and

$$S = S_{\mathrm{eq}} \exp(-\alpha\phi),$$

where S is the long-range order parameter and α represents the initial disordering rate of a fully ordered alloy. For doses measured in displacements per atom (dpa), $\alpha = 24$ for the data shown in Fig. 9. One dpa is the dose required to create a Frenkel pair on every lattice site one time (see (1)); it is a convenient measure of dose since it is independent of the type of ion employed, and it provides physical insight into the damage level.

The behavior is very different at elevated temperatures. As shown in the lower inset of Fig. 10, a short pulse of irradiation (1 s) with He ions at 635.5 K or 607.7 K results in a rapid change in the order parameter, as monitored here by the change in electrical resistivity, ΔR. At the end of the pulse, reordering takes place. The time constants for reordering are shown in an Arrhenius plot in the same figure for various ion irradiations. At low temperatures the relaxation times show a linear behavior, whereas near the order–disorder temperature in equilibrium, T_{c}, the times become much longer owing to the reduced driving force for ordering. The curves fit well to an equation of the form,

$$\tau^{-1} \propto D \tanh\left[\frac{\varepsilon_{\mathrm{a}}}{2T}\left(\frac{T_0 - T}{T_0}\right)\right], \tag{11}$$

where $\varepsilon_{\mathrm{a}} \approx 1000\,\mathrm{K}$ and is related to the ordering energy, and $T_0 = T_C \pm 2\,\mathrm{K}$ [43], illustrating that recovery in this case is due primarily to equilibrium

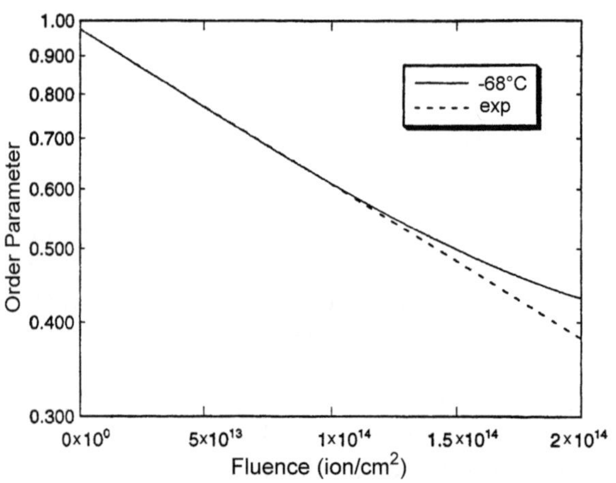

Fig. 9. Order parameter in Cu_3Au, following Ne irradiation at 205 K. *Dashed line* indicates an exponential fit to the data with $\alpha = 24\,(\mathrm{dpa})^{-1}$. After [41]

kinetics. The excess vacancies produced by the short pulse of irradiation find sinks in times far shorter than τ, and thus cause little reordering. Also shown in this plot is the feature that the time constants become significantly longer as the primary recoil is shifted to higher energies. From these times, the average number of jumps each atom undergoes during the reordering process may be deduced. For He irradiation, at the lowest temperatures, reordering can be achieved with only a few atomic jumps, whereas with Kr irradiation, also at low temperatures, the number increases by a factor of three. The low number for He shows that the disorder is comprised predominantly of isolated antisite defects and that no large volumes of disorder are present. For Kr, the number is increased, showing that small volumes of disorder are created by the cascades and these regions require more atomic motion to reorder.

Figure 11 illustrates the ordering kinetics under continuous irradiation. Initially, the order increases with He ion fluence until the ordering/disordering processes come into steady state. At temperatures well below T_c, the kinetic equations for this case can be approximated by [44],

$$\frac{\mathrm{d}S}{\mathrm{d}t} = -K\dot{\phi}S + K_1 D(S_{\mathrm{ss}} - S)^2. \tag{12}$$

The reordering process is mediated, as before, by vacancy diffusion, but under persistent irradiation the vacancy concentration is comprised of both equilibrium and excess vacancies. Notice in Fig. 11 that when the irradiation intensity is increased the balance between ordering and disordering is upset and the order parameter decreases and a new steady state is achieved. If the irradiation is reduced to its original intensity, the previous degree of order is restored. The process is reversible, illustrating that the steady state of the system is independent of the starting point and thus represents a state characterized by the temperature and ion flux, as will be described in fuller detail below in Sect. 4. When the irradiation is switched off, the equilibrium state of order is obtained. Notice, however, that the time constants for the system order to change from one state to another is longer when the irradiation is switched off compared to that when the irradiation is only reduced from $\phi + \Delta\phi$ to ϕ. The difference arises primarily from the excess vacancies in the system during irradiation. When the irradiation is switched off, the recovery is due only to equilibrium vacancies. By comparing these relaxation times, the concentrations of vacancies in the system can be accurately measured, relative to the equilibrium concentration, for any temperature and irradiation intensity [43]. In terms of materials processing, this study illustrates that the irradiation flux, ϕ, provides an independent control variable, like temperature and pressure, for materials processing.

It is noteworthy that in Cu_3Au, $T_c \approx 380°C$, or approximately 3/8 the melting temperature (T_m). Since diffusion is sluggish at this temperature, it is very difficult to reach high degrees of order without extensive thermal annealing. In such cases irradiation can prove very helpful, since it enhances diffusion. For example, the $L1_0$ ordered phase of equiatomic Fe–Ni, which has a

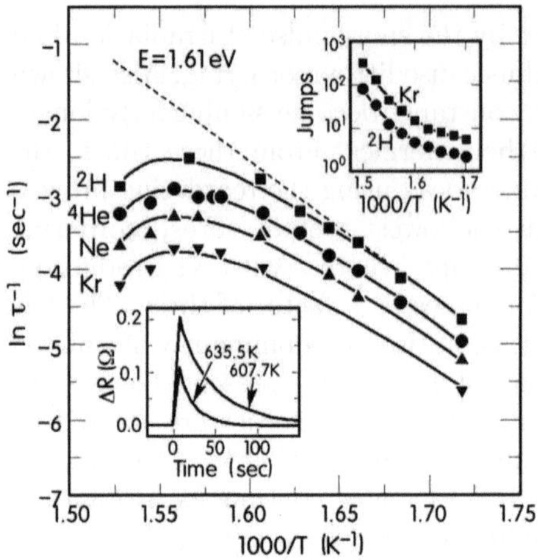

Fig. 10. An Arrhenius plot of inverse recovery time of Cu_3Au following a short pulse of He irradiation. The lines are fits to (7). *Lower inset*: Time dependence of recovery of order. *Upper inset*: number of jumps per atom to recover order [43]

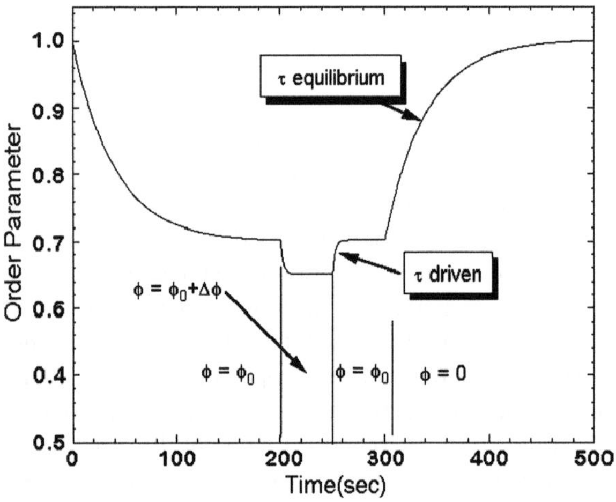

Fig. 11. Response of the order parameter to changes in the irradiation intensity [43]

critical temperature of 320°C, was discovered after irradiation with neutrons at 295°C [45]. More recently, interest has arisen in achieving high degrees of order in thin films such as FePt for high-density magnetic recording media, since the ordered phase has a high magnetic anisotropy. The required temperature for ordering is far too high for manufacturing thin-film devices, however, as discussed in chapter "Magnetic Properties and Ion Beams: Why and How" by Devolder and Bernas, ion irradiation can be employed to accelerate the ordering kinetics, as illustrated for Cu_3Au in Fig. 10 [46].

3.2 Phase-Separating Alloys: AgCu

Similar behavior is observed in phase-separating alloys, such as the eutectic alloy, AgCu, as shown in Fig. 12. Here, X-ray diffraction patterns from a multilayer thin-film sample are shown following irradiation with 1.8-MeV Kr ions to a dose of $1 \times 10^{16}\,\mathrm{cm}^{-2}$ at various temperatures. At this dose, the microstructure reaches a steady state and the diffraction patterns no longer change. The diffraction pattern obtained from the as deposited sample shows pure phases of Cu and Ag. The peak widths are quite broad since each layer is only 10 nm thick. The absence of (200) peaks indicates strong preferential alignment of the films. Curve F, obtained after irradiation at room temperature, shows that this immiscible alloys has been forced into a homogeneous AgCu alloy due to ion-beam mixing. The peak width has sharpened considerably, illustrating that the grain size is much larger than the initial layer thickness and that phase boundaries are no longer present. Irradiation at higher temperatures results in two-phase alloys, but now with the solubility limits in the steady state being greatly extended.

Similar to the results on ordered alloys, the solubility limits at steady state are independent of the sample history as illustrated in Fig. 13. Diffraction patterns D and A derive from Cu–Ag multilayers irradiated at room temperature and 368 K, respectively, to a dose of $1 \times 10^{16}\,\mathrm{cm}^{-2}$. Curves B and C are diffraction patterns from samples first irradiated at room temperature and subsequently reirradiated at 368 K to doses of 1×10^{16} and $7 \times 10^{16}\,\mathrm{cm}^{-2}$, respectively. Independent of the initial microstructure, therefore, the solubility limits of the alloy reach the same values for a given irradiation flux and temperature.

A large number of different alloy systems have been irradiated at room temperature to explore the extension of solubility limits in immiscible alloys. Similar to CuAg, several of these alloys form single-phase solid solutions, while many do not. For example, irradiation of AgNi at temperatures as low as $\approx$80 K shows a maximum solubility of 16 at.% Ag in Ni, but only 4 at.% Ni

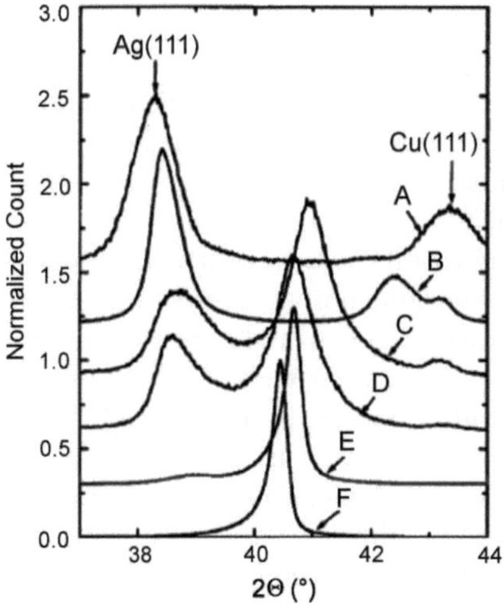

Fig. 12. X-ray diffraction patterns of multilayer Cu–Ag samples after irradiation to a dose of $1 \times 10^{16}\,\mathrm{cm}^{-2}$ with 1.0-MeV Kr ion at (A) as deposited; (B) 473 K; (C) 423 K; (D) 398 K; (E) 348 K; (F) 298 K. From [47]

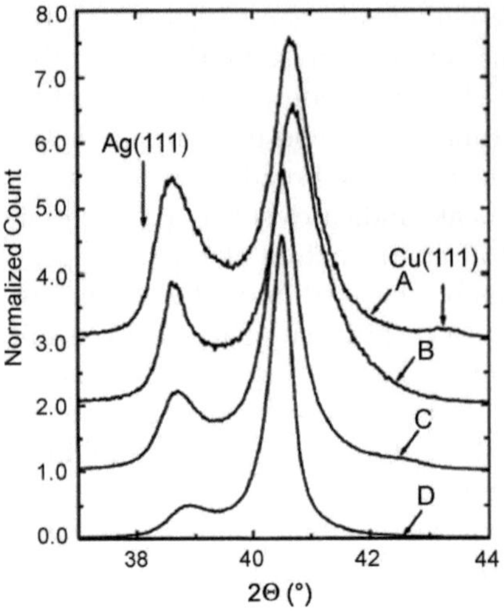

Fig. 13. X-ray diffraction patterns of multilayer Cu–Ag samples after irradiation to a dose of $1 \times 10^{16}\,\text{cm}^{-2}$ with 1.0-MeV Kr ion at (A) 398 K; (D) 298 K; (B) 398 K – reirradiation of the sample shown as (D); (C) 398 K – but irradiated to $7.4 \times 10^{16}\,\text{cm}^{-2}$ K. From [47]

in Ag [48]. This result may appear surprising at first since radiation-enhanced diffusion is negligible at this low temperature. This example illustrates, however, that atomic mixing in cascades is comprised of both purely ballistic mixing, where solutes flow down their concentration gradients, and thermal spike mixing where atoms move in the liquid state flow down gradients in their chemical potentials. Noteworthy is that the solubility limit of Ag in Ni just above $T_m(\text{Ni}) = 1453°\text{C}$ is ≈5 at.%, while that for Ni in Ag is 1 at.% just above $T_m(\text{Ag}) = 961°\text{C}$, suggesting that equilibrium is not attained during the short lifetime of the thermal spike, but see Sect. 4, below, for further details.

3.3 Amorphization

Many irradiated materials have been shown to undergo a crystalline to amorphous transition during irradiation at temperatures sufficiently far below the crystallization temperature of the amorphous phase. Covalently bonded systems are particularly conducive to amorphization, with even pure elements Si and Ge undergoing amorphization during irradiation. In contrast, no pure metal undergoes amorphization under irradiation, nor do alloys that form solid solutions, even at irradiation temperatures below 10 K. Several intermetallic compounds, on the other hand, have been amorphized [49]. A review of the models for amorphization under irradiation can be found in [50]. In the present work, we concern ourselves only with how this phase transition depends on the conditions of the irradiation, and therefore we will simply assume that the accumulation of defects and disorder eventually leads to amorphization. We note, however, that in pure metals the largest concentration of point defects and defect clusters that can be accumulated is ≈0.1 at.%, even at temperatures where point defects are immobile. At higher concentrations, the Frenkel pairs recombine and defect clusters collapse into dislocation loops, which have much lower energies. Owing to these relaxation mechanisms, amorphization is prevented. In many intermetallic compounds both defect accumulation and chemical disorder are introduced, enabling amorphization. Simulations suggest that in some intermetallic compounds either chemical disorder or point defects are sufficient for amorphization, but not always [50].

Figure 14 schematically illustrates the effects of primary recoil spectrum and temperature on the critical ion dose required for amorphization, while Fig. 15 shows a similar plot obtained from experiments on CuTi [51]. At low temperatures, amorphization is achieved at approximately the same irradiation dose (measured in dpa), regardless of primary recoil spectrum. At higher temperatures, however, the critical dose for amorphization increases

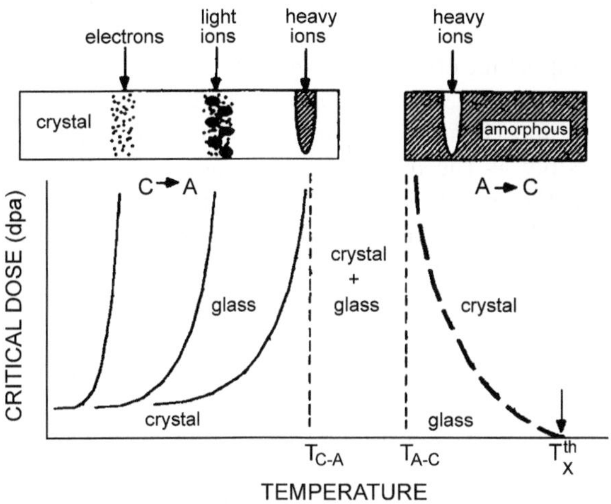

Fig. 14. Critical dose required for amorphization as a function of temperature, schematically shown for different types of irradiation. After [50]

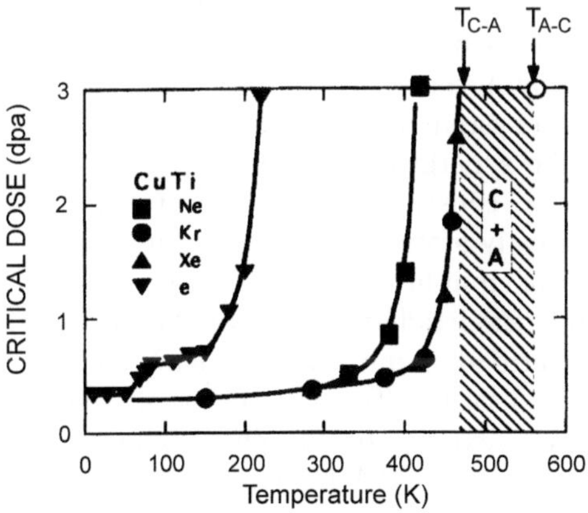

Fig. 15. Critical dose required for amorphization of CuTi as a function of temperature for different types of irradiation. After [51]

owing to the onset of thermally activated recovery mechanisms. Presumably recovery of order and defect annealing is mediated predominantly by vacancy migration. The situation for amorphization thus very much resembles the behavior shown in Fig. 10, for the order–disorder alloy, Cu_3Au. For Cu_3Au it was pointed out that the relaxation time of the order parameter increases markedly on increasing the mass of the irradiation ion. If we assume that similar mechanisms control amorphization, the data shown in Fig. 13 follow directly, but see [50] for details.

4 Phase Transformations: Effective Temperature Model

Sustained irradiation can lead to the dynamical stabilization of nonequilibrium phases at steady state, as illustrated in Sect. 3. In order to assess the radiation resistance of materials, for instance in nuclear reactors or in matrices for radioactive waste immobilization, it is of high practical interest to be able to rationalize or even predict the phases eventually stabilized by a given irradiation environment. An important observation in that respect is that, in almost all experiments, these steady states are observed to be independent of the initial state of the alloy, and that the transformation from one steady state to another occurs reversibly as the irradiation parameters are varied. Furthermore, small changes in the irradiation conditions can lead to drastically different steady states. For instance, during 1-MeV electron irradiation of the ordered alloy Ni_4Mo, a temperature drop from 470 K to 450 K results in a transition from a chemically long-range ordered to a disordered steady state [52].

This general behavior suggests that it may be possible to recast the problem of phase stability under irradiation into a framework resembling equilibrium thermodynamics, and thereby place ion-beam processing on the same footing as more conventional processing. Earlier attempts were made to rationalize radiation-induced phase transformations using free energies constrained by high point defect supersaturations. This approach, however, fails to reproduce experimental results, in particular transitions from one steady state to another. As initially proposed by Adda et al. [53], one should instead consider an alloy under irradiation as a system subjected to several dynamical processes in parallel, which can be synergistic or competing. For such a dynamical system, one could envision constructing a steady-state phase diagram that yields the most stable steady state under specified irradiation conditions, thus extending the concept of equilibrium phase diagram. Clearly, axes in such a diagram need to include the irradiation flux, or the displacement rate, in addition to common thermodynamic variables such as composition and temperature. The experimental results reviewed above indicate that, in these steady-state diagrams, one expects to find phase boundaries, which correspond to dynamical phase transitions. A fundamental and practical question is then to determine the location of these dynamical phase boundaries. Although there is no general proof that an effective free energy can be derived for dissipative systems [54], a simplified but powerful approach was introduced by Martin in the mid-1980s [55], leading to the so-called effective temperature criterion. We now review this criterion and illustrate its application to various phase transformations.

4.1 Phase Decomposition

Consider the case of an alloy such as Cu–Ag subjected to sustained irradiation, as illustrated in Figs. 12 and 13. In order to distinguish solid solution from phase-separated steady states, Martin proposed to write the evolution of the composition profile in such an alloy as

$$\frac{\partial c(r,t)}{\partial t} = -M'\nabla\mu + D_{\mathrm{b}}\nabla c, \tag{13}$$

where M' is the atomic mobility, enhanced by the point-defect supersaturation, μ the equilibrium chemical potential of the alloy, and D_{b} a ballistic diffusion coefficient that takes into account the random mixing forced by the nuclear collisions discussed in Sect. 2. Two important approximations are made in writing (13). First, the forced mixing is assumed to be random, and second, this mixing is assumed to be short-range and thus it can be described by a diffusive process. Using a regular-solution model for the chemical potential of the alloy, including a Cahn–Hilliard inhomogeneity term, Martin showed that the stable steady state reached under irradiation at the temperature T, corresponds to the equilibrium state that the same alloy system would have reached at an effective temperature,

$$T_{\mathrm{eff}} = T(1 + D_{\mathrm{b}}/D'_{\mathrm{th}}), \tag{14}$$

where D'_{th} is the radiation-enhanced interdiffusion coefficient due to thermally activated atomic transport.

Figure 16 illustrates how the effective temperature T_{eff} varies with the actual irradiation temperature T and the irradiation flux ϕ. At elevated temperatures, $T_{\mathrm{eff}} \to T$, irradiation accelerates the thermal kinetics but without much affecting the alloy state. At low temperatures, T_{eff} becomes very large since ballistic effects dominate. In fact, if one retains only the ballistic term on the RHS of (14), the alloy reaches an infinite temperature state. At a given intermediate temperature, the higher the irradiation flux, the higher the effective temperature. For a model alloy simply displaying a miscibility gap at equilibrium, the application of the effective temperature criterion leads to the dynamical phase diagrams shown in Fig. 17 [56]. Experimental results discussed in Sect. 3.2 for the Cu–Ag system under irradiation are in good agreement with such diagrams.

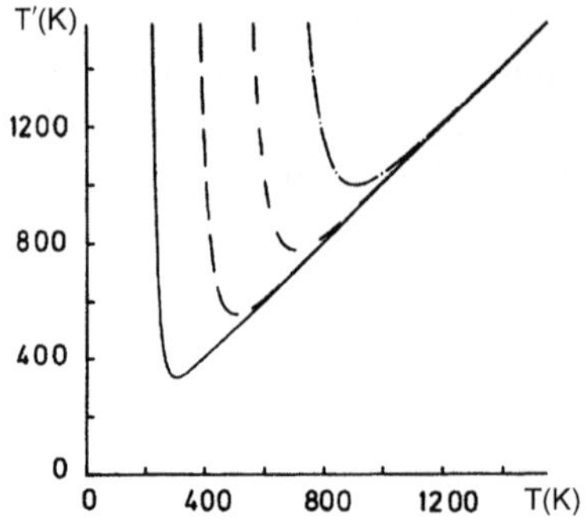

Fig. 16. Schematic plots of the effective temperature T_{eff} as a function of the actual irradiation temperature and flux (from [55])

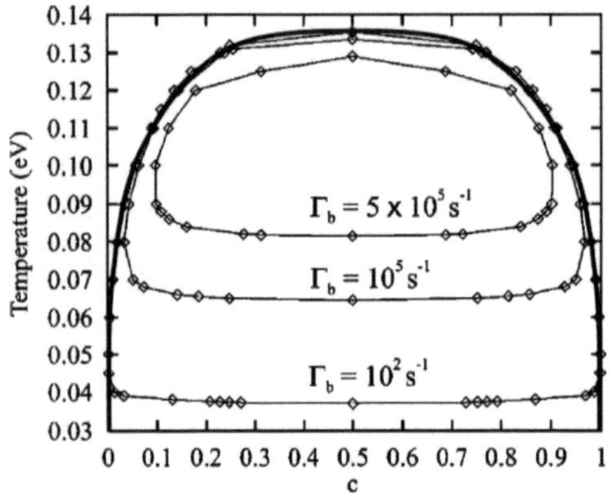

Fig. 17. Steady-state dynamical phase diagrams at three constant frequencies of nearest-neighbor ballistic exchanges Γ_{b}. The various miscibility loops are calculated using the effective temperature criterion. The *bold line* corresponds to the equilibrium phase diagram (after [56])

4.2 Order–Disorder

Although it is not possible to derive an exact expression for an alloy undergoing an order–disorder transition under irradiation, the effective temperature criterion provides a very good approximation of the steady-state degree of order parameter reached under irradiation [55]. From a qualitative perspective, it reproduces very well the features discussed in Sect. 2 for irradiated Cu_3Au. For ordered phases that remain ordered up to their melting point, T_{M}, such as NiTi, by extension of the T_{eff} criterion, one could suggest that, when $T_{\mathrm{eff}} > T_{\mathrm{M}}$, the alloy would reach an amorphous steady state. Certain compounds, however, such as Ni_3Al, while fully disordered by low-temperature irradiation, do not transform to amorphous phases, even at cryogenic temperatures [57]. It should be kept in mind, however, that the effective temperature model refers to diffusion and chemical compositions, it does not consider free energies of competing structures.

4.3 Beyond the Effective Temperature Criterion

While the effective-temperature criterion captures the dynamical competition between the rates of irradiation-induced mixing or disordering and thermally activated relaxation toward a low free-energy state, it does not take into account the fact that these processes may operate at different length scales. Irradiation with heavy and energetic ions in particular leads to the formation of dense displacement cascades, thus introducing two new length scales in the problem, the cascade size, ≈1 to 10 nm, and the average relocation distance of atoms within the cascade, ≈1 to 10 Å. Recent analytical and simulation works have shown that a general property of systems where dynamical processes compete at different length scales is the propensity to self-organize in space into patterns. Self-organization of composition fields in the bulk and patterning of surface morphology, for instance, are covered in detail in chapters "Precipitate and Microstructural Stability in Alloys Subjected to Sustained Irradiation" by Bellon and "Spontaneous Patterning of Surfaces by Low-Energy Ion Beams" by Chason and Chan.

5 Conclusions

The underlying principles of ion-beam processing are now well understood, and it is possible to design ion-beam processing schemes to achieve desired structures. Many challenges, however, remain in predicting the response of more complex materials that are of interest for engineering applications. The difficulties arise because ion irradiation drives materials far from their equilibrium states and this opens many pathways for the material to respond. For example, prolonged irradiation of metals creates point defects that can form defect clusters, and these clusters may be immobile or mobile, they may trap other defects and solute, and they alter the properties of the material, such as conductivity and strength. Accounting for these defect interactions is difficult, particularly in concentrated, multiphase alloys. Another example where ion-beam processing is proving a promising processing tool concerns the synthesis of nanostructured materials. As was noted in this chapter, the dimensions of the displacement cascades, or diameter of ion tracks are ideal for forming nanostructures both in the interior and at the surfaces of materials. The irradiation process, however, is stochastic and it remains challenging to create nanostructures that are spatially organized in patterns. Progress in achieving self-organization and patterning in irradiated materials is discussed in chapters "Precipitate and Microstructural Stability in Alloys Subjected to Sustained Irradiation" by Bellon and "Spontaneous Patterning of Surfaces by Low-Energy Ion Beams" by Chason and Chan.

A promising new direction in treating the complexity involved in ion-beam processing of engineering materials involves multiscale computer modeling. Several examples in this article concerning the displacement process in irradiated materials illustrate the power of molecular-dynamics computer simulations for this purpose. With current computing power, and progress in developing accurate yet tractable interatomic potentials, it is indeed now possible to reliably calculate the displacement process in most materials, although materials where charge exchange is significant remains difficult. The more challenging task of computing the evolution of microstructures at elevated temperatures during prolonged irradiation, or during postirradiation annealing still lies ahead. Advances in such methods as kinetic Monte Carlo [58, 59], accelerated molecular dynamics [60, 61], and phase-field modeling [62, 63], however, appear very promising for this purpose and the field of ion-beam processing.

Acknowledgements

The present work was supported by the U.S. Department of Energy U.S.DOE Basic Energy Sciences, under Grant No. DEFG02-05ER46217, and the National Science Foundation, under grant DMR 04-07958.

References

1. P. Sigmund, *Particle Penetration and Radiation Effects* (Springer, Heidelberg, 2006), Chap. 2 2
2. M. Toulemonde, C. Trautmann, E. Balanzat, K. Hjort, A. Weidinger, Nucl. Instrum. Methods B **216**, 1 (2004) 2
3. L. Civale et al., Phys. Rev. Lett. **67**, 648 (1991) 2
4. J.F. Ziegler, J.P. Biersack, U. Littmark, *The Stopping and Range of Ions in Solids*. Stopping and Ranges of Ions in Matter, vol. 1 (Pergamon, New York, 1984) 3, 4, 11
5. R.S. Averback, R. Benedek, K.L. Merkle, Phys. Rev. B **18**, 4156 (1978) 4
6. M.J. Norgett, M.T. Robinson, I.M. Torrens, Nucl. Eng. Des. **33**, 500 (1975) 4
7. R.S. Averback, R. Benedek, K.L. Merkle, Phys. Rev. B **18**, 4156 (1978) 4
8. D.J. Bacon, in *Computer Simulation of Materials*, ed. by H.O. Kirchner et al. (Kluwer Academic, Dordrecht, 1996), p. 198 4
9. H.H. Andersen, in *Ion Implantation and Beam Processing*, ed. by J.S. Williams, J.M. Poate (Academic Press, New York, 1984), Chap. 6 4
10. P. Sigmund, N.Q. Lam, in *Fundamentals Processes in Sputtering of Atoms and Molecules (SPUT'92)*, ed. by P. Sigmund, Medd. Kgl. Dan. Vindensk. **43** (1992), 255 4
11. H.H. Andersen, Appl. Phys. **18**, 131 (1979) 5
12. R.S. Averback, T. Diaz de la Rubia, in *Solid State Physics*, vol. 51, ed. by H. Ehrenreich, F. Spaepen (Academic Press, New York, 1998), p. 282 5, 12
13. H.L. Zhu, R.S. Averback, M. Nastasi, Philos. Mag. A **71**, 735 (1995) 6, 7

14. C.P. Flynn, R.S. Averback, Phys. Rev. B **38**, 7118 (1988) 6
15. A. Caro, M. Victoria, Phys. Rev. A **40**, 2287 (1989) 6
16. D.M. Duffy, A.M. Rutherford, J. Phys. Condens. Matter **19**, 016207 (2007) 6
17. R. Sizmann, J. Nucl. Mater. **69/70**, 386 (1978) 8
18. J. Philibert, *Atomic movements Diffusion and Mass Transport in Solids* (Les Editions de Physique, Les Ulis Cedex A, 1991), p. 497 10
19. J.L. Bocquet, N.V. Doan, G. Martin, Philos. Mag. **85**, 559 (2005) 9
20. N.V. Doan, G. Martin, Phys. Rev. B **67**, 134107 (2003) 9
21. K. Cho et al., Appl. Phys. Lett. **47**, 1321 (1985) 9
22. J.F. Ziegler, J.P. Biersack, U. Littmack, *The Stopping and Ranges of Ions in Solids* (Pergamon, New York, 1985) 11
23. P. Ehrhart, in *Interactions of Atomic Defects in Metals and Alloys*, ed. by H. Ullmaier. Landolt-Bornstein, New Series III, vol. 25 (Springer, Berlin, 1991), p. 88, Chap. 2 12
24. P. Ehrhart, J. Nucl. Mater. **216**, 170 (1994) 13
25. E.P. EerNisse, Appl. Phys. Lett. **18**, 581 (1971) 13
26. M. Bruel, Electron Lett. **31**, 1201 (1995) 13
27. L. Cartz, Radiat. Eff. Defects Solids **54**, 57 (1981) 13
28. S. Klaumunzer, G. Schuhmacher, Phys. Rev. Lett. **51**, 1987 (1983) 13
29. C.A. Volkert, J. Appl. Phys. **74**, 7107 (1983) 13
30. S.G. Mayr, R.S. Averback, Phys. Rev. Lett. **87**, 196106 (2001) 13
31. E. Snoeks, T. Weber, A. Cacciato, A. Polman, J. Appl. Phys. **78**, 4723 (1995) 13
32. H. Trinkaus, J. Nucl. Mater. **223**, 196 (1995) 13
33. H. Trinkaus, J. Nucl. Mater. **246**, 244 (1997) 13
34. S.G. Mayr, Y. Ashkenazy, K. Albe, R.S. Averback, Phys. Rev. Lett. **90**, 055505 (2003) 13
35. M. Ghaly, R.S. Averback, Phys. Rev. Lett. **72**, 364 (1994) 14
36. S.G. Mayr, R.S. Averback, Phys. Rev. B **68**, 214105 (2003) 14
37. M. Morgenstern, T. Michely, G. Cosma, Philos. Mag. **79**, 775 (1999) 14
38. S.G. Mayr, R.S. Averback, Phys. Rev. Lett. **87**, 6106 (2001) 14
39. T.G. Bifano, H.T. Johnson, P. Bierden, R. Mali, J. Microelectromech. Syst. **11**, 592 (2002) 15
40. H. Trinkaus, A.I. Ryazanov, Phys. Rev. Lett. **74**, 5072 (1995) 15
41. Y.S. Lee, Ph.D. thesis, University of Illinois at Urbana-Champaign 15, 16
42. S. Siegel, Phys. Rev. **75**, 1823 (1949) 16
43. L. Wei, Y.S. Lee, R.S. Averback, C.P. Flynn, Phys. Rev. Lett. **84**, 6046 (2000) 16, 17, 18
44. G.J. Dienes, Acta Metall. **3**, 549 (1955) 17
45. L. Néel, J. Paulevé, R. Pauthenet, J. Laugier, D. Dautreppe, J. Appl. Phys. **35**, 873 (1964) 18
46. H. Bernas, J.-Ph. Attané, K.-H. Heinig, D. Halley, D. Ravelosona, A. Marty, P. Auric, C. Chappert, Y. Samson, Phys. Rev. Lett. **91**, 077203 (2003) 18

47. L.C. Wei, R.S. Averback, J. Appl. Phys. **81**, 613 (1997) 19, 20
48. B.Y. Tsaur, J.W. Mayer, Appl. Phys. Lett. **37**, 389 (1980) 20
49. J.L. Brimhall, E.P. Simonen, Nucl. Instrum. Methods B **16**, 187 (1986) 20
50. P.R. Okamoto, N.Q. Lam, L.E. Rehn, in *Solid State Physics*, vol. 52, ed. by H. Ehrenreich, F. Spaepen (Academic Press, New York, 1999), p. 1 20, 21, 22
51. G. Xu, J. Koike, M. Meshii, P.R. Okamoto, in *The 47th Annual Meeting of the Electron Microscopy Society of America* (San Francisco Press, San Francisco, 1989), p. 658 21
52. S. Banerjee, K. Urban, M. Wilkens, Acta Metall. **32**, 299 (1984) 22
53. Y. Adda, M. Beyeler, G. Brebec, Thin Solid Films **25**, 107 (1975) 22
54. G. Martin, P. Bellon, Solid State Phys. **50**, 189 (1997) 23
55. G. Martin, Phys. Rev. B **30**, 1424–1436 (1984) 23, 24
56. R. Enrique, P. Bellon, Phys. Rev. B **60**, 14649 (1999) 23, 24
57. S. Müller, C. Abromeit, S. Matsumura, N. Wanderka, H. Wollengberger, J. Nucl. Mater. **271–272**, 241 (1999) 24
58. O. Trushin, A. Karim, A. Kara, T.S. Rahman, Phys. Rev. B **72**, 115401 (2005) 26
59. K. Sastry, D.D. Johnson, D.E. Goldberg, P. Bellon, Phys. Rev. B **72**, 085438 (2005) 26
60. M.R. Sorensen, A.F. Voter, J. Chem. Phys. **112**, 9599–9606 (2000) 26
61. Y. Shim, J.G. Amar, B.P. Uberuaga, A.F. Voter, Phys. Rev. B **76**, 205439 (2007) 26
62. L.Q. Chen, Annu. Rev. Mater. Res. **32**, 113 (2002) 26
63. Q. Bronchart, Y. Le Bouar, A. Finel, Phys. Rev. Lett. **100**, 015702 (2008) 26

Index

Precipitate and Microstructural Stability in Alloys Subjected to Sustained Irradiation

P. Bellon

Department of Materials Science and Engineering, University of Illinois at Urbana-Champaign, Urbana, IL 61801, USA, e-mail: bellon@uiuc.edu

Abstract. The sustained irradiation of a material by energetic particles leads to the continuous production of damage in the form of point defects, point-defect clusters, and forced atomic relocations, as reviewed in Chap. 1. These elementary processes lead to an acceleration of thermally activated diffusion owing to point-defect supersaturation, as well as a forced mixing of chemical species due to atomic replacements. In materials with precipitates or ordered phases, this forced mixing alone would lead to dissolution and chemical disordering, respectively. At high enough temperatures, however, these dynamical processes compete with thermally activated diffusion, which tends to restore an equilibrium state. The outcome of this competition depends of course on the relative intensity, or rates, of these processes, but also on their characteristic length scales. We review in some detail the evolution of pre-existing precipitates under irradiation to illustrate the complex material's response to these dynamical processes, including the potential self-organization of the microstructure. Similar effects are anticipated in materials undergoing order–disorder transformations. In addition, the kinetic coupling between point defects and chemical fluxes can lead to radiation-induced segregation and precipitation. Finally, we discuss the contribution of point-defect evolution to microstructural changes, which can produce dimensional changes and alter mechanical properties.

1 Introduction

Energetic projectiles, such as electrons, ions, and neutrons, are progressively slowed down as they propagate through a solid material. This slowing down can originate from interactions with the nuclei of the target atoms, with the electrons of the target atoms in the case of charged projectile, and from resonant nuclear reactions. These processes result in the continuous introduction of defects and disorder in the target material. Such a material is thus maintained in a nonequilibrium state, and, as a consequence, the driving force for its evolution does not solely originate from equilibrium thermodynamics [1]. There are indeed numerous experimental reports demonstrating that irradiation and implantation can induce nonequilibrium phase transformations

H. Bernas (Ed.): Materials Science with Ion Beams,
Topics Appl. Physics **116**, 29–52 (2010)
DOI: 10.1007/978-3-540-88789-8_2,

and microstructural evolutions. We review some of these nonequilibrium evolutions in this chapter. While these nonequilibrium evolutions often lead to a degradation of service properties, for instance embrittlement of the steels used in nuclear-reactor pressure vessels, these driven material systems display a propensity to undergo self-organization, as briefly indicated at the end of chapter "Fundamental Concepts of Ion-Beam Processing" by Averback and Bellon. There are indications that such self-organized microstructures, in fact nanostructures, may offer very desirable properties, for instance for optical and magnetic applications, as well as for designing radiation-resistant materials. In order to understand or to predict material evolutions under irradiation, it is first necessary to identify all the dynamical processes that are active under given irradiation conditions. As emphasized by Martin in his seminal paper in 1984 [2], these various dynamical processes may compete with one another, often driving the material into a nonequilibrium steady state. As the irradiation parameters are varied, for instance the irradiation temperature or the mass of the energetic projectiles, a new steady state may be reached, and it is therefore of prime interest, both fundamentally and practically, to establish a map of these steady states as a function of the relevant irradiation parameters. Various theoretical tools, computer simulations, as well as experiments have been employed to construct these dynamical equilibrium phase diagrams. They represent an extension of equilibrium phase diagrams, however, for materials subjected to sustained external forcing [2–4]. In Sect. 2 we briefly review the key features of the radiation-induced elementary processes relevant to this chapter, and, in Sect. 3, we discuss in some detail how the synergies and competition between these dynamical processes can affect the stability of precipitates; we then review more succinctly in the following sections other microstructural evolutions, such order–disorder transformations, segregation and precipitation, and defect clustering.

2 Elementary Processes in Metallic Alloys Subjected to Irradiation

The elementary effects produced in a host material by irradiation or implantation have been introduced in chapter "Fundamental Concepts of Ion-Beam Processing" by Averback and Bellon, and a comprehensive account of these effects can be found in [5]. Here, we focus on the case of crystalline metallic targets irradiated by relatively low projectile energy, much less than 1 MeV/amu. The damage created by irradiation is then solely due to nuclear collisions between the energetic projectiles, and the target atoms set in motion by prior collisions, with the target atoms at rest. The recoil energy, T, defined as the amount of kinetic energy transferred from an energetic projectile to an atom at rest, plays a determinant role in defect formation. A binary collision approximation is sufficient to calculate the number of Frenkel pairs created in one recoil event at low T, but not for $T \geq 1\,\text{keV}$, due to the many-body

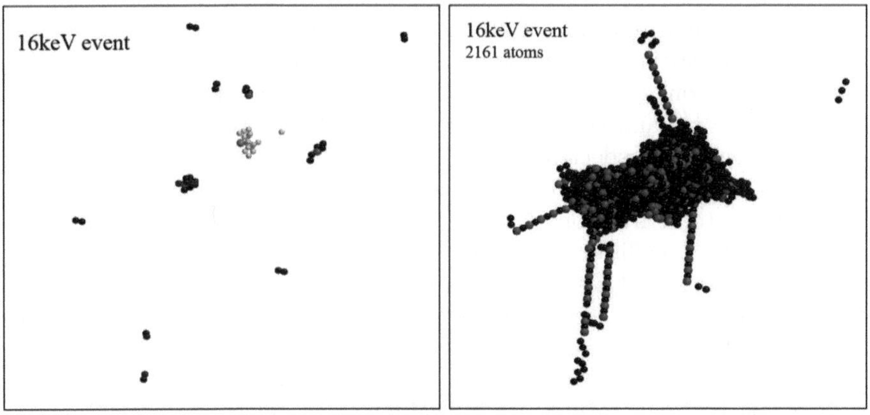

Fig. 1. MD simulation of the damage produced by a 16-keV displacement cascade in Ni_3Al: *at left*, isolated vacancies (*white spheres*), interstitials dumbbells (*dark and light gray spheres* for Ni and Al, respectively), and their clusters; *at right*, the 2161 atoms that have been replaced in that cascade (from [6])

interactions taking place in displacement cascades. These cascades extend from 1 to about 10 nm, and intracascade processes take place over a few picoseconds. Molecular-dynamics (MD) simulations have revealed that a large fraction of the Frenkel pairs produced in displacement cascades agglomerate into clusters. In particular, for materials with high atomic number, and based on close-packed lattices, dense and energetic cascades lead to the clustering of up to 80% the Frenkel pairs, with vacancy clusters at the center of the cascade and interstitial clusters at the periphery [5], as illustrated in Fig. 1. This point defect clustering is at the origin of the efficiency function introduced in (1.a) in chapter "Fundamental Concepts of Ion-Beam Processing" by Averback and Bellon, with $\xi(T) \approx 1/3$ for high recoil energies.

Since the displacements produced by irradiation are at the origin of the various nonequilibrium dynamical processes that may be found in an irradiated material, following Norgett et al. [7], it is common to define the intensity of an irradiation by the rate of production of these displacements, in units of NRT displacements per atom per second (dpa/s). Note that NRT displacements and displacement rates are calculated while ignoring defect clustering in displacement cascades, that is with $\xi(T) = 1$ in (1.a) in chapter "Fundamental Concepts of Ion-Beam Processing" by Averback and Bellon. Typical values for this displacement rate ϕ range from 10^{-8} dpa s^{-1} for components in nuclear reactors that are weakly irradiated, such as the pressure vessel in pressurized water reactors (PWR), up to 10^{-3} dpa s^{-1} during electron irradiation in a high-voltage electron microscope or during high-flux ion irradiation and implantation.

The continuous production of point defects and point-defect clusters under irradiation lead to a supersaturation, and thus to an acceleration of the thermally activated diffusion of atoms. Depending upon the relative weight of vacancy-interstitial recombination and point-defect elimination on sinks such as dislocations, grain boundaries, and surfaces, the defect supersaturation scales as $\phi^{1/2}$ and ϕ in the recombination-dominated and elimination-dominated regimes, respectively [8, 9].

Besides displacing atoms from lattice sites, nuclear collisions also force the relocation of atoms from one lattice site to another. These atomic replacements result from collisions involving energies that are usually much larger than the typical enthalpies involved in phase transformations, and, in a first approximation, one can assume that these relocations are decoupled from the chemical interactions between atoms. For this reason, Martin [2] proposed to use the term ballistic to describe this forced atomic mixing. Irradiation at low homologous temperatures are typically carried out to isolate and quantify this ballistic mixing, so as to suppress any possible contribution from thermally activated unmixing processes [3, 10].

We note, however, that in alloy systems with large and positive heat of mixing, typically $\Delta H_{\mathrm{m}} > 15\,\mathrm{kJ\,mol^{-1}}$, such as Ag–Fe and Cu–Mo, ballistic mixing appears to be negligible [11–14], even at cryogenic temperatures. Following Averback et al. [11, 12], this effect can be rationalized by noting that, in such alloy systems, the elements are typically immiscible both in the solid and in the liquid state. Any mixing produced during the ballistic phase of a displacement cascade can therefore be undone during the thermal spike phase, since the cascade region is essentially liquid-like during that phase.

Returning now to the case where irradiation produces forced (ballistic) mixing, the frequency of these relocations per atom, and the distance of these relocations are the two important characteristics of this ballistic mixing. The ballistic mixing rate can be expressed in terms of the displacement rate, multiplied by the number of replacements per displacement. This ratio ranges from a few for electron and light-ion irradiation, up to several hundred for heavy-ion irradiation. The range of the ballistic mixing has been the subject of confusion for quite a while. This range was sometimes taken as large, of the order of 100 Å, based on early simplistic pictures of displacement cascades [15]. Haff [16] and Andersen [17] proposed later to describe ballistic mixing as a random walk, with a jump distance ranging from 5 to 10 Å. With the widespread use of MD simulations, it became clear that indeed most of the ballistic mixing takes place between nearest-neighbor atoms [5]. Nevertheless, on top of this short-range mixing, the histogram of relocation distances display a tail of longer distances, as illustrated in Fig. 2. This tail can be well fitted by an exponential decay, and we will show in the following section that R, the decay length of this exponential tail, which ranges from $\approx$1 Å to a few Å, plays a key role in precipitate stability and composition patterning under irradiation.

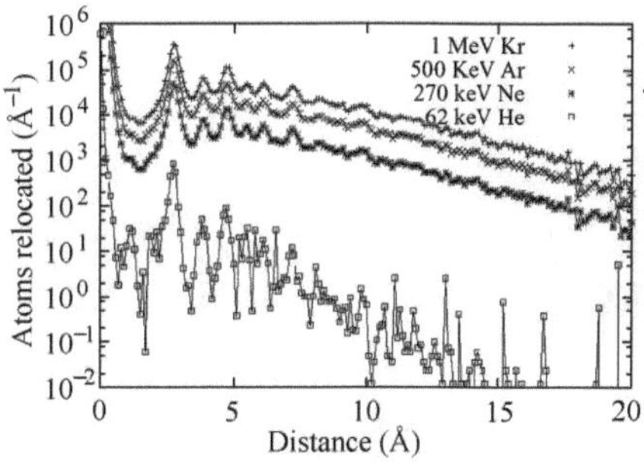

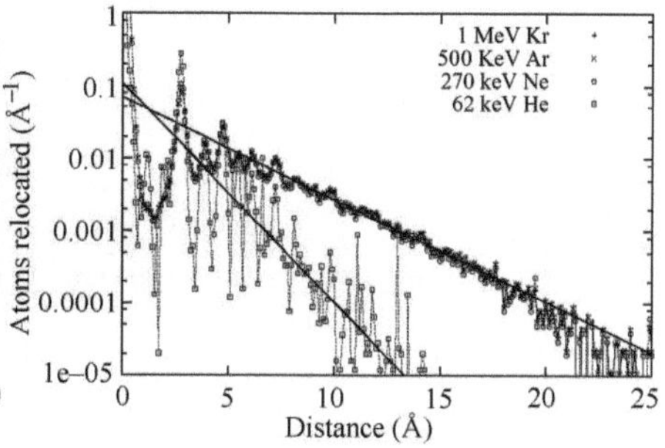

Fig. 2. Histograms of relocation distances produced by one projectile in $Ag_{50}Cu_{50}$ measured from MD simulations: (*left*) absolute values; (*right*) normalized values. The *solid lines* represent fits by exponential decay with a decay length of 1.44 Å for He and 3.08 Å for the three other ions. All four ions have the same projected range in the host material (from [18])

3 Precipitate Evolution in Irradiated Alloys

In this section, we review the evolution under irradiation of materials with pre-existing precipitates. This choice is partly motivated by the fact that precipitates often play a critical role in conferring optimized mechanical properties in engineering alloys, as for instance in the new generation of ODS ferritic/martensitic steels [19, 20]. The study of the stability of precipitates under irradiation also provides an instructive example of the challenges faced in modeling microstructural evolutions in nonequilibrium systems. After a short review of some key experimental findings, in particular on the self-organization of precipitates, we show how successive models have contributed to elucidate these puzzling findings.

3.1 Experimental Observations

Early on, it was recognized that neutron irradiation could lead to the dissolution of existing precipitates, as reported by Boltax for Cu–Fe alloys [21], or to the homogenization of a two-phase mixtures, as reported by Berman for ZrO_2–UO_2 [22]. Nelson et al. [23] studied systematically the effect of the irradiation temperature on the stability of $L1_2$ ordered precipitates in Ni–Al alloys. These rather large Ni_3Al precipitates were formed by heat treatment prior to irradiation, and were thus thermodynamically stable. These authors found, however, that, at low irradiation temperature, e.g., RT, the precipitates are disordered and then dissolved, as also confirmed later by Bourdeau, and coworkers [24, 25]. This result is well rationalized by the predominance of the ballistic mixing over thermally activated diffusion at low temperature. At higher temperatures, it is expected that the initial precipitates will instead remain stable or even grow, owing to radiation-enhanced thermal diffusion. At intermediate irradiation temperatures, however, the initial large

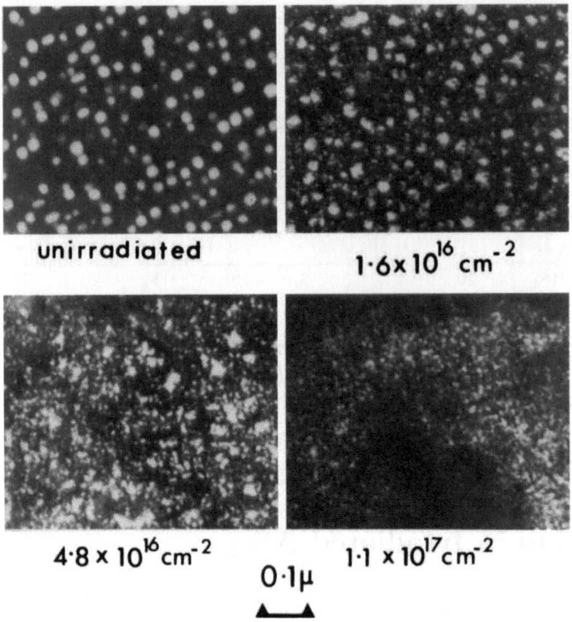

Fig. 3. Dark-field TEM imaging of $L1_2$-ordered Ni-rich precipitates in Ni-13.5 at.% Al irradiated at 550°C with 100-keV Ni ions; notice the refinement of the precipitate microstructure and the stabilization of nanoscale precipitates at large irradiation dose, given here in Ni ions cm^{-2} (from [23])

precipitates were replaced by a dispersion of very small precipitates, ≈10 nm in diameter. A more recent study on the same Ni–Al system reported that Ni_3Al precipitates, whose initial diameter was 5 nm after thermal annealing, shrank and stabilized at 2 nm during Ni ion irradiation. The dynamical stabilization of precipitates to a finite average size is clearly a nonequilibrium phenomenon since, if the alloy system were to be evolving toward its thermodynamic equilibrium state, the average size of pre-existing precipitates should either increase continuously, through growth and coarsening, or go to zero, i.e., the precipitates would dissolve. In contrast, the above observations show that the stationary microstructure of these irradiated alloys is organized at the mesoscale, that is a scale that is neither macroscopic nor atomic. This self-organization, which is characteristic of dissipative systems [26–28], has intrigued the community for several decades, and we will discuss in detail in Sects. 3.2 and 3.3 the models that have been proposed to account for such nonequilibrium evolutions. Beyond the case of metallic alloys, we note that Jones [29] reported that large ThO_2 precipitates in Ni became decorated by a halo of small thoria precipitates after high-dose Ni irradiation. Very recently, Rizza et al. [30, 31] reported a similar precipitate refinement, but this time for metallic precipitates, Au and Ag, in an oxide matrix, SiO_2. It thus appears that self-organization of a second phase in irradiated alloys may be a quite general phenomena.

3.2 Models with Unidirectional Ballistic Mixing

Many models on the effect of ballistic mixing on the stability of precipitate under irradiation have relied on a "unidirectional" ballistic mixing in the sense that this mixing is restricted to the resolution of solute atoms from the precipitates to the matrix. The first of these models was introduced by Nelson et al. to rationalize their puzzling experimental observations on irradiated two-phase Ni–Al alloys [23]. They proposed that the evolution of thermodynamically stable and pre-existing precipitates should result from a competition between the forced atomic mixing produced by nuclear collisions, leading to the so-called recoil dissolution, and thermally activated chemical diffusion, which tends to restore the precipitates since they have been assumed to be thermodynamically stable. This latter dynamics is accelerated by irradiation, in proportion to the supersaturation of point defects that is reached in the irradiated material. This model contains in fact two variants, depending as to whether the forced mixing induces direct dissolution or a dissolution mediated by the disordering of chemically ordered precipitates. From a general perspective, however, these two variants lead to similar predicted behaviors, and, for simplicity, we will only summarize here the variant for direct dissolution.

The Nelson, Hudson, and Mazey (NHM) model rests on the following approximations for the thermally activated contribution. The solute concentration in the matrix c, is assumed to be small, all the precipitates are assumed to have the same radius r_{p}, the number density of precipitates N, is low enough so that each precipitate can be treated separately, and their total volume fraction f_{p}, is small. Defining p as the atomic fraction of solute atoms constituting the precipitate phase, the conservation of the total solute concentration, c_0, imposes that

$$c_0 = f_{\mathrm{p}} + (1 - f)c \approx \frac{4}{3}\pi r_{\mathrm{p}}^3 pN + c. \tag{1}$$

Note that it is also assumed that atomic volumes are the same in both phases. NHM then write the growth rate of the precipitates due to the thermally activated diffusion as

$$\frac{\mathrm{d}r_{\mathrm{p}}}{\mathrm{d}t} = \frac{3D_{\mathrm{s}}^{\mathrm{irr}}c}{pr_{\mathrm{p}}}, \tag{2}$$

where $D_{\mathrm{s}}^{\mathrm{irr}}$ is the irradiation-enhanced solute diffusion coefficient in the matrix. Equation (2) assumes that the local equilibrium solute concentration at the matrix/precipitate interface is negligible compared to c, the average matrix solute concentration.

As regards the forced atomic relocation, NHM made two critical assumptions. The first one is that a solute atom initially in the precipitate, when forced to relocate into the matrix by a nuclear collision, is redistributed instantaneously anywhere in the matrix. This is equivalent to assuming that the average relocation distance is large compared to the separation distance between two precipitates. The second critical assumption is that the forced relocation events are not transporting solute atoms from the matrix to the precipitate. This assumption is common to all the models discussed in this section, and it will be shown that it results in serious deficiency regarding the conditions required for irradiation to trigger compositional patterning. With these two assumptions, NHM write the rate of precipitate shrinkage due to the recoil events as

$$\frac{\mathrm{d}r_\mathrm{p}}{\mathrm{d}t} = -\Omega\phi, \tag{3}$$

where Ω is the atomic volume, and ϕ is the flux of solute atoms recoiled into the matrix (per unit area), and it is thus proportional to the displacement rate. The resulting evolution of the precipitate radius is obtained by combining (2), (3)

$$\frac{\mathrm{d}r_\mathrm{p}}{\mathrm{d}t} = -\Omega\phi + \frac{3D_\mathrm{s}^\mathrm{irr}c_0}{pr_\mathrm{p}} - D_\mathrm{s}^\mathrm{irr}r_\mathrm{p}^2 N. \tag{4}$$

The third term in the right-hand side of (4) is to guarantee that, in the absence of irradiation, the equilibrium precipitate size is compatible with solute conservation, (1). The main outcome of the NHM model is that, for a given radiation flux ϕ, there is a threshold precipitate radius such that precipitates smaller than this value grow, while larger ones shrink. The model thus predicts that precipitates always reach a finite size for long enough irradiations. The higher the irradiation flux or the lower the temperature, the smaller is this steady-state size.

While this result appears to offer a rationalization of the experimental results of NHM on Ni–Al, the limitations of the NHM model are such that this rationalization is not justifiable. Indeed, neither nucleation nor coarsening is included in the NHM model, and, even in the absence of irradiation, it thus predicts that precipitates will reach a finite size. Furthermore, the modeling of ballistic mixing is clearly incorrect, since the relocation distance is unphysically large, and since it ignores the transport of solute atoms from the matrix to the precipitates.

Brailsford [32] improved the NHM model by taking into account the fact that the relocation distance R is finite, but, as noted by Brailsford, this improvement did not fix the main deficiencies of the model, in particular its unphysical behavior as $R \to 0$, that is when ballistic mixing takes place between nearest-neighbor atomic sites.

The next significant model was introduced by Frost and Russell [33, 34]. Ballistic mixing is assumed to displace solute atoms initially located in a precipitate by a vector of random magnitude and direction within a sphere a radius R. This mixing is modeled by a source term G, originally derived by Gelles and Garner [35]. The steady-state solute concentration profile is obtained by solving the diffusion equation

$$\frac{\partial c(r)}{\partial t} = D_\mathrm{s}^\mathrm{irr}\nabla^2 c(r) + G(r), \tag{5}$$

with $G(r) = 0$ for $r \geq r_\mathrm{p} + R$, and boundary conditions $c = c_\mathrm{eq}$ at $r = r_\mathrm{p}$, $(\partial c/\partial r) = 0$ at $r = r_\mathrm{p} + R$. Global solute conservation is also enforced. The steady-state solute concentration profile rises from $c = c_\mathrm{eq}$ at $r = r_\mathrm{p}$ to a maximum value at $r = r_\mathrm{p} + R$, and remains constant beyond that distance. This maximum solute concentration is given by

$$c(r \geq r_\mathrm{p} + R) = c_\mathrm{eq} + \frac{\phi_\mathrm{FR} R^2}{12 D_\mathrm{s}^\mathrm{irr}} \left(1 - \frac{R}{4 r_\mathrm{p}}\right), \tag{6}$$

where ϕ_FR is the creation rate of solute recoil per atom, thus proportional to the irradiation flux. Frost and Russell then included coarsening by using the Gibbs–Thomson equation $c_\mathrm{eq} = c_\infty(1 + r_\mathrm{cap}/r_\mathrm{p})$, where c_∞ is the equilibrium interface solute concentration for a planar interface, and r_cap is the capillary length. Equation (6) then becomes

$$c(r \geq r_\mathrm{p} + R) = c_\infty \left(1 + \frac{\phi_\mathrm{FR} R^2}{12 D_s^\mathrm{irr} c_\infty}\right) + \frac{1}{r_\mathrm{p}} \left(c_\infty r_\mathrm{cap} - \frac{\phi_\mathrm{FR} R^3}{48 D_\mathrm{s}^\mathrm{irr}}\right). \tag{7}$$

Frost and Russell interpreted (7) by noting that irradiation could lead to a change of sign of the factor in front of the $(1/r_\mathrm{p})$ dependence, and thus lead to an inverse coarsening, as illustrated in Fig. 4.

The critical irradiation flux for triggering inverse coarsening is given by

$$\phi_\mathrm{FR}^\mathrm{c} = \frac{48 D_\mathrm{s}^\mathrm{irr} c_\infty r_\mathrm{cap}}{R^3}. \tag{8}$$

Note that this critical flux is independent of the precipitate radius, and scales as $1/R^3$. Frost and Russell also solved the kinetics for precipitates to reach their finite steady-state size, r_p^ss, and showed that the characteristic time for reaching this steady state scales as $(r_\mathrm{p}^\mathrm{ss})^3 \phi_\mathrm{FR}^{-1} R^{-3}$. A significant improvement over the NHM model is that precipitates would coarsen continuously in the absence of irradiation, and that there exists a critical irradiation flux, or

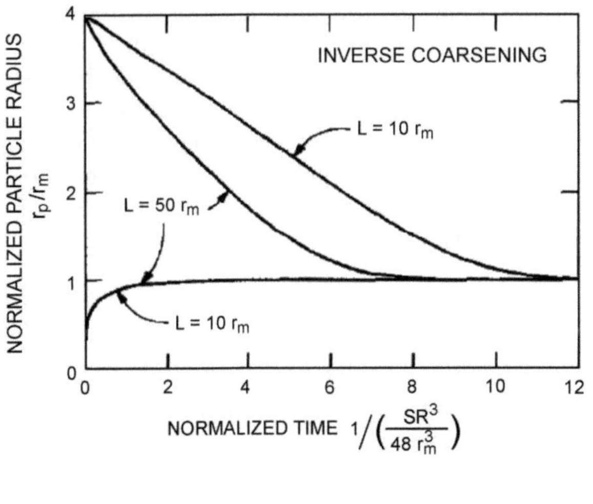

Fig. 4. Evolution of precipitate radius r_p, normalized to the steady-state size r_m, as a function of irradiation time in the case where inverse coarsening takes place in the Frost–Russell model. R is the relocation distance and S the recoil generation rate (from [34])

irradiation temperature if the flux is kept constant, for irradiation to trigger the instability that leads to compositional patterning.

Heinig and Strobel [36] later derived an expression for the solute concentration under irradiation using a better description of the source term by choosing an exponential distribution of relocation distances, in agreement with recent computer simulation results (see Sect. 2 and Fig. 2). Their expression for the solute concentration is, however, identical to (7), up to some numerical factors. Heinig and Strobel proposed to use the concept of effective interfacial energy to rationalize inverse coarsening when the irradiation flux exceeds the critical value given by (8).

3.3 Models Including Full Account of Forced Mixing

All the models reviewed so far made the key assumption that the forced mixing can only transport solute atoms from the precipitates to the matrix. The rationale behind this might have been that since the matrix is diluted in solute, one could neglect the transport of solute atoms from the matrix to the precipitates. This assumption is, however, clearly incorrect and inconsistent with the treatment of the forced mixing as a forced diffusive process, as proposed by Haff [16] and Andersen [17]. In particular, it yields an incorrect dependence of the rate of dissolution of precipitates due to this ballistic mixing. Comparing (2) and (3), one sees that the restrictive assumption made on the forced mixing leads to a rate $(\mathrm{d}r_\mathrm{p}/\mathrm{d}t)$ that is independent of r_p, whereas a diffusive process leads to a rate that is inversely proportional to r_p. This has significant consequences on the conditions required for irradiation to trigger a patterning or inverse coarsening reaction.

This point is made very clear by using the model introduced by Martin in 1984 [2]. Martin showed that, if one can neglect the medium- and long-range relocations forced by nuclear collisions, this short-range ballistic mixing is akin to a forced diffusion, and, in a crystal with cubic symmetry, Martin defined the ballistic diffusion coefficient as $D_\mathrm{b} = R^2 \Gamma_\mathrm{b}/6$. The frequency of ballistic replacements per atom, Γ_b, is proportional to the irradiation flux $\Gamma_\mathrm{b} = \sigma_\mathrm{r}\phi$, where σ_r is the replacement cross section. Martin's model is the first one that treats correctly the competition between this ballistic diffusion and thermally activated diffusion. An important result is the so-called effective temperature criterion: under irradiation at a temperature T an alloy reaches a steady state that is equivalent to the equilibrium state that it would have reached at a higher, effective temperature given by

$$T_\mathrm{eff} = T\left(1 + \frac{D_\mathrm{b}}{\tilde{D}_\mathrm{irr}}\right), \tag{9}$$

where $\tilde{D}_\mathrm{irr}$ is the interdiffusion coefficient, accelerated by irradiation. Let us apply (9) to assess the stability of precipitates under irradiation. Consider an alloy system with a positive heat of mixing, yielding a miscibility gap with a critical temperature T_c, which is reached for the composition $c = 1/2$. Equation (9) predicts that precipitates will not be stable under irradiation if the effective temperature exceeds the solvus temperature T_{solvus}. For simplicity, we approximate here that temperature by the instability temperature in a Bragg–Williams mean-field approximation, which yields $T_{\text{solvus}} = 4c(1-c)T_c$. Equation (9) predicts therefore that precipitates will be dissolved when the irradiation flux exceeds a critical value given by

$$\phi_{\text{M}}^{\text{c}} = \frac{\tilde{D}_{\text{irr}}}{\sigma_{\text{r}}} \frac{[4c(1-c)T_{\text{c}} - T]}{TR^2}. \tag{10}$$

We note that this critical flux is given by an expression that is very similar to the one obtained from Frost and Russell or Hening and Strobel, (8), especially since the capillary length displays a temperature dependence that is essentially identical to the one in the second fraction on the right-hand side of (10). The main difference, however, is that the critical flux for dissolution scales as $1/R^2$, and this result is consistent with a correct treatment of ballistic diffusion. Figure 5 shows a schematic plot of the critical fluxes for inverse coarsening, as predicted by (8) and for precipitate dissolution as predicted by (10). It is clear that, as $R \to 0$ in a continuum description, or $R \to a_{\text{nn}}$ in a discrete description – a_{nn} is the nearest-neighbor distance, inverse coarsening will not take place since the precipitates will already be dissolved at a flux lower than the one required for inverse coarsening. Inverse coarsening can take place only when the characteristic relocation distance for ballistic jumps exceeds some critical value R_c. This important result is absent from the models of Frost and Russell, and of Heinig and Strobel because they neglect the ballistic transport of solute atoms from the matrix to the precipitates. This contribution plays a key role when R is small.

While Fig. 5 clearly indicates that patterning and dissolution compete with one another, with the models presented so far it is not possible to pre-

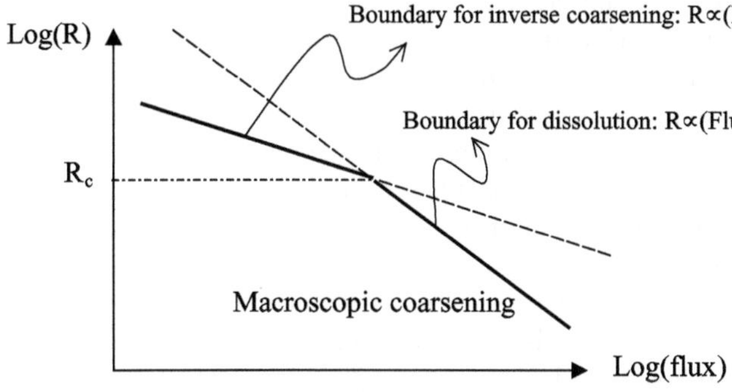

Fig. 5. Schematic plot of the dynamical boundary separating macroscopic coarsening from inverse coarsening and from precipitate dissolution. The different dependences of these boundaries with the irradiation flux lead to the existence of a threshold value R_c for the relocation distance for inverse coarsening to take place

dict the boundary between these two possible regimes. Enrique and Bellon [37, 38] introduced a model that overcomes these limitations and that makes it possible to determine the boundaries between macroscopic precipitation, compositional patterning, and dissolution of precipitates into a solid solution. This model is based on a phase-field-type description, thus belonging to the class of diffuse interface models. Considering a one-dimensional system for simplicity, the evolution of the local deviation from the nominal concentration $\psi(x) = c(x) - \bar{c}$ results again from the competition between forced mixing and thermally activated diffusion

$$\frac{\partial \psi}{\partial t} = M_{\mathrm{irr}} \nabla^2 \left(\frac{\delta \Omega F}{\delta \psi} \right) - \Gamma_{\mathrm{b}} \left(\psi - \langle \psi \rangle_{\mathrm{R}} \right), \tag{11}$$

where M_{irr} is the thermally activated atomic mobility, enhanced by irradiation. In the absence of irradiation, (11) reduces to Cahn's diffusion equation [39], and a Cahn–Hilliard expression is chosen for the free-energy functional $F\{c(x)\}$

$$F = \frac{1}{\Omega} \int \left(-A\psi^2 + B\psi^4 + C|\nabla\psi|^2 \right). \tag{12}$$

The relocation distances of ballistic jumps in (11) is given by a distribution w_{R}, with an average distance R, and $\langle \psi \rangle_{\mathrm{R}} = \int w_{\mathrm{R}}(x - x')\psi(x')\,\mathrm{d}x'$ is a local average of ψ, as sampled by w_{R}. Using a variational analysis, one can plot a map of the stable steady states predicted by (11). Figure 6 displays a cut of this map, which we refer to as a dynamical equilibrium phase diagram, for a $\mathrm{A}_{50}\mathrm{B}_{50}$ binary alloy with a positive heat of mixing at a given irradiation temperature ($T < T_{\mathrm{c}}$). This dynamical phase diagram possesses the features expected from Fig. 5: at small R values, as the reduced irradiation intensity $\gamma = \Gamma_{\mathrm{b}}/M_{\mathrm{irr}}$ increases, the alloy undergoes a transition from macroscopic phase separation to a solid solution, and the irradiation intensity at this boundary scales as R^{-2}, as expected from (10). When the characteristic relocation distance R exceeds the critical value of $R_{\mathrm{c}} = \sqrt{C/A}$, there is a range of irradiation intensities that drive the alloy into a compositional patterning steady state. In the large R limit, $R \to \infty$, the boundary between macroscopic phase separation and patterning scales as $\gamma_1 \propto \sqrt{AC}/R^{-3}$, in agreement with (8). Furthermore, this model yields the boundary (labeled γ_2 in Fig. 6) that separates compositional patterning from solid solution.

The predicted dynamical phase diagram shown in Fig. 6 has been confirmed by kinetic Monte Carlo simulations [37, 38]. In particular, when the replacement distance R is smaller than a critical distance R_{c}, no patterning is ever observed in the atomistic simulations. For a binary alloy system on an fcc lattice, with a positive heat of mixing producing a critical temperature of 1300°C, e.g., close to that of the Ag–Cu system, the simulations suggest that $1.5\,\text{Å} < R_{\mathrm{c}} < 3.0\,\text{Å}$, while the continuum model predicts $R_{\mathrm{c}} \approx 1.38\,\text{Å}$ [18]. Even though the critical value is small, it is experimentally relevant since, as reviewed in Sect. 2, recent MD simulations indicate that, while most the

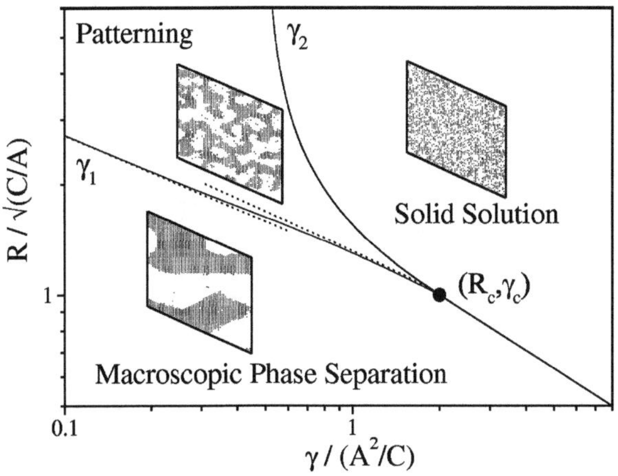

Fig. 6. Dynamical equilibrium phase diagram for an immiscible $A_{50}B_{50}$ alloy. $\gamma = \Gamma_b/M_{irr}$ is a dimensionless forcing intensity, and R is the average atomic relocation distance (see text for definition of A, C). The transition lines are calculated from the effective free energy. Insets are typical KMC steady states in a (111) plane (from [37])

ballistic mixing takes place between the first few nearest neighbors, there is a longer-range tail that is well described by a decaying exponential for ion irradiations. For light ions, e.g., He, this exponential tail is decaying quickly, with $R \approx 1.44$ Å, and KMC simulations predict that irradiation cannot trigger patterning. A similar conclusion should also apply to electron and proton irradiations. In contrast, displacement cascades with heavier ion, e.g., Ne and beyond, lead to $R \approx 3.08$ Å, a value large enough in KMC simulations to induce compositional patterning.

Enrique et al. [40] reported indirect evidence that 1-MeV Kr irradiations of Ag–Cu multilayers at temperatures ≈100°C to 200°C lead to compositional patterning. Krasnochtchekov et al. [41] carried out a systematic study, by performing 1-MeV Kr irradiations of $Cu_{1-x}Co_x$ thin films with $10\% \leq x \leq 20\%$, using SQUID magnetometry and the superparamagnetic character of the small Co clusters to determine the evolution of the precipitate size under irradiation. Three different initial states were studied: as-deposited, which is partly decomposed, preirradiated at room temperature, leading to randomization of the composition, and annealed, which produced coarse-scale decomposed microstructures. As in the KMC simulations, the same steady state was reached regardless of the initial state (see Fig. 7). Furthermore, while low-temperature irradiation produced random alloys, irradiation temperatures between RT and ≈300°C resulted in the stabilization of finite-size precipitates (see Fig. 7), the average size of which increased continuously with the irradiation temperature. For irradiation temperatures of 350°C and above, the Co precipitates appear to grow continuously with the irradiation dose, and this was interpreted as a clear indication of coarsening typical of a regime of macroscopic phase separation. Additional experiments performed on the Cu–Ag and Cu–Fe systems have led to similar conclusions [42].

While this section was focused on the effect of the forced mixing on precipitate stability, patterning reactions involving point-defect clusters, voids,

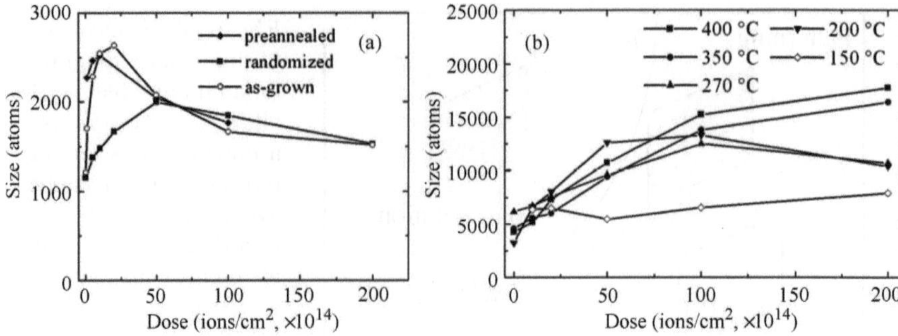

Fig. 7. Size of Co precipitates obtained from magnetic measurements (**a**) in $Cu_{90}Co_{10}$ irradiated at 200°C with 1-MeV Kr ions: the same steady state is reached regardless of the initial state; (**b**) in $Cu_{85}Co_{15}$ irradiated at various temperatures; note the finite precipitate size at steady state, i.e., patterning, for $T_{irr} \leq 270°C$, in contrast to continuous growth, indicative of macroscopic phase separation, for $T_{irr} \geq 350°C$ (from [41])

and gas bubbles have also been reported in irradiated alloys (see [43, 44] for reviews). The models developed to rationalize these patterning reactions have relied on bias in defect production in displacement cascades, bias in defect elimination on sinks, and anisotropy of defect migration. In light of the above review on compositional patterning under irradiation, however, it would be interesting to investigate whether certain length scales could also be relevant for these reactions. In particular, in the case of He bubbles in metallic matrices and of fission gas bubbles in nuclear fuel oxides, current models [45, 46] rely on a description of the resolution rate that parallels the approach used in the NHM model, and it would be prudent to include a full treatment of the forced mixing of He and fission gas atoms, whether they belong to bubbles or are in solution in the matrix.

From a fundamental perspective, it is interesting to draw a parallel between irradiation-induced patterning reactions and self-organization taking place in equilibrium systems. In particular, it has been observed that *competing interactions* with different characteristic length scales can lead to the formation of mesoscopic structures, for instance when short-range attractive chemical interactions compete with long-range repulsive electrostatic [47, 48] or elastic interactions [49]. In the case of an irradiated solid, we are, however, dealing with a dynamical system, and we propose then that a general criterion for self-organization is the competition between *dynamical processes* that operate at different length scales.

From a practical perspective, we want to stress that when one subjects an engineering material to an accelerated test of radiation resistance by using particles that create damage faster than in service conditions, a direct extrapolation of the results of such accelerated tests to service conditions can be quite misleading. In particular, this extrapolation will be wrong if the representative points in the (R, γ) control parameter space (Fig. 6) lie in different domains, since these alloys evolve toward different steady states.

4 Order–Disorder Transformations

In the case of alloys with an ordering tendency, ballistic mixing and radiation-enhanced thermally activated diffusion may also compete and produce results similar to the ones reviewed in the previous section (see chapter "Fundamental Concepts of Ion-Beam Processing" by Averback and Bellon and [3, 50] for reviews). In particular, irradiation at low temperature usually leads to the chemical disordering of pre-existing ordered phases. In fact, at these low temperatures, where thermal diffusion is so sluggish that it can be neglected, the measurement of the disordering rates provides an experimental way to determine the rate of replacements per atom per second [50, 51] (see also (10) in chapter "Fundamental Concepts of Ion-Beam Processing" by Averback and Bellon). Under electron irradiation, random recombination of Frenkel pairs can also contribute to the chemical disordering [52]. At elevated temperatures, owing to radiation-enhanced diffusion, irradiation can in fact be used to achieve high degree of chemical order over times much shorter than during conventional thermal annealing [50]. This effect, combined with thin-film technology and masking techniques, can be used to tailor the regions of the materials that are chemically ordered in functional materials [53].

For irradiation conditions leading to the formation of displacement cascades, the chemical disorder resulting from these displacement cascades can be imaged by dark-field transmission electron microscopy [54], and this in turn provides an experimental evaluation of the size L of these displacement cascades. L ranges typically from about 1 nm to 10 nm. Similarly to the case of compositional patterning driven by the finite relocation distances, which requires $R > R_c$, atomistic simulations suggest that when the cascade size exceeds a critical value L_c, irradiation may trigger a patterning of the chemical order [55, 56]. This patterning takes place when the reordering of large disordered zones lead to the renucleation of ordered domains, which are not necessarily in phase with the ordered matrix. Figure 8 gives an example of the resulting dynamical phase diagram for an $L1_0$ compound. The similarities with the dynamical phase diagram shown in Fig. 6 are striking. This phenomenon of irradiation-induced patterning of order awaits experimental validation.

Irradiation may also induce the amorphization of crystalline compounds, or the crystallization of amorphous phases but, due to lack of space, the reader is referred to recent reviews [3, 57] on this topic.

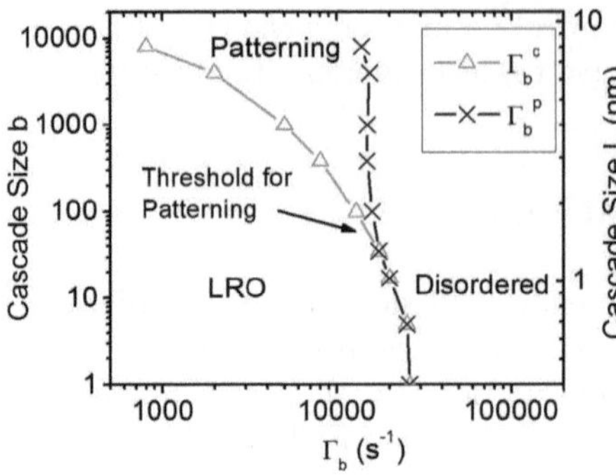

Fig. 8. KMC dynamical phase diagram for an $A_{50}B_{50}$ alloy with $L1_0$ ordering tendency irradiated with heavy ions at $T \approx 1000\,\mathrm{K}$. The threshold for patterning of order is for a cascade size $L \approx 2\,\mathrm{nm}$ using FePt data (from [56])

5 Radiation-Induced Segregation and Precipitation

The continuous production of point defects homogeneously in an irradiated material, combined with the preferential elimination or recombination at localized extended defects or regions lead to the build-up of permanent nonzero net fluxes of point defects. These fluxes may couple preferentially with chemical species, thus setting nonzero chemical fluxes, a process referred to as the inverse Kirkendall effect when induced by vacancy fluxes. A first result of these nonzero chemical fluxes is to produce radiation-induced segregation (RIS) [58–65], an effect that is of particular technological relevance in stainless steels since it is experimentally observed that RIS often leads to Cr depletion at grain boundaries, and it is thus suspected of contributing to stress corrosion cracking (SCC) in these irradiated materials. The amount of segregation can be so large that precipitates form. Heterogeneous radiation-induced precipitation (RIP) was first observed in dilute Ni–Si alloys [66, 67]; it resulted from the preferential transport of Ni atoms by vacancies, thus increasing the Si concentration at sinks until Ni_3Si precipitates formed. When the dominant sink is a free surface, this RIP can lead to the formation of a thick Ni_3Si layer [68]. In contrast, in dilute Ni–Al alloys, the sinks become depleted in Al, and irradiation can induce the precipitation of Ni_3Al phase between the sinks [69, 70]. Radiation-induced homogeneous precipitation has also been observed experimentally and explained by the irreversible effect of vacancy–interstitial recombination on solute transport [71–73].

A particular challenge in the modeling of RIS and RIP is the determination of the coupling between the point defects and the chemical species. In the case of infinitely dilute alloys, there is a small set of independent defect jump frequencies, and the coefficients describing the kinetic coupling between the various species, the so-called Onsager coefficients, can be expressed directly in terms of these few frequencies [74–77]. In the case of concentrated alloys, however, this approach is no longer possible. Manning introduced a two-frequency model that made it possible to obtain approximate analytical expressions for these coefficients [78, 79]. These expressions, which have been a key ingredient in RIS models [59–61, 63, 65], however, could not reproduce important effects, such as negative off-diagonal Onsager coefficients and percolation effects. A recent breakthrough has been made by Barbe and Nastar [80, 81], who solved a Master Equation with an effective Hamiltonian under an imposed chemical potential gradient. Using various mean-field approximations, Barbe and Nastar have derived transport coefficients for vacancy and interstitial diffusion that are in remarkable agreement with KMC simulations [82, 83]. It remains to integrate these expressions with RIS models.

For simplicity, solute–defect coupling was neglected in our discussion of the stability of pre-existing precipitates in Sects. 3.2 and 3.3, but this coupling should of course be included as it can contribute significantly to the evolution of the precipitates (see [84] for a review). Recent atomistic modeling work by Krasnochtchekov et al. [85] indicates that, when interstitials couple preferentially with solute atoms, and solute clusters are effective trap for interstitials, irradiation may induce the formation of "mushy" precipitates. This coupling effect can in fact be strong enough to induce precipitation in alloys that form ideal solid solutions at equilibrium.

6 Defect Clustering and Related Microstructural Evolutions

While point defects produced by irradiation annihilate by recombination and by elimination on permanent sinks, they also form defect clusters, either directly in displacement cascades [5] or by nucleation due to point-defect supersaturation. These clusters then act as sinks, and thus influence the build-up and the fluxes of point defects in irradiated materials. Simple rate theory models [8, 9] distinguish a steady state dominated by recombination at low temperature and/or high irradiation flux, and a steady state dominated by the elimination on sinks at elevated temperatures and/or low irradiation flux. From a practical perspective, it would be desirable that service conditions maximize recombination, thus suppressing the long-range transport of point defects and chemical species and its potentially deleterious consequences. For instance, a preferential elimination of interstitials and interstitial clusters on sinks results in an excess of vacancies, which may form voids, or bubbles, leading to the swelling of the irradiated material.

An important factor in the study of point-defect evolution during irradiation is that there may exist an asymmetry, often referred to as bias, in the kinetics of interstitials and vacancies. There are several effects that can induce such a bias. First, there may be an elimination bias, owing to the typically larger relaxation volume of interstitials compared to vacancies, thus leading to larger elastic interaction between interstitials and the hydrostatic component of the stress field of dislocations [86]. Diffusion anisotropy in noncubic materials, and applied stresses, provide other sources for this elimination bias, potentially leading to macroscopic shape changes, which are referred to as growth and creep, respectively. The dimensional changes brought about by swelling, creep, and growth may limit the lifetime of materials in service and lead to premature failure, and therefore a large amount of research has been devoted to the understanding of these phenomena, and to the design of materials that are resistant to these effects. Due to lack of space, these topics are not developed here but the interested reader can find several review articles covering these topics in [87–91]. Another source of point-defect bias comes directly from their production. Molecular-dynamics simulations reveal clearly this so-called production bias [92, 93]. A larger fraction of interstitials form stable clusters at the end of the lifetime of a displacement cascades, leaving a net excess of free vacancies in the matrix. The production bias can thus contribute significantly to microstructural evolutions, such as swelling [94].

An asymmetry has also been reported in the transport mechanism of small defect clusters. While small vacancy clusters migrate in three dimensions through a mechanism analogous to Brownian motion, small interstitial clusters, which often take planar shapes to minimize strain energy, are observed to glide unidimensionally at short timescales, along directions constrained by the Burgers vector and the plane of the clusters. This glide is almost athermal since MD results indicate effective activation energies of the order of 0.05 to 0.2 eV (thermal activation in rate theory requires activation barriers 3 to 5 times $k_{\mathrm{B}}T$). At longer times, clusters may reorientate through Burgers-vector changes, thus allowing the defect clusters to migrate in three dimensions. It is been proposed that this particular 1D–3D migration is the origin of the self-organization of defect clusters [95].

In addition to dimensional and shape changes, point-defect clustering is also at the origin of important changes in the mechanical properties of irradiated alloys. The interaction between defect clusters with dislocations is a complex function of the cluster nature, size, geometry, and distance to a given dislocation. While MD simulations provide a convenient way to investigate systematically these complex interactions [96], more work is still needed to uncover these effects, to test the resulting predictions experimentally, and finally to include this new information into continuum models. Overall, it is observed experimentally that these interactions lead to hardening and embrittlement of irradiated materials. In addition, it was recognized quite early by Wechsler that plastic deformation may not be homogeneous in irradiated solids [97]. It has been observed that irradiation can lead to the formation of dislocation-free channels, and to the localization of plastic flow in these channels (see Fig. 9). This localization is at the origin of the softening that is sometimes observed at larger doses. The modeling of the mechanical response of alloys subjected to irradiation is a complex task since it involves a large spectrum of effects and interactions, which cover timescales ranging from the picosecond to the hours or years, and length scales ranging from the atomic scale to tens or hundreds of micrometers. This task may benefit from the use of multiscale modeling, as illustrated recently in [98].

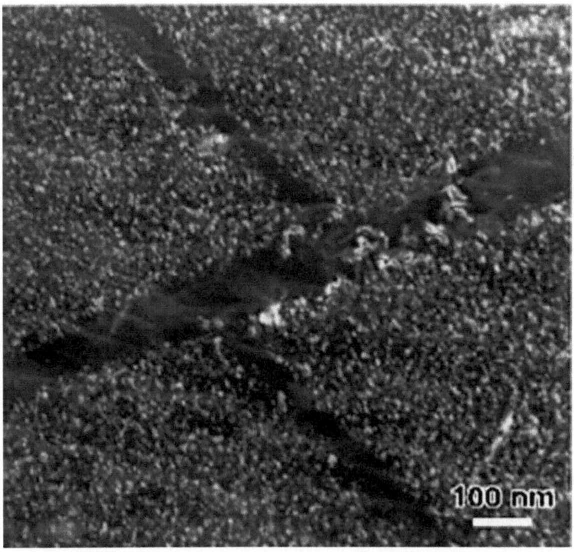

Fig. 9. Defect-free channels observed in Pd irradiated with protons (from [99])

7 Conclusion

The continuous production of damage in alloys under irradiation drives these material systems into nonequilibrium states. While the physics of damage creation, through point defects, point-defect clusters, and ballistic mixing, is rather well understood, predicting the resulting macroscopic and long-term evolution of irradiated alloys remains a challenge owing to the dynamical competition between various elementary processes, which cover a wide range of length and time-scales. We have, nevertheless, illustrated in this chapter that, as far as microstructural evolutions are concerned, significant advances have been made through the use of physics-based kinetic models that integrate the relevant characteristics of the dynamical processes. Besides the rate of these processes, e.g., the mixing rate for ballistic effects, we have shown that it is also necessary to take into account the characteristic length scales of these processes. In particular, the scale-dependent competition between ballistic mixing and thermally activated reordering can lead to the self-organization of the composition and of the degree of chemical order. Microstructural evolutions induced by irradiation can in turn lead to dimensional or shape changes, as well as to modifications of mechanical properties.

Acknowledgements

Stimulating discussions with G. Martin and R.S. Averback are gratefully acknowledged. The research was supported by the National Science Foundation, under grant DMR 04-07958, and the U.S. Department of Energy U.S.DOE Basic Energy Sciences, under Grant No. DEFG02-05ER46217.

References

1. Y. Adda, M. Beyeler, G. Brebec, Thin Solid Films **25**, 107 (1975) 29
2. G. Martin, Phys. Rev. B **30**, 1424 (1984) 30, 32, 38
3. G. Martin, P. Bellon, Solid State Phys. **50**, 189 (1997) 30, 32, 43
4. P. Bellon, G. Martin, in *Alloy Physics*, ed. by W. Pfeiler (Wiley VCH, Weinheim, 2007), p. 423 30
5. R.S. Averback, T.D. de la Rubia, Solid State Phys. **51**, 281 (1998) 30, 31, 32, 45
6. J. Ye, Y. Li, R.S. Averback, P. Bellon, to be published 31
7. M.J. Norgett, M.T. Robinson, I.M. Torrens, Nucl. Eng. Des. **33**, 50 (1975) 31
8. R. Sizmann, J. Nucl. Mater. **69&70**, 386 (1978) 32, 45
9. H. Wiedersich, J. Nucl. Mater. **205**, 40 (1993) 32, 45
10. P. Bellon, R.S. Averback, Scr. Mater. **49**, 921 (2003) 32
11. R.S. Averback, D. Peak, L.J. Thompson, Appl. Phys. A **39**, 59 (1986) 32
12. R.S. Averback, Nucl. Instrum. Methods B **15**, 675 (1986) 32
13. S.J. Kim, M.A. Nicolet, R.S. Averback, D. Peak, Phys. Rev. B **37**, 38 (1988) 32
14. K. Nordlund, M. Ghaly, R.S. Averback, J. Appl. Phys. **83**, 1238 (1998) 32
15. J.A. Brinkman, Am. J. Phys. **24**, 246 (1956) 32
16. P.K. Haff, Z.E. Switkowski, J. Appl. Phys. **48**, 3383 (1977) 32, 38
17. H.H. Andersen, Appl. Phys. **18**, 131 (1979) 32, 38
18. R. Enrique, K. Nordlund, R.S. Averback, P. Bellon, J. Appl. Phys. **93**, 2917 (2003) 33, 40
19. R.L. Klueh, J.P. Shingledecker, R.W. Swindeman, D.T. Hoelzer, J. Nucl. Mater. **341**, 103–114 (2005) 33
20. M.S. El-Genk, J.-M. Tournier, J. Nucl. Mater. **340**, 93–112 (2005) 33
21. A. Boltax, in *Symposium on Radiation Effects on Materials.* ASTM, vol. 208 (1957), p. 183 33
22. R.M. Berman, J. Nucl. Mater. **17**, 313 (1965) 33
23. R.S. Nelson, J.A. Hudson, D.J. Mazey, J. Nucl. Mater. **44**, 318 (1972) 33, 34, 35
24. F. Bourdeau, E. Camus, Ch. Abromeit, H. Wollenberger, Phys. Rev. B **50**, 16205 (1994) 33
25. E. Camus, Ch. Abromeit, F. Bourdeau, N. Wanderka, H. Wollenberger, Phys. Rev. B **54**, 3142 (1996) 33
26. G. Nicolis, I. Progogine, *Self-organization in Nonequilibrium Systems: From Dissipative Structures to Order Through Fluctuations* (Wiley, New York, 1977) 34
27. H. Haken, *Advanced Synergetics* (Springer, Berlin, 1983) 34
28. M.C. Cross, P.C. Hohenberg, Rev. Mod. Phys. **65**, 851 (1993) 34
29. R.H. Jones, J. Nucl. Mater. **74**, 163 (1978) 34

30. G.C. Rizza, M. Strobel, K.H. Heinig, H. Bernas, Nucl. Instrum. Methods B **178**, 78 (2001) 34
31. G.C. Rizza, H. Cheverry, T. Gacoin, A. Lamasson, S. Henry, J. Appl. Phys. **101**, 014321 (2007) 34
32. A.D. Brailsford, J. Nucl. Mater. **91**, 221 (1980) 36
33. H.J. Frost, K.C. Russell, J. Nucl. Mater. **103–104**, 1427 (1981) 36
34. H.J. Frost, K.C. Russell, Acta Metall. **30**, 953 (1982) 36, 37
35. D.S. Gelles, F.A. Garner, J. Nucl. Mater. **85–86**, 689 (1979) 36
36. K.H. Heinig, T. Müller, B. Schmidt, M. Strobel, W. Möller, Appl. Phys. A **77**, 17 (2003) 38
37. R.A. Enrique, P. Bellon, Phys. Rev. Lett. **84**, 2885 (2000) 40, 41
38. R.A. Enrique, P. Bellon, Phys. Rev. B **63**, 134111 (2001) 40
39. J.W. Cahn, Acta Metall. **9**, 795 (1961) 40
40. R. Enrique, P. Bellon, Appl. Phys. Lett. **78**, 4178 (2001) 41
41. P. Krasnochtchekov, R.S. Averback, P. Bellon, Phys. Rev. B **72**, 174102 (2005) 41, 42
42. S.W. Chee, Ph.D. thesis, University of Illinois at Urbana-Champaign, 2008 41
43. W. Jäger, H. Trinkaus, J. Nucl. Mater. **205**, 394 (1993) 42
44. N.M. Ghoniem, D. Walgraef, S.J. Zinkle, J. Comput.-Aided Mater. Des. **8**, 1 (2002) 42
45. H. Trinkaus, J. Nucl. Mater. **318**, 234 (2003) 42
46. D.R. Olander, D. Wongsawaeng, J. Nucl. Mater. **354**, 94 (2006) 42
47. M. Seul, V.S. Chen, Phys. Rev. Lett. **70**, 1658 (1993) 42
48. S.L. Keller, H.M. McConnell, Phys. Rev. Lett. **82**, 1602 (1999) 42
49. T. Miyazaki, M. Doi, T. Kozakai, Solid State Phenom. **3–4**, 227 (1988) 42
50. E.M. Schulson, J. Nucl. Mater. **83**, 239 (1979) 43
51. M.A. Kirk, T.H. Blewitt, T.L. Scott, Phys. Rev. B **15**, 2914 (1977) 43
52. T. Mukai, C. Kinoshita, S. Kitajima, Philos. Mag. A **47**, 255 (1983) 43
53. J.-Ph. Attané, K.-H. Heinig, D. Halley, D. Ravelosona, A. Marty, P. Auric, C. Chappert, Y. Samson, Phys. Rev. Lett. **91**, 077203 (2003) 43
54. M.L. Jenkins, C.A. English, J. Nucl. Mater. **108–109**, 46 (1982) 43
55. J. Ye, P. Bellon, Phys. Rev. B **70**, 094104 (2004) 43
56. J. Ye, P. Bellon, Phys. Rev. B **73**, 224121 (2006) 43, 44
57. P.R. Okamoto, N.Q. Lam, L.E. Rehn, Solid State Phys. **52**, 1 (1999) 43
58. T.R. Anthony, in *Radiation-induced Voids in Metals*, ed. by C.W. Corbett, L.C. Ianniello, US Atomic Energy Commission, Conf. 710601 (1971), p. 630 44
59. P.R. Okamoto, H. Wiedersich, J. Nucl. Mater. **53**, 336 (1974) 44
60. R.A. Johnson, N.Q. Lam, Phys. Rev. B **13**, 4364 (1976) 44
61. W. Wagner, L.E. Rehn, H. Wiedersich, V. Naundorf, Phys. Rev. B **28**, 6780 (1983) 44
62. G. Martin, R. Cauvin, A. Barbu, in *Phase Transformations During Irradiation*, ed. by F.V. Nolfi (Applied Science, London, 1983), p. 47 44

63. G.S. Was, Prog. Mater. Sci. **32**, 211 (1990) 44
64. Y. Grandjean, P. Bellon, G. Martin, Phys. Rev. B **50**, 4228 (1994) 44
65. T.R. Allen, J.T. Busby, G.S. Was, E.A. Kenink, J. Nucl. Mater. **255**, 44 (1998) 44
66. A. Barbu, A.J. Ardell, Scr. Metall. **9**, 1233 (1975) 44
67. A. Barbu, G. Martin, Scr. Metall. **11**, 771 (1977) 44
68. R.S. Averback, L.E. Rehn, W. Wagner, H. Wiedersich, P.R. Okamoto, Phys. Rev. B **28**, 3100 (1983) 44
69. D.I. Potter, H.A. Hoff, Acta Metall. **24**, 1155 (1976) 44
70. D.I. Potter, D.G. Ryding, J. Nucl. Mater. **71**, 14 (1977) 44
71. R. Cauvin, G. Martin, Phys. Rev. B **23**, 3322 (1981) 44
72. R. Cauvin, G. Martin, Phys. Rev. B **23**, 3333 (1981) 44
73. R. Cauvin, G. Martin, Phys. Rev. B **25**, 3385 (1982) 44
74. A.B. Lidiard, Philos. Mag. **46**, 1218 (1955) 44
75. A.B. Lidiard, Acta Metall. **34**, 1487 (1986) 44
76. A. Barbu, Acta Metall. **28**, 499 (1980) 44
77. A.R. Allnatt, A.B. Lidiard, *Atomic Transport in Solids* (Cambridge University Press, Cambridge, 1993), pp. 380–537 44
78. J.R. Manning, Phys. Rev. **124**, 470 (1961) 44
79. J.R. Manning, Phys. Rev. B **4**, 1111 (1971) 44
80. V. Barbe, M. Nastar, Philos. Mag. **86**, 1513 (2006) 45
81. V. Barbe, M. Nastar, Philos. Mag. **86**, 3503 (2006) 45
82. M. Nastar, V. Barbe, Faraday Discuss. **134**, 331 (2007) 45
83. V. Barbe, Ph.D. thesis, Université de Paris-Orsay, 2006 45
84. A.D. Marwick, Nucl. Instrum. Methods **182–183**, 827 (1981) 45
85. P. Krasnochtchekov, R.S. Averback, P. Bellon, Phys. Rev. B **75**, 144107 (2007) 45
86. P.T. Heald, M.V. Speight, Philos. Mag. **29**, 1075 (1974) 45
87. *Proceedings of the International Workshop on Mechanisms of Irradiation Creep and Growth*, Hecla Island, Manitoba, Canada, June 22–25, 1987, published in J. Nucl. Mater. **159** (1988) 46
88. *Proceedings of the International Workshop on Defect Production, Accumulation and Materials Performance in an Irradiation Environment*, Davos, Switzerland, 2–8 October, 1996, published in J. Nucl. Mater. **251** (1997) 46
89. *Proceedings of the Symposium on Microstructural Processes in Irradiated Materials*, 2005 Annual TMS meeting, San Francisco, CA, USA, 14–17 February 2005, published in J. Nucl. Mater. **351** (2006) 46
90. *Proceedings of the Symposium on Radiation Effects, Deformation and Phase Transformations in Metals and Ceramics*, TMS Annual Meeting, San Antonio, Texas, USA, 12–16 March 2006, published in J. Nucl. Mater. **362** (2007) 46
91. *Proceedings of the E-MRS 2006 Spring Meeting: Symposium N on Nuclear Materials and Materials for Fusion*, Nice, France, 29 May–2 June 2006, published in J. Nucl. Mater. **362** (2007) 46

92. C.H. Woo, B.N. Singh, Phys. Status Solidi B **159**, 609 (1990) 46
93. C.H. Woo, B.N. Singh, Philos. Mag. A **65**, 889 (1992) 46
94. H. Trinkaus, B.N. Singh, S.I. Golubov, J. Nucl. Mater. **283–287**, 89 (2000) 46
95. H.L. Heinisch, B.N. Singh, Philos. Mag. **83**, 3661 (2003) 46
96. D.J. Bacon, Y.N. Osetsky, JOM, April 2007, p. 40 46
97. M.S. Wechsler, in *The Inhomogeneity of Plastic Deformation* (Amer. Society for Metals, Metals Park, 1972), p. 19 46
98. T. Diaz de la Rubia, H.M. Zbib, T.A. Khraishi, B.D. Wirth, M. Victoria, M.J. Caturla, Nature **406**, 871 (2000) 46
99. M. Victoria, N. Baluc, C. Bailat, Y. Dai, M.I. Luppo, R. Schäublin, B.N. Singh, J. Nucl. Mater. **276**, 114 (2000) 47

Index

Spontaneous Patterning of Surfaces by Low-Energy Ion Beams

Eric Chason[1] and Wai Lun Chan[2]

[1] Brown University, Division of Engineering, Providence, RI 02912, USA, e-mail: eric_chason@brown.edu
[2] University of Illinois at Urbana Champaign, Department of Materials Science and Engineering, Urbana, IL 61801, USA

Abstract. Pattern formation by low-energy ion beams results from a balance among different kinetic processes on the surface. Some increase the roughness of the surface (e.g., sputtering) while others tend to smoothen the surface (e.g., diffusion) and the interaction between them leads to the development of a characteristic periodicity on the surface. In this chapter, we describe the different physical mechanisms that contribute to sputter ripple formation and their dependence on the processing parameters of flux and temperature. This is used to develop a linear instability model that can be applied to understanding the different features of patterning that are found under different processing conditions.

1 Introduction

The phenomenon of pattern formation by ion beams (also known as sputter ripples) is fairly easily described – a collimated low-energy ion beam is used to bombard a surface and the initially flat morphology spontaneously develops a well-ordered periodicity over a large area. An example of a sputter ripple formed on an amorphous SiO_2 surface is shown in the atomic force micrograph (AFM) shown in Fig. 1. In this case, a 1-keV Xe ion beam was used to irradiate the surface at an angle of 54° from the normal direction in the direction shown by the arrow. After sputtering, the surface developed a one-dimensional sinusoidal periodicity with a wavelength of approximately 30 nm. The height of the ripples was approximately one tenth of the wavelength, which is typical of the aspect ratios often found. In this material, the direction of the pattern on the surface is determined by the direction of the ion-beam, i.e., if the ion beam direction is changed by rotating azimuthally around the surface normal, the pattern rotates as well.

The patterning does not require any masking or rastering of the beam, so its origin must reside in the physics of the interaction between the ion and surface. Therefore, understanding ripple formation can provide tremendous insight into the fundamental kinetic mechanisms operating on a surface during the highly nonequilibrium conditions of sputter bombardment. In addition, because the pattern typically has small dimensions, it represents a potential method for inexpensive self-organized patterning of nanoscale features on surfaces. However, although they were first observed over 45 years

H. Bernas (Ed.): Materials Science with Ion Beams,
Topics Appl. Physics **116**, 53–71 (2010)
DOI: 10.1007/978-3-540-88789-8_3,

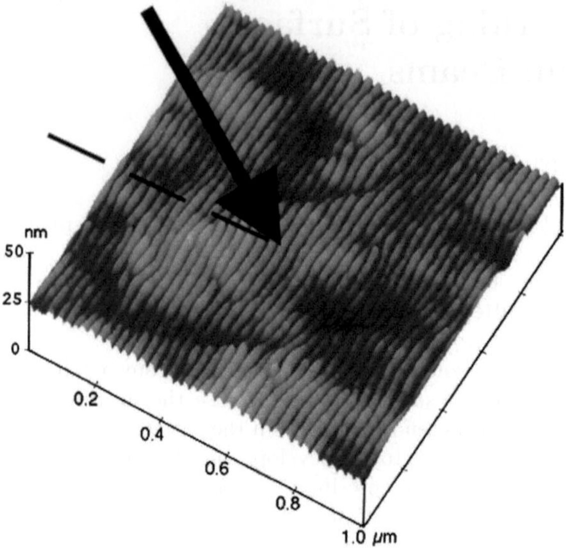

Fig. 1. AFM image of SiO_2 surface after sputtering with 1-keV Xe ions. The ion beam is incident at an angle of 54° relative to the surface normal in the direction indicated by the *arrow* on the figure. The *dashed line* is the projection of the arrow onto the surface. Figure reprinted from [1]

ago [2], understanding the rich variety of morphologies that develop under different conditions and the physical mechanisms controlling them still represents a significant challenge.

In a general sense, this type of ripple formation can be understood as the result of a dynamic balance among different kinetic processes induced by the ion bombardment. Some of the processes, such as sputter removal of atoms, lead to roughening of the surface. Others, such as surface diffusion of point defects, operate to make the surface smoother. The competition among these simultaneous processes leads to the selection of a preferred length scale on the surface which appears as a characteristic ripple pattern. Shifting the balance among the processes (e.g., by changing the flux, temperature or other processing condition) can be used to induce a transition from one type of behavior to another.

The goal of this chapter is to describe the different physical mechanisms that are at the heart of sputter rippling and to relate them to different types of patterning. We start by describing a range of patterning phenomena that have been observed and the processing regimes in which they are found. In the following section, we describe a number of kinetic mechanisms active during sputtering and use them to develop a linear instability model that accounts for many features of ripple format. We use this model to tie together various experimental observations under different processing conditions. We end by summarizing where our understanding of ripple behavior is most lacking and avenues for future developments. For further detail than can be provided in this chapter, the reader is guided to several valuable review articles [3–6].

2 Varieties of Ion-Induced Pattern Formation

Over the years, ion-induced pattern formation has been observed in many classes of material (metals, semiconductors, oxides) with different types of surface structures (crystalline, amorphous, polycrystalline). The morphology can vary from 1-dimensional waves to 2-dimensional fields of pits and mounds to quantum-dot-like individual islands. Under some conditions, no roughening is observed or no coherent pattern emerges. The direction of the pattern can be determined by the orientation of the ion beam, while in others it is determined by the surface and not the beam. In addition, different types of behavior can be observed in the same system under different processing conditions.

Understanding this wide variety of patterning can present a daunting challenge to a reader who is not familiar with the field. However, much of the behavior that has been observed can be organized into several classes of pattern formation. In the following section, we describe the characteristic features of these classes and the materials systems and conditions under which they are typically found. Note that these categories are described in order to organize a wide variety of behaviors and may not strictly account for all the observed behaviors that have been seen. They are presented to reflect our current understanding and illustrate the types of behavior that can occur in a coherent manner. A linear instability model that enables us to associate the pattern formation with corresponding kinetic processes is described in the following section.

Bradley–Harper Ripples (Ion-Induced Orientation)

This form of patterning behavior is named in reference to an instability model that was developed by Bradley and Harper (BH) [7] to explain their behavior. Ripples of this type occur when the beam is oriented in a direction off normal to the surface and have several characteristic features. They are typically 1-dimensional with a surface wavevector that is oriented either parallel or perpendicular to the direction of the ion beam projected onto the surface. The orientation of the pattern can change with the incident angle of the beam relative to the surface normal. At angles near to normal incidence, the surface wavevector is parallel to the ion direction projected on the surface, while at higher incidence angles (nearer to grazing) the wavevector rotates to be perpendicular to the ion direction. For the parallel direction, the ripples have been observed to travel along the surface [8] while they are believed to be stationary for the perpendicular direction. In the early stages of ripple formation, the amplitude is observed to grow exponentially as a function of sputtering time. This rapid growth is observed only in the early stages of sputtering and is found to saturate as the sputtering is continued. Over the range of sputtering during which the amplitude grows rapidly, the ripple wavelength is typically observed to be roughly constant. An example of BH

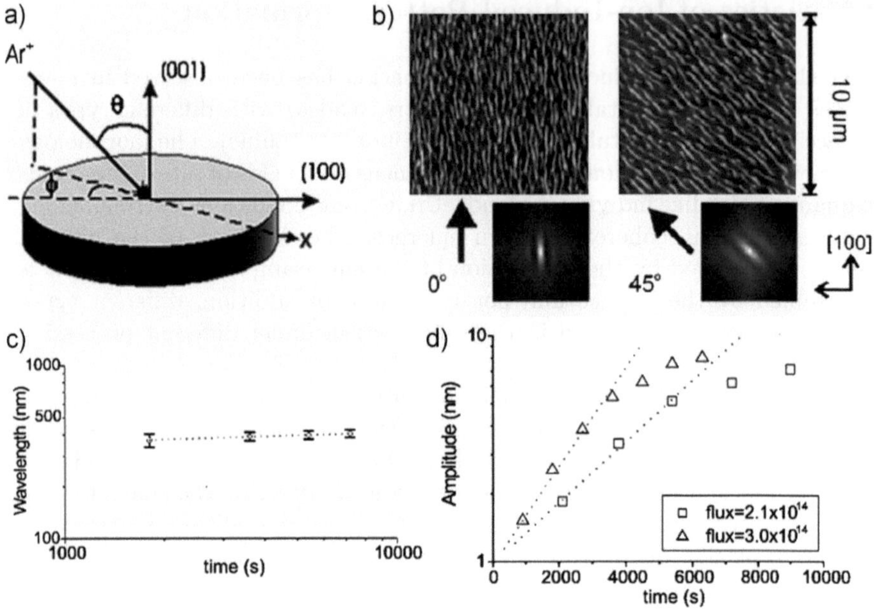

Fig. 2. Features of BH ripples formed on Cu(001) surfaces. (**a**) Schematic of ion-beam orientation relative to pattern. (**b**) AFM of surface after sputtering with 800-eV Ar ions. Alignment of the pattern with the ion orientation is shown by the arrow indicating the ion direction in each figure. *Inset* is the autocorrelation function of the surface. Evolution of (**c**) ripple wavelength and (**d**) ripple amplitude with sputtering time. Figure reprinted from [9]

ripples induced on Cu(001) surfaces, together with the evolution of wavelength and amplitude in time, is shown in Fig. 2 [9].

This type of behavior was first observed on SiO_2 surfaces but was then seen on many amorphous oxide and semiconductor surfaces (e.g., Si [10–12], Ge [13], C [14], GaAs [15]). It has also been observed more recently on metal surfaces as well [9]. Arrays of well-ordered ripples with long-range order have been produced by careful control of the ion source [16, 17].

The BH instability theory (described below) relates the ripple-formation kinetics in this regime to different fundamental kinetic processes occurring on the surface, such as sputtering and surface diffusion. This permits us to calculate the flux and temperature dependence of characteristic features such as the ripple wavelength and ripple growth rate and directly compare the prediction of the continuum theory with experiments and simulations [18]. Because of different surface smoothing mechanisms occurring on the surface, the dependence of the patterning process on the processing conditions can be complex. For instance, the temperature and flux dependence of the ripple wavelength for ripples produced at high temperature can be very different from those produced at low temperature [19].

Ehrlich–Schwoebel Patterns (Diffusion-Controlled Orientation)

Sputtering of many metal surfaces results in a pattern in which the orientation of the pattern is aligned with the crystallographic axes on the surface rather than the direction of the ion beam. When the sample is rotated azimuthally around the surface normal, while keeping the ion beam fixed, the pattern rotates with the sample [20]. This indicates that the pattern orientation is not determined by the ion beam directly and that surface kinetics play an important role. Similar pattern formation has been observed during deposition on various surfaces and has been attributed to barriers to diffusion between different layers on the surface (known as Ehrlich–Schwoebel (ES) barriers [21, 22]).

Unlike the BH ripples, this type of pattern can occur for ion incidence angles normal to the surface. The pattern can consist of mounds or vacancy islands (pits) on the surface, corresponding to the agglomeration of different types of defects on the surface produced by the ion beam (i.e., ion bombardment produces both adatom and vacancy-type defects on the surface). The resulting pattern has symmetry related to the surface so that hexagonal or triangular patterns develop on (111) surfaces [23, 24], square arrays on (100) surfaces [25–27] and 1-d ripples on (110) surfaces [20]. For ES ripples, the wavelength of the pattern is not constant in time and often grows with a power-law dependence [28, 29]. The amplitude of the roughness can also exhibit a power-law behavior.

Because it is diffusion controlled, the magnitude of the surface roughness in this regime can be strongly temperature dependent. This can be seen in the degree of roughness that develops on a Ag(001) surface for the same amount of ion fluence at different temperatures (shown in Fig. 3) [25]. The surface roughening due to the ES mechanism has a maximum contribution over a limited temperature range. At higher and lower temperatures away

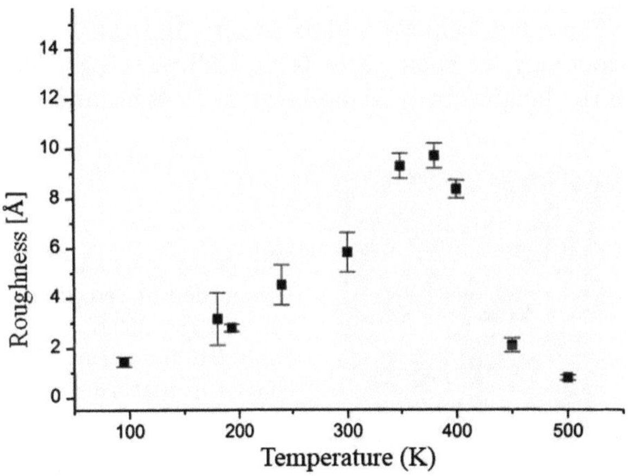

Fig. 3. RMS roughness vs. temperature of a Ag(001) surface bombarded with 1-keV Ar ions. For the same ion fluence, the roughness has a maximum as a function of temperature. Figure reprinted from [25]

from this maximum, the ES mechanism no longer dominates and there may be a transition to other types of patterning behavior such as BH-type ripples.

Low-Temperature or Athermal BH Behavior

At low temperatures, metals have been observed to change from the ES behavior described above to a behavior more like BH [6, 30]. For example, sputtering of a Ag(001) surface at 350 K produces a square pattern aligned with the surface axes while the pattern becomes a 1-dimensional ripple aligned with the ion-beam direction when the sputtering temperature is reduced to 180 K [6] (see Fig. 4). In this low-temperature regime, the wavelength of the ripple is generally only weakly dependent on the temperature (as compared to the high-temperature BH behavior). As discussed below, this type of behavior is attributed to a smoothing mechanism related to the ion bombardment (proposed by Makeev and Barabasi [31]) that does not depend on thermally activated processes. It is important to note that not all surfaces exhibit BH-type behavior at low temperatures; instead they roughen without developing a well-defined periodicity.

Nonroughening Behavior

Under some conditions, the surface is observed not to roughen even after prolonged amounts of sputtering [23, 32, 33, 30]. This type of behavior is generally observed for processing conditions that promote surface-smoothing behavior, for instance when the temperature is high (so that surface diffusion is rapid) or the ion flux is low (so that ion-induced roughening is slow).

Other Types of Patterning (Quantum Dots, Kinetic Roughening)

On some semiconductor surfaces (GaSb [15], InP [34], Si [35]), the formation of nanoscale islands or quantum-dot (QD) structures due to ion bombardment have been observed. The QDs typically form at normal or near-normal incidence of the ion beam. These can achieve a high density and take on a hexagonal close-packed morphology. In some cases (e.g., GaSb), the aspect ratio can be quite high, with the height of the island being at least as large as

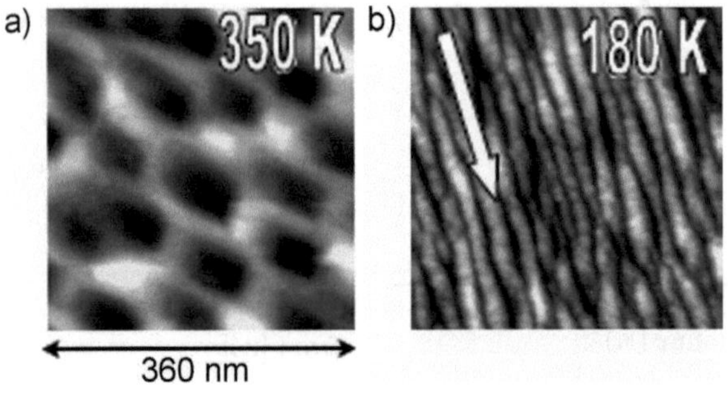

Fig. 4. Ag(001) surfaces bombarded at temperatures of (**a**) 350 K and (**b**) 180 K indicating the change in pattern morphology with temperature. Figure reprinted from [6]

the base. In other cases (Si), the islands have a much smaller aspect ratio. At the other extreme, in many systems the surface never develops a well-defined periodicity [36, 37]. In this case, the roughness often follows a power-law behavior and we refer to it as kinetic roughening. Even in systems that do show patterning behavior, after prolonged periods of sputtering the surface pattern may be replaced by this more random form of roughening.

Kinetic Phase Diagram for Cu(001)

Depending on the processing conditions, different types of patterning can be observed on the same surface. Many different processing parameters (flux, temperature, ion energy, incidence angle, etc.) have been shown to affect the pattern forming behavior. One way to visualize this balance among parameters is through the use of a kinetic phase diagram. As shown in Fig. 5, we delineate the regimes of different behavior observed on a Cu(001) surface for different values of ion flux and temperature [3, 39]. The different symbols correspond to studies performed by different groups as explained in the figure caption. Note that the diagram should not be interpreted strictly quantitatively since not all the experiments were performed under identical conditions (different ion energies, incidence angles, etc. were used in some of the studies). However, the broad dependence of the different types of patterning behavior can be easily seen on the diagram.

Examining the regimes in which different types of patterning emerge is useful for understanding the dominant mechanisms in each case. As described above, one of the characteristics of the BH-type ripples is the pattern direction determined by the ion beam. This can occur at high ion fluxes when the patterning effects of the ion beam are dominant. On the other hand, non-roughening behavior occurs when surface-smoothing effects are dominant.

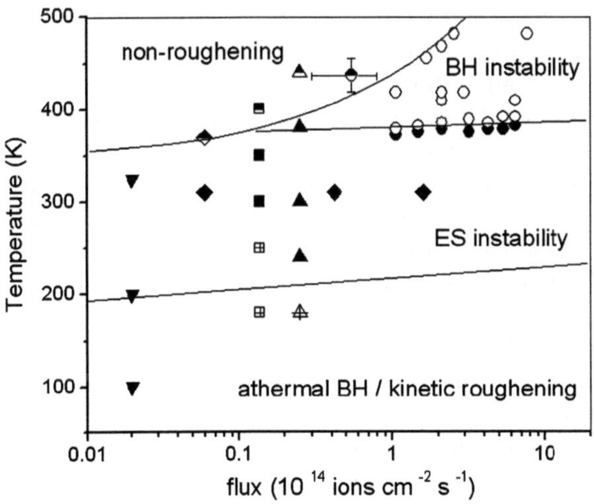

Fig. 5. Kinetic phase diagram on ion-induced patterns on Cu(001) and Ag(001) surfaces. The different symbols correspond to the following references: ○ [19]; ◇ [27]; ▽ [38]; △ [25]; □ [6]. Figure reprinted from [3]

This is seen to occur when the ion flux is small or when the temperature is high. ES behavior is determined by surface-diffusion effects and occurs only over a limited temperature range on the diagram. At high temperatures, the effects of diffusion barriers decrease so that the ES behavior changes to non-roughening behavior (if the ion flux is small) or to BH behavior if the ion flux is large. At low temperature, surface diffusion also decreases so that a transition to low-temperature BH behavior or kinetic roughening is observed.

Although the Cu(001) represents many types of pattern formation, there are other forms of patterning (e.g., quantum dots) that are not observed on it. In addition, the Cu surface remains crystalline during sputtering, while other surfaces such as semiconductors can become amorphous and therefore other relaxation mechanisms such as ion-induced viscous flow can be active [1, 40]. Nonetheless, it is useful as a way to see the relationships among a wide range of pattern formation.

3 Competing Kinetic Mechanisms and the Linear Instability Model

The kinetic phase diagram is a useful qualitative way to describe the competition among different kinetic processes. However, to develop a more rigorous approach to the patterning, we can use continuum equations to describe the evolution of the surface in terms of different kinetic processes on the surface. This approach was first used by Bradley and Harper [7] and resulted in the instability model that is often named after them. In the intervening years, additional terms corresponding to additional kinetic mechanisms have been included to form a more complete model.

3.1 BH Instability Model

To understand this model, it is first useful to consider the different kinetic processes occurring during ion bombardment that should be included in it. In the first place, we can consider the effect that the ion bombardment has directly on the surface. When the ion impinges on the surface, it gives up its kinetic energy in a series of collisions with atoms in the near-surface region. As shown in Fig. 6a, the sequence of collisions creates point defects (vacancies and interstitials) that may be mobile and recombine, diffuse to form clusters or diffuse to the surface to form adatoms and surface vacancies. Since the defects are created by atomic displacements, they primarily form Frenkel pairs with equal amounts of vacancy and interstitial defects. However, some of these collisions may occur near the surface and result in an atom being knocked off the surface (sputtering), creating a vacancy without a corresponding adatom or interstitial defect. Also, the different diffusion kinetics of the defects may result in a different number of adatoms and vacancies being created on the surface after the initial displacements occur [41, 42].

For low-mass ions, the individual collisions can be modeled within the binary collision approximation using a Monte Carlo simulation such as the well-known SRIM program [43]. For higher ion masses, the collisions occur in a denser cascade and can be modeled using molecular-dynamics simulations [44]. However, these calculations are stochastic in nature and difficult to incorporate into a model of the surface evolution. Sigmund [45, 46] described the sputtering process within a continuum framework by approximating the energy deposited by an ion into the near surface region as a Gaussian ellipsoid, as shown schematically in Fig. 6b. The energy per unit volume deposited by collisions at each point in the material is described by

$$\varepsilon(\rho, z) = \frac{\varepsilon_0}{(2\pi)^{3/2}} \exp\left(-\frac{(z-a)^2}{2\sigma^2} - \frac{\rho^2}{2\mu^2} \right), \tag{1}$$

where ε_0 is a normalization factor and ρ and z are defined in a cylindrical coordinate system with z parallel to the initial ion trajectory, ρ perpendicular to the trajectory and the origin at the point where the ion enters the surface. The center of the energy deposition occurs at a distance a from the point of impact with a spread of σ and μ in the parallel and perpendicular directions.

The analytical expression in (1) is meant to represent the average energy deposition over a large number of incident ions, not for an individual ion trajectory. The sputter yield at each point on the surface is taken to be proportional to the average amount of energy deposited there by the ions. Bradley and Harper [7] used this form for the sputter yield to calculate the effect on the surface morphology of a flux of collimated ions striking the surface uniformly. By integrating the effect of multiple ions over the surface, they calculated the change in the surface height h to be

$$\frac{\partial h(x,y)}{\partial t} = -v_0 + \frac{\partial v_0}{\partial \theta}\frac{\partial h}{\partial x} + v_x \frac{\partial^2 h}{\partial x^2} + v_y \frac{\partial^2 h}{\partial y^2}, \tag{2}$$

where x, y are surface coordinates in the direction parallel and perpendicular to the direction of the incident ion projected onto the surface and θ is the angle of the ion relative to the surface normal. The parameters v_x and v_y relate the sputter yield to the surface curvature and depend on the ion-beam parameters (a, σ and μ) of the Sigmund sputtering model.

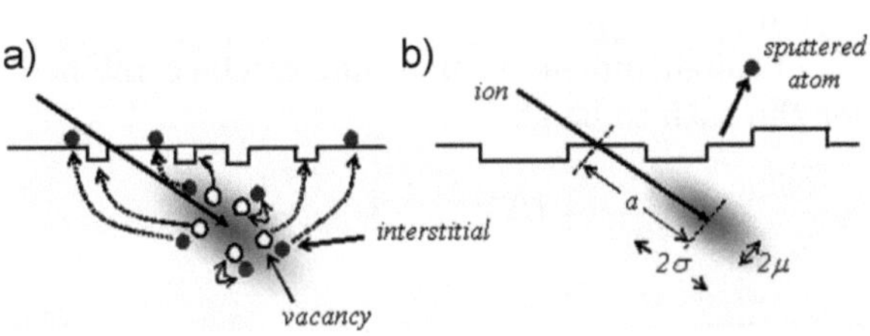

Fig. 6. (**a**) Schematic of defect generation, diffusion and recombination processes occurring during ion bombardment. (**b**) Schematic of sputtering process following mechanism in Sigmund model

In opposition to the roughening induced by sputtering, there is also smoothing of the surface due to the diffusion of defects on the surface. As described by Mullins [47] and Herring [48], the driving force for the smoothing process is reduction of the surface energy. Based on this approach, the change in the surface height is given by

$$\frac{\partial h}{\partial t} = -B\nabla^4 h, \tag{3}$$

where B is equal to $\gamma D_\mathrm{s} C/n^2 k_\mathrm{B} T$ and γ is the surface energy, D_s is the diffusivity of a mobile surface defect, C is the average concentration of mobile defects on the surface (number per unit area) and n is the number of atoms per unit volume. The negative sign in this expression indicates that the surface height is driven to decrease due to divergence in the surface curvature. Note that the rate of surface smoothing depends on both the mobile-defect diffusivity and the defect concentration. Therefore, the nonequilibrium concentration of surface defects induced by the ion–atom collisions can strongly affect the relaxation rate during sputtering.

An additional component of surface smoothing due to the ion bombardment itself can also decrease the surface roughness. The curvature-dependent roughening due to sputtering described in (2) was derived assuming that the local radius of curvature is large relative to the depth of the ion penetration. Makeev and Barabasi [31] carried the calculation of the ion–solid interaction out to higher order and recognized that there are additional terms in the expression for the change in surface height of the form

$$-B_{I,x}\frac{\partial^4 h}{\partial x^4} - B_{I,xy}\frac{\partial^4 h}{\partial x^2 \partial x^2} - B_{I,y}\frac{\partial^4 h}{\partial y^4}. \tag{4}$$

The negative sign in front of these terms leads to smoothing of the surface in a similar way to surface diffusion. However, the coefficients in (4) depend only on the ion–solid interaction and not on any surface transport. This effect was originally called "ion-induced effective surface diffusion" by Makeev et al., but we prefer to refer to it as athermal smoothing since it is unrelated to surface diffusion. In addition to athermal smoothing, additional terms in the expansion of the surface height lead to dispersion so that the velocity of he ripple along the surface depends on the wavelength. These effects have been seen in experiments [38, 8] and in kinetic Monte Carlo simulations [18, 49] of ripple formation. These effects of roughening and smoothing can be combined into a single rate equation for the surface height:

$$\begin{aligned}\frac{\partial h}{\partial t} = -v_0 + \frac{\partial v_0}{\partial \theta}\frac{\partial h}{\partial x} + v_x\frac{\partial^2 h}{\partial x^2} + v_y\frac{\partial^2 h}{\partial y^2} - B\nabla^4 h - B_{I,x}\frac{\partial^4 h}{\partial x^4} \\ - B_{I,xy}\frac{\partial^4 h}{\partial x^2 \partial x^2} - B_{I,y}\frac{\partial^4 h}{\partial y^4}.\end{aligned} \tag{5}$$

Because it is linear, the solution can be found by Fourier transforming the equation and considering the evolution of individual Fourier components with wavevector (k_x, k_y) on the surface. The amplitude of each Fourier component $h_k(k_x, k_y, t)$ increases or decreases independently with time at an exponential rate determined by the amplification $r(k_x, k_y)$:

$$h_k(k_x, k_y, t) = h_k(k_x, k_y, 0)\mathrm{e}^{r(k_x, k_y)}, \tag{6a}$$

where

$$\begin{aligned} r(k_x, k_y) = &-v_x k_x^2 - v_y k_y^2 - B\left(k_x^2 + k_y^2\right)^2 \\ &- B_{I,x} k_x^4 - B_{I,xy} k_x^2 k_y^2 - B_{I,y} k_y^4 \end{aligned} \tag{6b}$$

and $h_k(k_x, k_y, 0)$ is the initial amplitude. In addition to amplifying existing roughness, it is also possible to include the effect of the sputtering to create fluctuations that grow [13, 50], but we have not included this here.

The different spatial dependences of the roughening and smoothing processes determine the different wavelength dependences of the rates in (6b). Since the roughening rate depends on the surface curvature $(\partial^2 h/\partial x^2)$, the amplitude of a component with wavevector k will increase with a rate that depends on k^2. At the same time, since the smoothing depends on $\partial^4 h/\partial x^4$, this contributes to a decrease in the amplitude with a rate that depends on k^4. As shown schematically in Fig. 7, the simultaneous action of the roughening and smoothing leads to an amplification rate that depends on the wavevector. For large wavevectors (i.e., high spatial frequency or short wavelength), the amplification factor is negative. For these Fourier components, the surface diffusion dominates over the sputter-induced roughening and the surface height decreases with time. At small wavevectors, the sputter roughening dominates and the surface height at this spatial frequency grows with time. The amplification factor has a maximum growth rate r^* at the wavevector k^*. These values are determined in terms of the different kinetic processes by computing the maximum of the growth rate in (6b):

$$r^* = \left(\frac{v_{\max}^2}{4(B + B_{I,\max})}\right), \tag{7a}$$

and

$$k^* = \left(\frac{v_{\max}}{2(B + B_{I,\max})}\right)^{1/2}. \tag{7b}$$

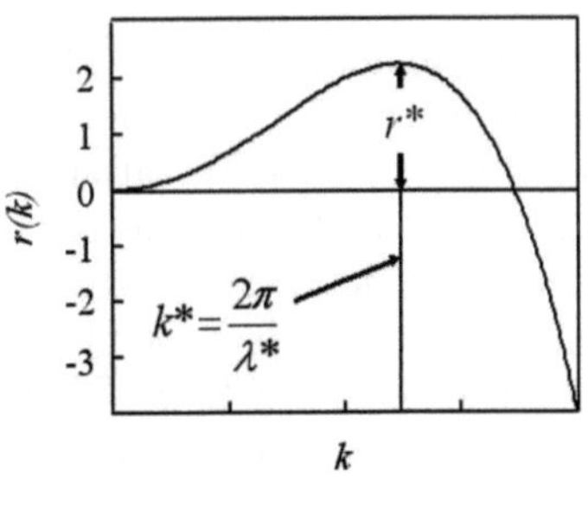

Fig. 7. Schematic of dependence of amplification factor on ripple wavevector in instability model. The maximum corresponds to the fastest growing surface wavevector that will appear as characteristic periodicity on surface

The subscript max in the expression above refers to the value of x or y for which the parameter v is larger.

The instability model derived here is consistent with many of the features observed in the BH type of pattern formation. The amplitude of the ripple is predicted to grow exponentially in time with a wavelength that is constant, as seen experimentally. The alignment of the pattern is predicted to be determined by the ion beam; the surface wavevector is either parallel or perpendicular to the ion-beam direction projected onto the surface depending on the values of v_x and v_y. These parameters depend on the incidence angle so that the ripple wavevector can change from the x- to the y-direction as the angle increases from near-normal to grazing incidence, also in agreement with experiment.

To relate these parameters to the processing conditions, we can consider the flux (f) and temperature (T) dependence of the different parameters in the model. The BH sputter roughening parameters (v_x, v_y) and the athermal smoothing parameters (B_I) depend linearly on the ion flux and are independent of the temperature. The diffusional parameter B depends on the flux and temperature through the individual dependences of the defect concentration $C(f,T)$ and the diffusivity $D_s(T)$. The wavelength therefore depends on the flux and temperature as:

$$\lambda^* = 2\pi\sqrt{\frac{2(B + B_{I,\max})}{v_{\max}}} \sim \sqrt{\frac{C(f,T)D(T)}{fT} + A_I}, \tag{8}$$

where $A_I = B_{I,\max}/v_{\max}$ is independent of temperature and flux.

Measurements of the temperature and flux dependence of the ripple wavelength [19] on Cu(001) (shown as the symbols in Fig. 8) are consistent with this picture. For example, the non-Arrhenius temperature dependence of the wavelength shown in Fig. 8a can be explained by the changing balance among the different processes. At high temperature (where $CD/ft \gg A_I$), thermal effects dominate and the wavelength has a strong temperature dependence. At low temperature, the athermal smoothing dominates and wavelength becomes independent of temperature. Using a simple model for $C(f,T)$ that includes thermal and ion-induced defect generation, the different flux dependence of the wavelength at high and low temperature (seen in Figs. 8b and c) can also be explained. The results from using (8) are shown as the solid lines in the figures.

In addition to experiments, kinetic Monte Carlo simulations have also recently been performed that include the Sigmund mechanism for sputter removal combined with surface diffusion of adatoms and vacancies by atomic hopping [18]. The results of the simulations agree very well with the predictions of the BH theory for the temperature and flux dependence of the ripple wavevector and growth rate. This suggests that the BH theory is a good continuum approximation of surface evolution under the action of the kinetic processes that are put into the model. Discrepancies between experi-

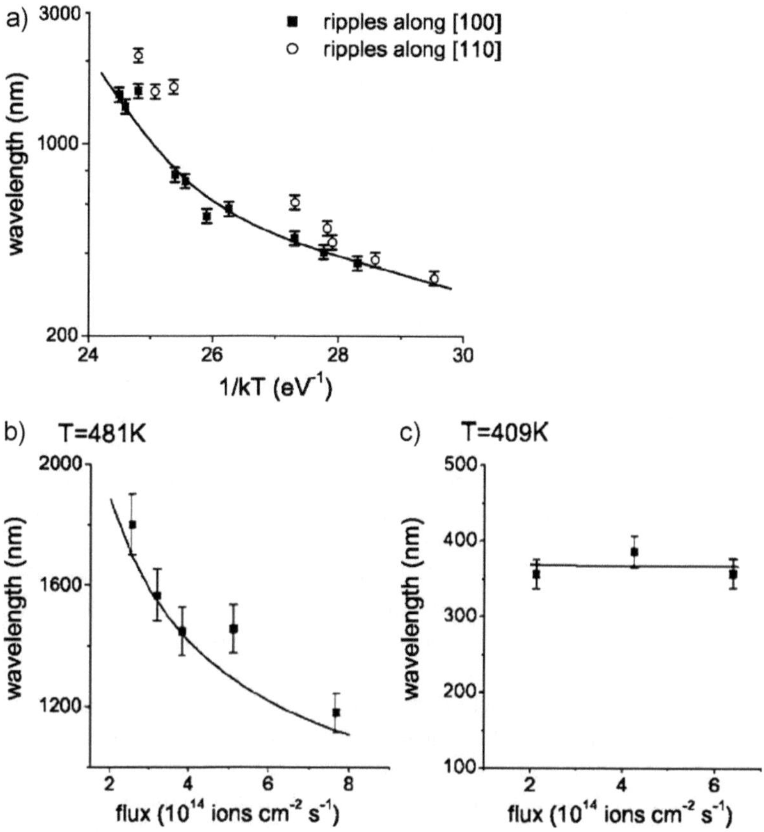

Fig. 8. Dependence of BH ripple wavelength on (**a**) temperature and (**b**), (**c**) ion flux measured on a Cu(001) surface. Figure reprinted from [19]

ments and the BH model are therefore likely due to differences in the physical assumptions of the BH model (e.g., sputter yields that are different from the Sigmund model or mechanisms that have not been included) rather than errors in the continuum approximation.

3.2 Diffusional Roughening and the ES Instability

As discussed above, in the ES regime of pattern formation the alignment is determined by the surface crystallography. This type of patterning has been attributed to the presence of ES barriers to diffusion from one level on the surface to another. Villain [51] proposed that these barriers lead to an instability in surface morphology during vapor deposition. Valbusa et al. [6] extended this mechanism to surfaces during ion bombardment. Although the mechanism is nonlinear, in the early stages it can be linearized. In this regime, the surface-height evolution is proportional to the surface curvature, similar to the BH mechanism.

Combining the BH model with the effect of diffusional roughening leads to:

$$\frac{\partial h}{\partial t} = -v_0 + \frac{\partial v_0}{\partial \theta}\frac{\partial h}{\partial x} + v_x\frac{\partial^2 h}{\partial x^2} + v_y\frac{\partial^2 h}{\partial y^2} + \sum_{i=1,2}\left[S_i\nabla_i^2 h - B_i\nabla_i^2(\nabla^2 h)\right]$$
$$- B_{I,x}\frac{\partial^4 h}{\partial x^4} - B_{I,xy}\frac{\partial^4 h}{\partial x^2 \partial x^2} - B_{I,y}\frac{\partial^4 h}{\partial y^4}, \qquad (9)$$

where the S_i parameters depend on the diffusion barriers and are aligned with respect to the crystallographic directions, as denoted by the subscript i.

Within the linear regime, the addition of the ES barriers does not change the form of the equation and the instability picture is still valid. However, the presence of the diffusional roughening can change which wavevectors are the fastest growing. If the ES barrier terms dominate, then the pattern will be aligned with the crystallographic directions instead of the ion beam direction.

The shifting balance between the ES and BH roughening terms can explain a large amount of the patterning behavior on metal surfaces where ES barriers to interlayer diffusion are significant. The temperature dependence of the S_i parameters is complex, but in general it will decrease at high temperature (where the diffusion barriers over step edges become small compared to kT) and at low temperature (when the diffusivity itself becomes small). On the other hand, the BH roughening term is not expected to depend on temperature. Therefore, at high temperature the ES-type pattering is predicted to transition to BH-type patterning as diffusive roughening effects become weaker than ion-induced effects. Alternatively, at low temperature the athermal smoothing becomes dominant and low-temperature BH ripples are predicted to occur. This is consistent with the temperature dependence seen in the kinetic phase diagram.

The balance between ES and BH patterning can also be adjusted by changing the ion parameters. As shown by Rusponi et al. [20], ripples on Cu(110) surface are aligned along crystallographic directions when the incidence angle is 45°. When the angle is increased to 70°, the pattern alignment is determined by the ion-beam direction. This is consistent with the fact that the BH parameters become increasingly large as the incidence angle increases, causing a transition from diffusional patterning to ion-induced patterning.

3.3 Other Regimes of Patterning – Beyond the Instability Model

The instability model described above is not sufficient to explain all the observed patterning behavior. In this section, we describe some alternative mechanisms that may explain behavior that is not explained by the extended BH model.

Systematic measurements of BH ripple formation under different conditions have enabled the BH theory to be compared quantitatively with experiments. Several important discrepancies have emerged, e.g., the measured roughening rate of ripples on several surfaces is much faster than predicted by the BH theory [3] and the observed velocity of the ripples is opposite to that predicted by the model [38]. One explanation for the difference in roughening rate from the calculations focuses on the assumptions in the BH model for the morphology dependence of the sputter yield. Sputtering simulations based on binary collision approximations [52] and MD simulations [53] indicate that the morphology of the sputtered crater may be significantly different from that predicted by the Sigmund model. Incorporating a more advanced form for the sputter yield into the BH theory may still result in an instability, but with a rate that is significantly higher than predicted by the current theory. Alternatively, the apparently rapid ion-induced roughening may be due to additional mechanisms that are not included in the BH model, such as stress in the surface region due to the ion-bombardment process [54].

The regime of nonroughening behavior observed on many surfaces is also not predicted by instability models and its origin is still not certain. One explanation is based on a mechanism that was proposed by Tersoff for the transition from smooth to rough growth during epitaxial growth [55]. On crystalline surfaces there is a kinetic barrier to nucleation of new steps, so if the roughening rate is low relative to the diffusion rate it is not possible to create a new terrace on the existing surface. This mechanism seems to be consistent with measurements on Cu(001) and kinetic Monte Carlo simulations [3] but there may be other causes for nonroughening behavior. For Si surfaces sputtered at room temperature, the amorphization of the surface may induce additional smoothing mechanisms that keep the surface from developing roughness [56]. Alternatively, it has been proposed that using a more realistic morphology-dependent sputtering process than the Sigmund model may lead to regimes of stable behavior in which the surface does not roughen [57].

In our discussion of the instability model, we restricted our consideration to only the linear terms so the model is only valid for the early stages of roughening. At larger degrees of roughening, other mechanisms such as shadowing and redeposition become important. These effects modify the surface evolution equations by adding nonlinear terms as described by Castor et al. [58]. Even in systems that initially form BH ripples, these effects can cause the ripple amplitude to saturate and transition to other forms of roughening. These nonlinear terms dominate at higher degrees of surface roughness and lead to asymptotic behavior such as power-law roughening and coarsening of the characteristic surface periodicity. As pointed out by Makeev et al. [4], even within the BH model there are sputtering conditions under which the surface will not develop a periodic pattern even in the early stages. Therefore, kinetic roughening without pattern formation can occur immediately without an initial period of ripple formation. Because the nonlinear equations are far more complex, the long-term evolution of the surface morphology is not as easily understood as the linear instability regime.

Other patterning phenomena still remain as a persistent challenge to our understanding. Quantum-dot formation by ion bombardment is an exciting potential application, but many features of their formation are difficult to understand within the current models. Similarly, extremely well-ordered ripple patterns have been produced on Si [59] and Ge [60] surfaces by careful control of the sputtering parameters. Such a high degree of uniformity is not expected within the instability model and suggests additional nonlinear effects are involved that enhance the sharpness of the pattern. On surfaces produced by focused ion-beam micromachining, instabilities limit the surface finish that can be achieved [61, 62]. The morphology suggests that the change in the sputtering yield at near-grazing orientation of the beam may lead the surface to break up into a saw-toothed pattern, with some regions nearly parallel to the beam and others nearly normal.

In summary, the instability model provides a useful framework for understanding many types of ion-induced patterning in terms of a balance between different surface kinetic processes during sputtering. However, it can not explain all the phenomena that are observed. The emergence of new ion-induced phenomena and a continued interest in fabrication on the micro- and nanoscale fabrication suggests that further advances in our understanding and control of patterning are likely to continue.

Acknowledgements

The authors gratefully acknowledge many helpful discussions with Vivek Shenoy. The authors also gratefully acknowledge the support of the US Department of Energy under contract DE-FG02-01ER45913.

References

1. T.M. Mayer, E. Chason, A.J. Howard, J. Appl. Phys. **76**, 1633 (1994) 54, 60
2. M. Navez, C. Sella, D. Chaperot, C. R. Acad. Sci. (Paris) **254**, 240 (1962) 54
3. W.L. Chan, E. Chason, J. Appl. Phys. **101**, 121301 (2007) 54, 59, 67
4. M.A. Makeev, R. Cuerno, A.L. Barabasi, Nucl. Instrum. Methods Phys. Res. B **197**, 185 (2002) 54, 67
5. J. Munoz-Garcia, L. Vazquez, R. Cuerno, J.A. Sanchez-Garcia, M. Castro, R. Cuerno, in *Lecture Notes on Nanoscale Science and Technology*, ed. by Z. Wang (Springer, Heidelberg, 2007) 54
6. U. Valbusa, C. Borangno, F.R. de Mongeot, J. Phys., Condens. Matter **14**, 8153 (2002) 54, 58, 59, 65

7. R.M. Bradley, J.M.E. Harper, J. Vac. Sci. Technol. A **6**, 2390 (1988) 55, 60, 61
8. S. Habenicht, K.P. Lieb, J. Koch, A.D. Wieck, Phys. Rev. B **65**, 115327 (2002) 55, 62
9. W.L. Chan, N. Pavenayotin, E. Chason, Phys. Rev. B **69**, 245413 (2004) 56
10. A.-D. Brown, J. Erlebacher, Phys. Rev. B **72**, 075350 (2005) 56
11. G. Carter, V. Vishnyakov, Phys. Rev. B **54**, 17647 (1996) 56
12. J. Erlebacher, M.J. Aziz, E. Chason, M.B. Sinclair, J.A. Floro, Phys. Rev. Lett. **82**, 2330 (1999) 56
13. E. Chason, T.M. Mayer, B.K. Kellerman, D.T. Mcllroy, A.J. Howard, Phys. Rev. Lett. **72**, 3040 (1994) 56, 63
14. S. Habenicht, W. Bolse, K.P. Lieb, K. Reimann, U. Geyer, Phys. Rev. B **60**, R2200 (1999) 56
15. S. Facsko, T. Dekorsy, C. Koerdt, C. Trappe, H. Kurz, A. Vogt, H.L. Hartnagel, Science **285**, 1551 (1999) 56, 58
16. F. Frost, B. Ziberi, T. Hoche, B. Rauschenbach, Nucl. Instrum. Methods Phys. Res. B **216**, 9 (2004) 56
17. B. Ziberi, F. Frost, B. Rauschenbach, Th. Hoche, Appl. Phys. Lett. **87**, 033113 (2005) 56
18. E. Chason, W.L. Chan, M.S. Bharathi, Phys. Rev. B **74**, 224103 (2006) 56, 62, 64
19. W.L. Chan, E. Chason, Phys. Rev. B **72**, 165418 (2005) 56, 59, 64, 65
20. S. Rusponi, G. Costantini, C. Boragno, U. Valbusa, Phys. Rev. Lett. **81**, 2735 (1998) 57, 66
21. G. Ehrlich, F.G. Hudda, J. Chem. Phys. **44**, 1039 (1966) 57
22. R.L. Schwoebel, J. Appl. Phys. **40**, 614 (1969) 57
23. M. Kalff, G. Cosma, T. Michely, Surf. Sci. **486**, 103–135 (2001) 57, 58
24. M.V.R. Murty, T. Curcic, A. Judy, B.H. Cooper, A.R. Woll, J.D. Brock, S. Kycia, R.L. Headrick, Phys. Rev. Lett. **80**, 4713 (1998) 57
25. G. Costantini, S. Rusponi, R. Gianotti, C. Boragno, U. Valbusa, Surf. Sci. **416**, 245 (1998) 57, 59
26. H.J. Ernst, Surf. Sci. **383**, L755 (1997) 57
27. M. Ritter, M. Stindtmann, M. Farle, K. Baberschke, Surf. Sci. **348**, 243 (1996) 57, 59
28. S. Rusponi, G. Costantini, C. Boragno, U. Valbusa, Phys. Rev. Lett. **81**, 4184 (1998) 57
29. T. Michely, M. Kalff, G. Cosma, M. Strobel, K.-H. Heinig, Phys. Rev. Lett. **86**, 2589 (2001) 57
30. S. van Dijken, D. de Bruin, B. Poelsema, Phys. Rev. Lett. **86**, 4608 (2001) 58
31. M.A. Makeev, A.L. Barabasi, Appl. Phys. Lett. **71**, 2800 (1997) 58, 62
32. M.V.R. Murty, A.J. Couture, B.H. Cooper, A.R. Woll, J.D. Brock, R.L. Headrick, J. Appl. Phys. **88**, 597 (2000) 58
33. B. Poelsema, L.K. Verheij, G. Comsa, Phys. Rev. Lett. **53**, 2500 (1984) 58

34. F. Frost, A. Schindler, F. Bigl, Phys. Rev. Lett. **85**, 4116 (2000) 58
35. R. Gago, L. Vazquez, R. Cuerno, M. Varela, C. Ballesteros, J.M. Albella, Appl. Phys. Lett. **78**, 3316 (2001) 58
36. E.A. Eklund, R. Bruinsma, J. Rudnick, R.S. Williams, Phys. Rev. Lett. **67**, 1759 (1991) 59
37. E.A. Eklund, E.J. Snyder, R.S. Williams, Surf. Sci. **285**, 157 (1993) 59
38. P.F.K. Alkemade, Phys. Rev. Lett. **96**, 107602 (2006) 59, 62, 67
39. E. Chason, W.L. Chan, Nucl. Instrum. Methods Phys. Res. B **242**, 232 (2006) 59
40. C.C. Umbach, R.L. Headrick, K.-C. Chang, Phys. Rev. Lett. **87**, 246104 (2001) 60
41. P. Bedrossian, Surf. Sci. **301**, 223 (1994) 60
42. M. Morgenstern, T. Michely, G. Cosma, Philos. Mag. A **79**, 775 (1999) 60
43. J.F. Ziegler, J.P. Biersack, in *SRIM-2000.40* (IBM Co., Yorktown, 2000) 61
44. M. Ghaly, K. Nordlund, R.S. Averback, Philos. Mag. A **79**, 795–820 (1999) 61
45. P. Sigmund, Phys. Rev. **184**, 383 (1969) 61
46. P. Sigmund, J. Mater. Sci. **8**, 1545 (1973) 61
47. W.W. Mullins, J. Appl. Phys. **30**, 77 (1959) 62
48. C. Herring, J. Appl. Phys. **21**, 301–303 (1950) 62
49. O.E. Yewande, A.K. Hartmann, R. Kree, Phys. Rev. B **71**, 195405 (2005) 62
50. G. Ozaydin, K.F. Ludwig, H. Zhou, R.L. Headrick, J. Vac. Sci. Technol. B **26**, 551 (2008) 63
51. J. Villain, J. Phys. (France) **I-1**, 19 (1991) 65
52. M. Feix, A.K. Hartmann, R. Kree, J. Munoz-Garcia, R. Cuerno, Phys. Rev. B **71**, 125407 (2005) 67
53. K. Kalyanasundaram, M. Ghazisaeidi, J.B. Freund, H.T. Johnson, Appl. Phys. Lett. **92**, 131909 (2008) 67
54. W.L. Chan, E. Chason, C. Iamsumang, Nucl. Instrum. Methods Phys. Res. B **257**, 428 (2007) 67
55. J. Tersoff, A.W. Denier van der Gon, R.M. Tromp, Phys. Rev. Lett. **72**, 266 (1994) 67
56. G. Ozaydin, K.F. Ludwig, H. Zhou, R.L. Headrick, J. Appl. Phys. **103**, 033512 (2008) 67
57. B. Davidovitch, M.J. Aziz, M.P. Brenner, Phys. Rev. B **76**, 205420 (2007) 67
58. M. Castro, R. Cuerno, L. Wázquez, R. Gago, Phys. Rev. Lett. **94**, 016102 (2005) 67
59. B. Ziberi, F. Frost, Th. Höche, B. Rauschenbach, Phys. Rev. B **72**, 235310 (2005) 68
60. B. Ziberi, F. Frost, B. Rauschenbach, Appl. Phys. Lett. **88**, 173115 (2006) 68

61. D.P. Adams, M.J. Vasile, T.M. Mayer, V.C. Hodges, J. Vac. Sci. Technol. B **21**, 2334 (2003) 68
62. H.H. Chen, O.A. Orquidez, S. Ichim, L.H. Rodriguez, M.P. Brenner, M.J. Aziz, Science **310**, 294 (2005) 68

Index

Ion-Beam-Induced Amorphization and Epitaxial Crystallization of Silicon

J.S. Williams[1], G. de M. Azevedo[1,2], H. Bernas[3] and F. Fortuna[3]

[1] Research School of Physical Sciences and Engineering, Australian National University, Canberra, 0200, Australia, e-mail: jsw109@rsphy1.anu.edu.au
[2] Present address: Brazilian Synchrotron Light Laboratory (LNLS), 6192 CEP, 13084-971, Campinas, SP, Brazil
[3] CSNSM-CNRS, University Paris-Sud 11, 91405 Orsay, France

Abstract. Ion-induced collisions produce athermal atomic movements at and around the surface or interface, inducing step formation and modifying growth conditions. The latter may be controlled by varying the temperature and ion-beam characteristics, guiding the system between nonequilibrium and quasiequilibrium states. Silicon is an ideal material to observe and understand such processes. For ion irradiation at or below room temperature, damage due to collision cascades leads to Si amorphization. At temperatures where defects are mobile and interact, irradiation can lead to layer-by-layer amorphization, whereas at higher temperatures irradiation can lead to the recrystallization of previously amorphized layers. This chapter focuses on the role of ion beams in the interface evolution. We first give an overview of ion beam-induced epitaxial crystallization (IBIEC) and ion-beam-induced amorphization as observed in silicon and identify unresolved issues. Similarities and differences with more familiar surface thermal growth processes are emphasized. Theories and computer simulations developed for surface relaxation help us to quantify several important aspects of IBIEC. Recent experiments provide insight into the influence of ion-induced defect interactions on IBIEC, and are also partly interpreted via computer simulations. The case of phase transformations and precipitation at interfaces is also considered.

1 Introduction

Possibly the most important feature of surfaces is their irregularity: crystals only grow, when matter is added, because steps form on the surface. These may be due to the deposited adatoms, or/and to sample heating – the step free energy is reduced as the configurational entropy term increases [1]. Another way to induce step formation and modify growth conditions involves charged-particle irradiation: ion-induced collisions produce athermal atomic movements at and around the surface or interface, and these (and hence the configurational entropy) may be controlled to some extent by varying the irradiation conditions. Performing the irradiation at different temperatures provides a potentially powerful means of guiding the system between nonequilibrium and quasiequilibrium states. Several such effects at surfaces are discussed in the chapter "Spontaneous Patterning of Surfaces by Low-Energy

H. Bernas (Ed.): Materials Science with Ion Beams,
Topics Appl. Physics **116**, 73–111 (2010)
DOI: 10.1007/978-3-540-88789-8_4,

Ion Beams" by Chason and Chan. In this chapter, we examine a situation in which, because they penetrate inside matter, ion beams play an even more specific role: that of growing interfaces. The latter display both similarities to and differences from surface evolution, notably as concerns phase transformations and precipitation at interfaces. We shall emphasize the cases of ion-beam-induced epitaxial crystallization (IBIEC) and ion-beam-induced amorphization. The topic emerged as a typical materials-science problem, in which basic and applied physics are totally intertwined. A near-surface amorphous layer and hence a buried crystal/amorphous interface can be produced in a silicon matrix by ion implantation (say, of a dopant). Recrystallization of this amorphous layer may be induced by high-temperature annealing (solid-phase epitaxial growth, SPEG), a quasiequilibrium technique, or by IBIEC. In the latter, the ion beam provides the atomic displacement energy, so that IBIEC occurs in a temperature range (typically around 200–400°C) some 200–400 degrees below SPEG. This can be advantageous for applications, since it is far more compatible with the preservation of prior microelectronics fabrication steps in an industrial environment.

Experimental studies show that under both SPEG and IBIEC the initially blurred, defected interface is first smoothened, then progressively moves towards the surface. The SPEG mechanism has been modeled [2] in thermodynamical terms as a sequence of bond rearrangements at the interface. This will be summarized later in order to provide a reference for specific ion-beam effects, but first we very briefly indicate some of the main ideas to which our discussion relates. When adatoms are deposited on surfaces at high temperatures, steps form and flow; in the intermediate temperature range where limited atomic motion occurs on the surface after deposition, nucleation, growth and island coalescence occur. Growth is the result of adatoms reaching step edges from above or below, with correspondingly different energy barriers (and activation energies). At temperatures sufficiently low to hinder any long-range atomic motion, deposition may lead either to formation of an amorphous deposit or to local epitaxy, depending on the latent heat and kinetic energy released by the arriving atom. It is known [3, 4] that atom deposition or ion bombardment at hyperthermal energies enhances the kinetic energy component, leading to an effective "local annealing" that favors short-range epitaxial growth by a well-defined bond rearrangement. On the other hand, the overall surface roughness increases progressively as inhomogeneities produced by random deposition of atoms or islanding interfere with each other's lateral expansion. Since this process multiplies the number of steps, it enhances the growth speed for a given deposition density. In the case of ion irradiation, how then do surface growth models relate to ion-irradiation-induced interfacial growth? The latter's evolution does not involve any increase in the amount of matter (no adatoms). However, the interface moves and its roughness is modified, its overall shape becoming "more planar". This signals the existence of interface relaxation, implying that matter has effectively moved along the interface. Do atoms move (by random or non-random diffusion) on the interface, or are the atomic movements limited to local bond restructuring and subsequent relaxation? If the latter is correct, can one specify the relaxation mode? This is where the theories developed for surface relaxation may help us to quantify several important aspects of SPEG and IBIEC, as we will see in Sect. 4.

Let us first consider some of the main experimental features. When materials are irradiated with energetic ions, the ion-induced disorder can lead to a number of structural transformations, including amorphization and crystallization. The behavior is particularly interesting when irradiation is carried out at temperatures where the mobility of defects produced by the ion beam is progressively increased. Silicon is an ideal material to observe and understand such processes but, despite extensive studies over the past two decades, there are still unanswered questions relating to ion-induced defects and their influence on amorphization and crystallization.

For ion irradiation at or below room temperature in silicon, the disorder produced is essentially stable since point defects are readily immobilized within disordered regions. Under such conditions, ion damage generated within collision cascades builds up with ion dose, leading to amorphization of the silicon. At higher implant temperatures, where defects begin to move and interact during ion bombardment, significant defect annihilation can occur and it can be difficult to induce amorphization. In this regime, preferential amorphization can be observed at regions where extended defects first form, for example, at nanocavities or at surfaces [5]. Continued irradiation can lead to layer-by-layer amorphization, but at higher temperatures ion irradiation may not cause amorphization. Incomplete defect annihilation during bombardment can lead to the formation of defect clusters and even extended defects in an otherwise crystalline matrix. In this elevated temperature regime, where defects are mobile, the understanding of the observed defect-mediated processes is far from complete. Irradiation at even higher temperatures can even lead to the recrystallization of previously amorphized layers. The latter IBIEC process occurs at temperatures well below those at which normal thermally induced crystallization of amorphous silicon occurs. IBIEC has been shown to be an activated process, dependent on the generation of mobile 'defects' through atomic displacement during ion irradiation. There has been considerable controversy as to the role of defects in IBIEC but recent experiments have partly clarified this issue. Indeed, studies of ion-beam-induced amorphization (IBIA) and IBIEC not only indicate much about the behavior of defects and defect-induced phase changes in silicon but also provide considerable insight into the fundamental physics of defect interactions and epitaxial crystallization at the atomic level.

This review first gives an overview of ion-induced amorphization and crystallization phenomena that have been observed in silicon and identifies some unresolved issues. More recent experiments, that provide insight into both ion-induced defect interactions and IBIEC, are then presented and partly interpreted with the aid of computer simulations. Finally, a summary of what is known and what is not known in these areas is presented.

2 Overview of Ion-Beam-Induced Amorphization

2.1 The Effect of Temperature on Defect Accumulation

At sufficiently low irradiation temperatures, residual lattice disorder in semiconductors is controlled by the energy deposited by swift ions in nuclear collisions with lattice atoms. Individual heavy ions can generate dense displacement cascades (Fig. 1a) that result directly in amorphous zones [6] and the overlap of such zones with increasing dose leads to a continuous amorphous layer [7, 8] as shown in the cross-sectional transmission electron micrograph (XTEM) in Fig. 1b. For light ions, cascades are less dense and the lattice can collapse to an amorphous phase when a sufficiently high defect density builds up and the local free energy of the defective lattice exceeds that of the amorphous phase [9–11]. These two extremes of damage build up at low temperatures can be successfully treated by heterogeneous (heavy ion) or homogeneous (light ion) models, such as those of Morehead and Crowder [7] and Vook and Stein [10], respectively.

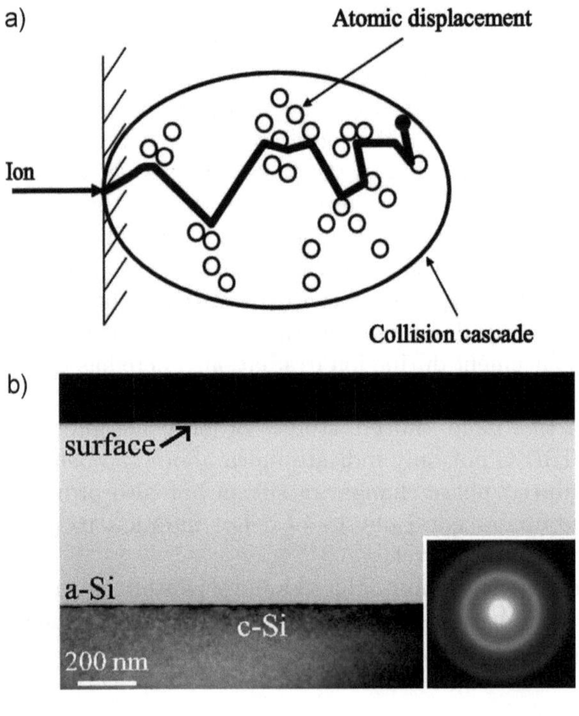

Fig. 1. (**a**) Schematic of displacements within a collision cascade. (**b**) A cross-sectional transmission electron microscope (XTEM) image of a continuous amorphous layer (a-Si) generated in silicon by 245-keV Si ion irradiation at room temperature to a dose of $3 \times 10^{15}\,\mathrm{cm}^{-2}$. The sample surface is indicated, as is the underlying crystalline silicon (c-Si)

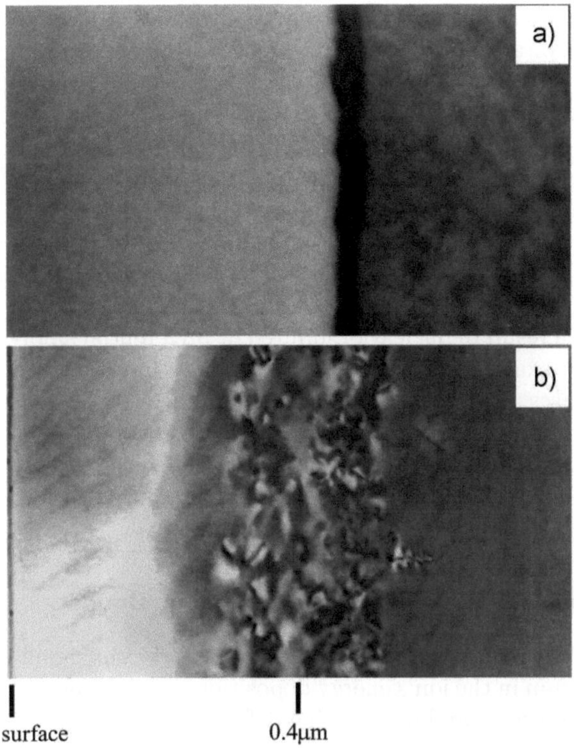

Fig. 2. XTEM images corresponding to 245-keV Si ion irradiation of silicon (100) to a dose of $4 \times 10^{15}\,\mathrm{cm}^{-2}$ at (**a**) room temperature, and (**b**) 350°C. After [12]

The implantation temperature can determine whether the defects generated within collision cascades are stable or whether they can migrate and annihilate. An example of temperature-dependent effects is shown in Fig. 2 [12]. Figure 2a is a XTEM micrograph depicting a continuous amorphous layer in silicon, produced by 245-keV Si ions at room temperature to a dose of $4 \times 10^{15}\,\mathrm{cm}^{-2}$. The ion range is around 3800 Å but, under these implant conditions, the amorphous layer is around 5000 Å thick.

Note that the boundary between the amorphous layer and the underlying silicon substrate is quite sharp. This may reflect the fact that defects produced in the tail of the Si implant distribution can annihilate quite effectively at this implant temperature and/or that there is an effective ordering correlation length or 'collective effect' operating on the crystalline side of the interface during irradiation. If the implant temperature is raised to 350°C, irradiation-produced defects are considerably more mobile and annihilate or cluster to effectively suppress amorphization [13], as shown in the XTEM micrograph in Fig. 2b. Here, there are clearly observed interstitial clusters that evolve into well-defined interstitial-based line defects such as {311} defects and dislocation loops [14] on annealing. It will be shown later that, at such implant temperatures where defects can annihilate, irradiation-induced displacements can induce crystallization of pre-existing amorphous layers.

Between the two extreme regimes illustrated in Fig. 2, the close balance between the rate of damage production within collision cascades and the rate of dynamic annealing (defect annihilation and clustering) can give rise to interesting defect-mediated phenomena, with strong dependencies on implantation temperature, dose and dose rate. Small changes in any of these parameters can result in dramatic differences in residual implantation damage from almost damage-free structures, as a result of efficient defect annihilation, to continuous amorphous layers. In this regime, amorphization can occur in an entirely different way, as a result of nucleation-limited or preferential amorphization processes [15]. For example, as the implantation dose increases and the density of defects increases, amorphous layers can spontaneously form at the depth of maximum disorder. Such layers can then grow to encompass the entire defective region [16]. Further examples of the critical balance between defect creation and defect annihilation, including preferential amorphization, are given below.

2.2 Preferential Amorphization at Surfaces and Defect Bands

Amorphous layers can be observed to nucleate preferentially at depths significantly away from the maximum in the ion's energy deposition distribution, at, for example, surfaces [17], interfaces and pre-existing defects [16, 18]. Figure 3 illustrates the case of preferential amorphization at a silicon surface or, more precisely, at a SiO_2–Si interface. Figure 3a [16] shows an RBS/channeling spectrum for an 80-keV Si implant into silicon at 160°C (dose $10^{16}\,cm^{-2}$ at a beam flux of 2.7×10^{13} ions $cm^{-2}\,s^{-1}$).

The spectrum shows a strong disorder peak at the surface and a buried peak around the end-of-ion-range at about 1200 Å. (The end-of-ion-range refers to the region in the tail of the ion-range distribution, about two standard deviations deeper than the projected ion range.) The corresponding XTEM micrograph in Fig. 3b [16] indicates that there are two amorphous layers present, one extending 300 Å from the surface and a buried layer from 500 to 1500 Å. Between these layers is a region of crystalline silicon containing few defects, but below the buried layer there is a region of crystalline silicon that is rich in (interstitial-type) defect clusters. This result shows not only the nucleation of an amorphous region around the maximum in the nuclear energy distribution at about 800 Å but nucleation of an amorphous layer well away from the maximum disorder depth, at the surface. When the evolution of this defect structure was examined as a function of ion dose [16] it was found that the deep disorder first accumulated by forming defect clusters of interstitial character at lower doses. This defective region then appeared to collapse into an amorphous layer as the dose increased. In addition, the surface amorphous layer was found to thicken with increasing dose. This behavior suggests that, in a regime where substantial dynamic annealing occurs

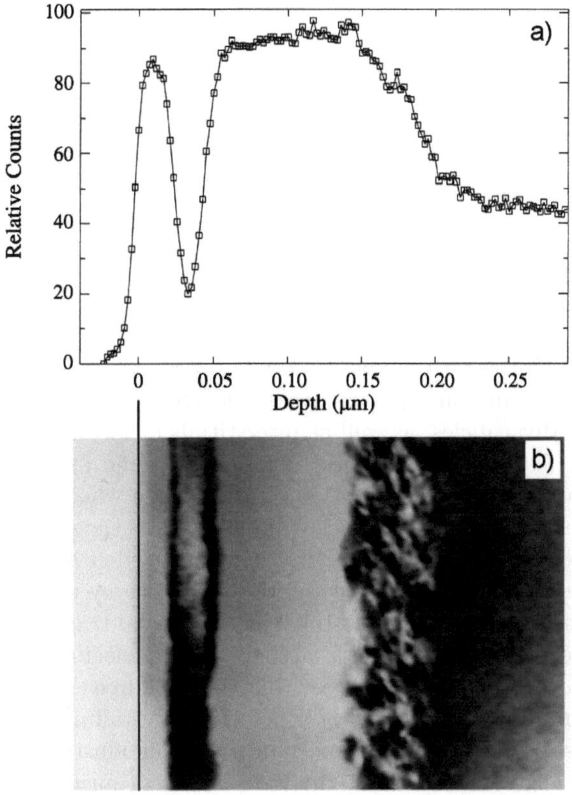

Fig. 3. (**a**) An RBS/channeling spectrum for an 80-keV Si implant into silicon at 160°C to a dose of $10^{16}\,\mathrm{cm}^{-2}$ at a beam flux of $2.7 \times 10^{13}\,\mathrm{ions\,cm}^{-2}\,\mathrm{s}^{-1}$. (**b**) XTEM image of the sample in (**a**). After [16]

during ion irradiation, mobile defects not only annihilate and locally form defect clusters, but can also migrate to and accumulate at SiO_2–Si interfaces. Collapse of such disorder to an amorphous phase can occur at a sufficiently high implantation dose. It has also been shown that a pre-existing dislocation band can act as a nucleation site for amorphization, even when it is situated well away from the disorder peak [15]. Furthermore, such dislocation bands have also been found to 'getter' interstitial-based defects formed deeper in the material during irradiation [15]. Thus, it would appear that both dislocation bands, surfaces (actually SiO_2–Si interfaces) and amorphous layers themselves are good trapping sites or sinks for mobile defects that may otherwise form stable clusters close to where they come to rest, in the absence of such sinks. When defect accumulation occurs at such interfaces, amorphous layers can be observed to "grow".

2.3 Mechanisms of Amorphization: The Role of Defects

The mechanism for the above defect trapping and preferential amorphization behavior deserves some comment. There has been considerable speculation

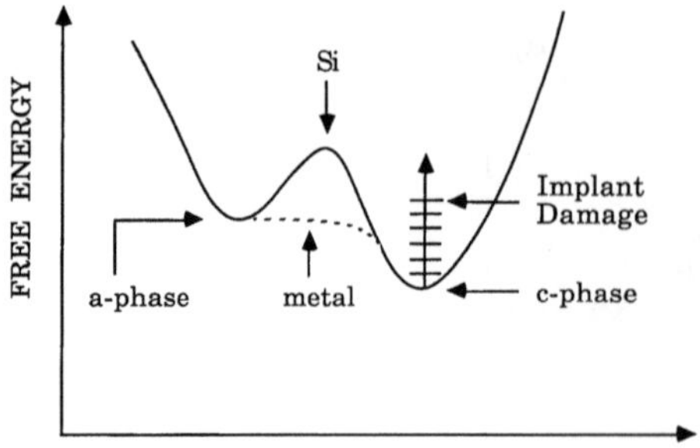

Fig. 4. Schematic illustrating free-energy differences and pathways between amorphous and crystalline materials. After [21]

in the literature [13, 15, 19, 20] as to the specific defects that are trapped at pre-existing defects, surfaces and amorphous layers. Clearly, open-volume defects such as vacancies or divacancies, as well as interstitials or interstitial complexes, are candidates. As we discuss more fully below, some experiments on the kinetics of amorphous layer formation, in the regime where the irradiation-induced amorphous phase is nucleation limited, have suggested that divacancies [19] may be the main defects preferentially trapped at amorphous layers. However, other experiments, where amorphous layers are nucleated at pre-existing dislocation bands, suggest [15] that interstitial trapping also may have a major role to play. Regardless of the specific defects that accumulate prior to amorphization, it would appear that the local free energy plays a major role in determining the collapse of a defective crystalline lattice to the amorphous phase. This free-energy mechanism [21] is illustrated schematically in Fig. 4. In the case of silicon, the free energy of an amorphous phase exceeds that of a crystalline phase and there is a strong driving force for amorphous regions to crystallize. However, the amorphous phase is metastable since there is a kinetic barrier that must be overcome before crystallization can occur. In contrast, for pure metals, an amorphous phase is unstable even at extremely low temperatures, since there is essentially no barrier to crystallization. Thus, under appropriate implantation conditions, implantation-induced disorder in silicon can build up until the local free energy exceeds that of the amorphous phase. It can then be energetically favorable for the defective crystalline lattice to collapse to the amorphous phase to achieve a local minimum in free energy. Such behavior suggests that, in cases where there is some defect mobility, defect annihilation and agglomeration occurs and the amorphous phase can preferentially form at sites that minimize the local free energy. Under such situations amorphization can be considered to be nucleation limited. This nucleation limited regime is not a general case and only occurs in a limited temperature range where defects are mobile enough to form dense networks of metastable defect clusters or extended defects but the temperature is not high enough to allow such defects to evolve into defect configurations that are in thermal equilibrium to minimize the free energy.

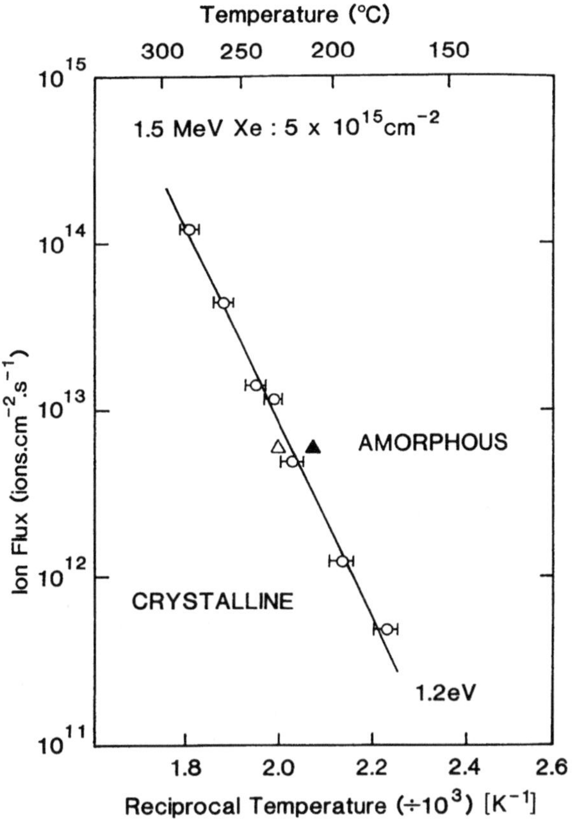

Fig. 5. Ion flux as a function of $1/T$ for ion-irradiation conditions (1.5-MeV Xe ions at a dose of $5 \times 10^{15}\,\mathrm{cm}^{-2}$) under which a buried amorphous layer is just formed in silicon. The *solid* and *open triangles* represent the cases in Figs. 7b and 8b, respectively. After [19]

In cases where there are no pre-existing nucleation sites for amorphization, the onset of amorphization (at elevated temperatures) usually occurs at the ion-end-of-range. Here, nucleation of the amorphous phase normally occurs where there is an interstitial excess and this corresponds roughly to the end-of-ion-range. In this regime, amorphization can exhibit interesting dependencies, including situations where the ion flux controls the critical amorphization temperature [19], as illustrated in Fig. 5. For a fixed dose of $5 \times 10^{15}\,\mathrm{cm}^{-2}$ for 1.5-MeV Xe ions irradiating silicon, amorphization at the end-of-ion-range can be observed only below 200°C if the average beam flux is kept below 10^{12} ions $\mathrm{cm}^{-2}\,\mathrm{s}^{-1}$, but up to 300°C if the ion flux is raised above 10^{14} ions $\mathrm{cm}^{-2}\,\mathrm{s}^{-1}$. This demonstrates the critical dependence of amorphization on the balance between the rate of disorder production (controlled by ion flux in the case of Fig. 5) and the extent of dynamic annealing, which is controlled by irradiation temperature. For implantation conditions on the left-hand side of the solid line in Fig. 5, no amorphous silicon was formed (only defect clusters in crystalline silicon), whereas buried amorphous layers are generated under conditions on the right. Note that the onset of amor-

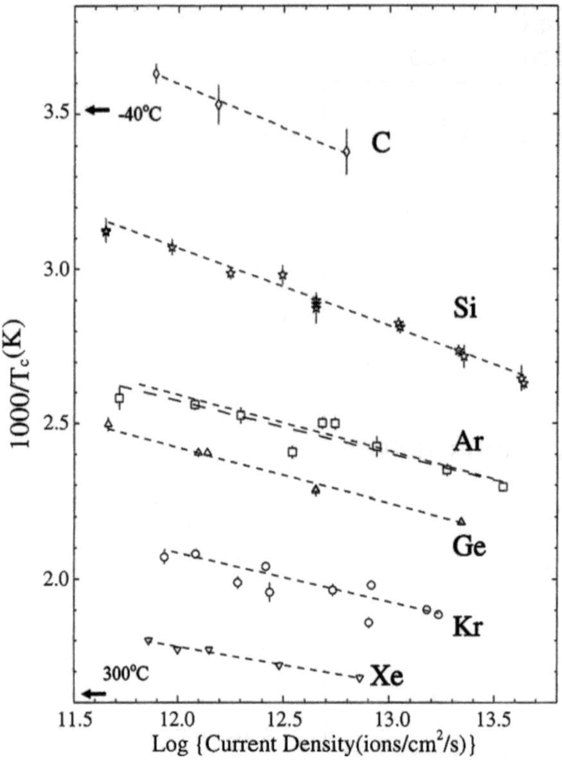

Fig. 6. Ion flux as a function of $1/T$ for ion-irradiation conditions under which a buried amorphous layer is just formed in silicon for a number of ions at a dose of $10^{15}\,\mathrm{cm}^{-2}$ except C where the dose was $2 \times 10^{15}\,\mathrm{cm}^{-2}$. After [22]

phization in Fig. 5 fits well to an activation energy of 1.2 eV. Elliman et al. [19] noted that this value corresponds to the dissociation energy of silicon divacancies and, consequently, suggested that the stability of divacancies may control amorphization in silicon. However, more recent studies, that use other ion beams to examine the dependence of the onset of amorphization on ion flux and temperature, have shown a range of apparent activation energies between 0.5 and 1.7 eV as shown in Fig. 6, taken from the work of Goldberg et al. [22]. The conclusion is that more complex defects and defect-interaction processes may control amorphization, depending on the implant conditions used, particularly the implantation temperature.

2.4 Layer-by-Layer Amorphization

Another intriguing case of preferential amorphization is layer-by-layer amorphization, which has been observed in some cases when silicon containing pre-existing amorphous layers is reirradiated at elevated temperatures [23]. An example of such behavior is illustrated by the XTEM micrographs in Fig. 7 [24]. Clearly, the near-surface amorphous layer in Fig. 7a has increased in thickness when irradiated with 1.5-MeV Xe ions at 208°C (Fig. 7b). It

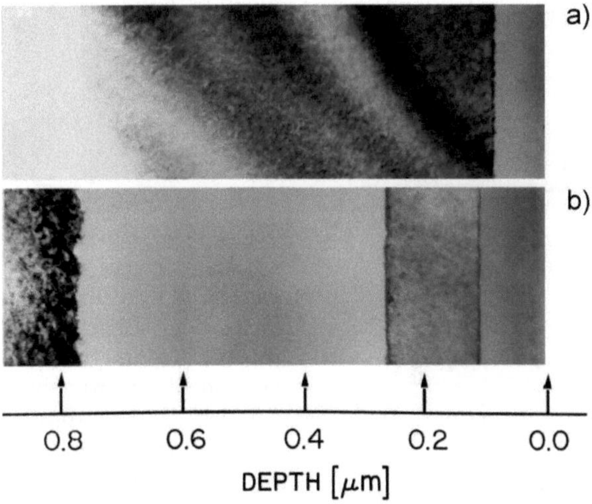

Fig. 7. XTEM images illustrating layer-by-layer amorphization of silicon by 1.5-MeV Xe ion irradiation to a dose of $5 \times 10^{15}\,\mathrm{cm}^{-2}$. (**a**) A pre-existing surface amorphous layer on silicon prior to Xe irradiation, and (**b**) following Xe irradiation at 208°C. After [24]

is also interesting to note that a buried amorphous layer has also formed at the Xe end-of-ion-range under these conditions. The region between the two amorphous layers is essentially free of defects, as a result of near-perfect defect annihilation in this region. Both amorphous layers are observed to extend layer-by-layer with increasing ion dose, presumably by the preferential trapping of mobile defects at the respective amorphous/crystalline interfaces. The degree of interface smoothness may be a function of the defect mobility and trapping at the interface but could also be related to cooperative effects associated with recrystallization coherence lengths in the crystalline side of the interface.

3 Overview of Ion-Beam-Induced Epitaxial Crystallization: Experiment and Modeling

3.1 IBIEC Temperature Dependence

The previous section illustrated implantation conditions where amorphization by ion irradiation is nucleation limited and can give rise to preferential amorphization and layer-by-layer amorphization phenomena. If the ion-irradiation conditions are changed to enhance the rate of dynamic annealing over defect production, by raising the temperature for example, pre-existing amorphous layers can be observed to crystallize epitaxially by the IBIEC process. IBIEC is illustrated for the case of 1.5-MeV Xe irradiation in Fig. 8 [24]. At an irradiation temperature of 227°C, the pre-existing surface amorphous layer is observed to shrink. Increasing the dose causes further epitaxial growth of

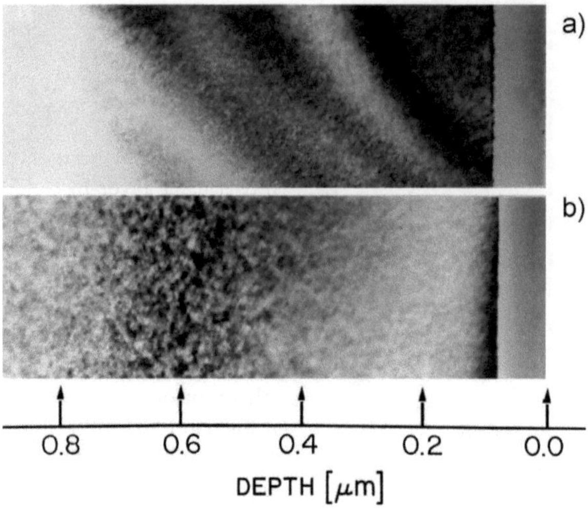

Fig. 8. XTEM images illustrating IBIEC of a pre-existing amorphous layer in silicon (**a**) using 1.5-MeV Xe ions to a dose of $5 \times 10^{15}\,\mathrm{cm}^{-2}$ at a temperature of 227°C. After [24]

the amorphous layer. It is interesting to note that a slight reduction in irradiation temperature to 208°C, keeping the other irradiation conditions the same, induces layer-by-layer amorphization, as previously shown in Fig. 7. If the temperature is increased further, above that corresponding to the data in Fig. 7b, the IBIEC rate speeds up. The temperature dependence of IBIEC is illustrated in Fig. 9 for the case of 600-keV Ne irradiation of silicon [25]. Note that a well-defined activation energy can be extracted from the data (0.24 eV), the magnitude of which is suggestive that defect-mediated processes control IBIEC, possibly vacancies [24, 25]. In Fig. 9, the kinetics of thermally induced epitaxial growth (SPEG) is also shown, with its activation energy of 2.8 eV [26]. It was accepted in early IBIEC studies [25, 26] that the low IBIEC activation energy arose as a result of athermally generated atomic displacements during ion irradiation. These displacements provide the defects for stimulating bonding rearrangements at the interface and hence crystallization. In the thermal (SPEG) case, the high activation energy was attributed [25] to two activation terms, nucleation of the defects influencing epitaxial crystallization and a second term involving migration and bond rearrangement. Hence, it has been suggested [25] that, in IBIEC, the first term can be eliminated by athermal defect generation and only the second activation term applies. This simple model does not take account of observations such as interface planarity and the processes involved in IBIEC may be decidedly more complex, as we discuss in later sections.

3.2 IBIEC Observations and Dependencies

Early studies [23–25, 27, 28] indicated that the IBIEC rate was proportional to ion dose and was controlled by nuclear-energy deposition. This demonstrates that atomic displacements are crucial for IBIEC. Indeed, experiments

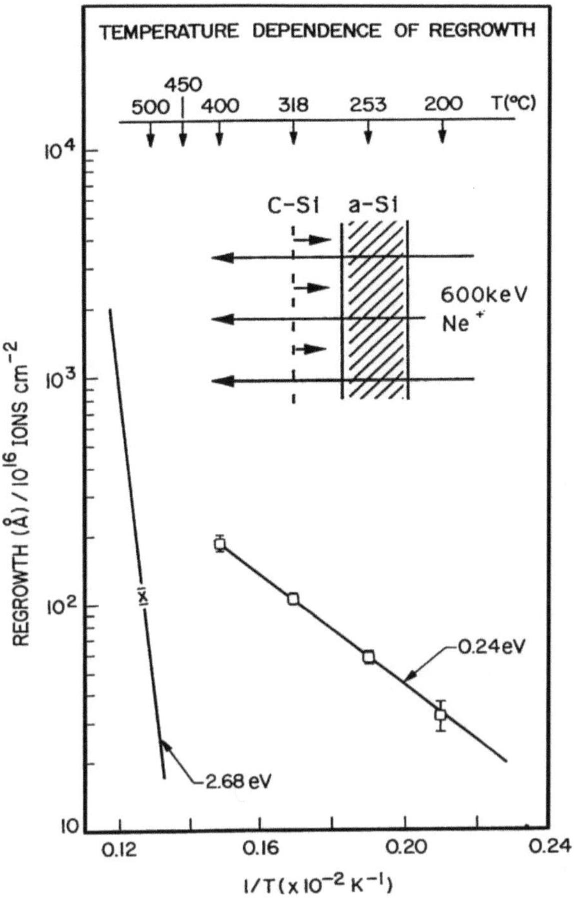

Fig. 9. IBIEC regrowth for a dose of 10^{16} Ne ions cm^{-2} (600 keV) as a function of $1/T$ (*open squares*) in silicon. The activation energy for thermally induced epitaxy (2.8 eV) is also shown. After [25]

with electron beams [29] have clearly shown that recrystallization only occurs if the energy of the electron beam is sufficient to produce atomic displacements in silicon in the region of the amorphous/crystalline interface. Several studies [23–25, 27–30] have suggested that atomic displacements generated by nuclear collisions very close to the amorphous/crystalline interface are responsible for IBIEC. For example, Fig. 10 from Williams et al. [31] shows the dependence of IBIEC on nuclear-energy deposition at the interface. In Fig. 10a, the RBS/channeling spectra show that for 1.5-MeV Ne ions at 318°C the extent of regrowth is linear with dose for this irradiation situation, where the nuclear-energy deposition is relatively constant at the interface as regrowth proceeds. In Fig. 10b, IBIEC growth is plotted as a function of nuclear-energy deposition at the interface (S_n) for Ne ion irradiation at 4 temperatures. Here, three Ne ion energies were used (0.6, 1.5 and 3 MeV) and the atomic displacements generated by the ion beam at the amorphous/crystalline interface (S_n) were obtained from simulations using the TRIM code [32]. The IBIEC rate

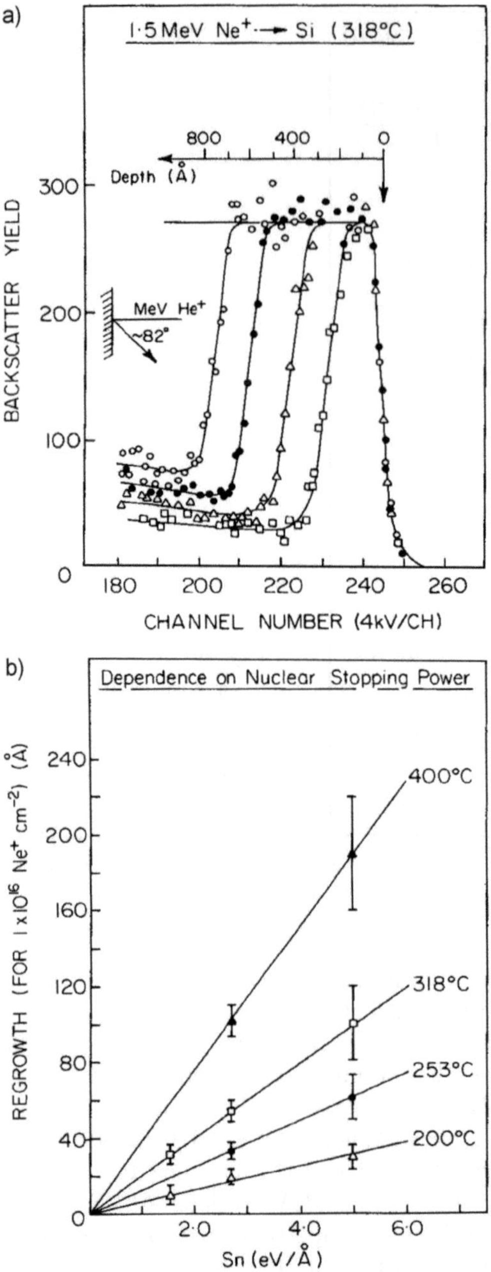

Fig. 10. (**a**) RBS/channeling of Si (with a preamorphized layer) at 318°C and irradiated sequentially by 1.5-MeV Ne ions (dose increments of $3 \times 10^{16}\,\mathrm{cm}^{-2}$). *Open circles*: data for pre-existing amorphous layer. (**b**) IBIEC growth normalized to Ne dose $10^{16}\,\mathrm{cm}^{-2}$ as a function of nuclear stopping power S_n (different substrate temperatures). After [31]

is observed to scale with the nuclear-energy deposition at the interface. This result strongly suggests that long-range diffusion of defects from the amorphous or crystalline sides of the interface do not contribute significantly to

IBIEC but does not rule out short-range diffusion, an issue we return to later.

The IBIEC growth rate is also found to be significantly different for different substrate orientations [28, 33, 34], where a 2–4 times slower rate is observed for ⟨111⟩ compared with ⟨100⟩ orientations. No difference between ⟨100⟩ and ⟨110⟩ orientations is observed for IBIEC. Compare this to the well-known thermally induced (SPEG) case [26], where the growth rate is 25 times slower in the ⟨111⟩ orientation than along ⟨100⟩, and 2.5 times slower along ⟨110⟩ than along ⟨100⟩. These SPEG results were accounted for by a model [2, 35] in which solid-phase epitaxial growth occurs by bond breaking and reforming processes at kinks and ledges on the amorphous/crystalline interface. Rate differences arise from the different concentrations of ledges depending on the sample orientation. Priolo et al. [36] suggested that similar processes may account for the IBIEC orientation dependence. Although it had to be corrected, this idea proved quite fruitful (see Sect. 4).

Impurity species also influence growth speeds along different orientations in thermally induced SPEG, as was discussed by Williams and Elliman [35]. The effects of impurity species on IBIEC are again qualitatively similar to those observed for thermally induced SPEG [26, 37, 38]. For example, slow-diffusing electrically active dopants, such as B and P, are observed to enhance the IBIEC growth rate [36], whereas species such as oxygen, that form strong bonds with silicon, are observed to retard the rate [39]. However, the magnitudes of the rate changes in IBIEC are considerably smaller than those observed for thermal epitaxy, again suggesting that the lower temperatures of IBIEC growth may not achieve thermal equilibrium behavior [30]. Priolo and Rimini [30] also reviewed the IBIEC behavior of fast-diffusing species such as Au and Ag, and noted the similar tendency for such impurities to strongly prefer to remain in the amorphous phase as growth proceeds. This leads to segregation at the moving amorphous/crystalline interface [40]. IBIEC allows such segregation phenomena to be studied at low temperatures, where the interface velocity can exceed the impurity diffusivity in the amorphous phase [41]. We return to this question in Sect. 4.

Although studies of the energy and depth dependence of IBIEC growth, such as that in Fig. 10, indicated that the IBIEC rate scales with nuclear-energy deposition, such scaling across widely different ion masses does not occur. Indeed, ion-mass effects were appreciated early [42], but only relatively recently have they been quantified in terms of an influence of cascade density on IBIEC rate [43]. Furthermore, a small ion-flux dependence of IBIEC [23, 42] was also found in early studies and has been examined over a wide flux range [43, 44]. Such mass effects, which illustrate the role of cascade density on IBIEC, and flux effects, which indicate the interaction times of defects contributing to IBIEC, are illustrated in Fig. 11, taken from the work of Kinomura et al. [43]. Figure 11a shows RBS/channeling spectra that illustrate the mass dependence of IBIEC. Here, 3-MeV Au, Ag, Ge and Si

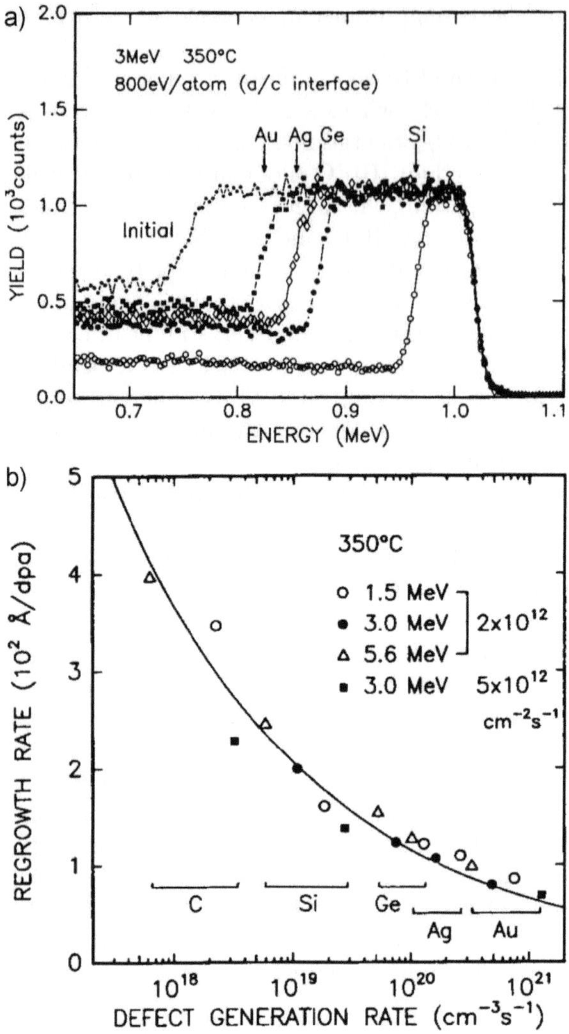

Fig. 11. (**a**) Channeling spectra showing the difference in regrowth thicknesses among four different ion species (Au, Ag, Ge, and Si) at 3.0 MeV. Irradiation doses were adjusted to provide the same total nuclear energy deposition (800 eV per atom) to the initial amorphous/crystalline interfaces. (**b**) Normalized regrowth rates as a function of defect-generation rate for five ion species (C, Si, Ge, Ag, and Au) at three energies (1.5, 3.0, and 5.6 MeV) with two dose rates (2×10^{12} and $5 \times 10^{12}\,\mathrm{cm^{-2}\,s^{-1}}$). After [43]

ions were used to irradiate an amorphous silicon layer of about 2000 Å in thickness on a silicon ⟨100⟩ substrate. Different doses were chosen to provide the same total nuclear-energy deposition at the amorphous/crystalline interface and MeV ions were chosen to provide a near-constant energy deposition at the interface during IBIEC growth. It is clear from Fig. 11a that the regrown thickness increases with decreasing ion mass, even though the total nuclear-energy deposition is similar for each ion within the range of the measured depth. This clearly shows that, at the same average ion flux, the *rate* of nuclear-energy deposition, or the cascade density, clearly influences IBIEC. Another effect observed by Kinomura et al. [43] was a flux dependence, whereby higher fluxes of the same ions under otherwise identical conditions resulted in less regrowth. This is again consistent with the observation that the rate of nuclear-energy deposition influences IBIEC. Figure 11b plots the IBIEC regrowth rate (normalized to constant nuclear-energy deposition at the interface) as a function of defect (i.e. vacancy) generation rate at the interface for five ion masses, four ion energies and two fluxes at 350°C. The defect generation was calculated using TRIM [32]. Note that the defect-generation rate varies over more than 4 orders of magnitude from C to Au and the normalized growth rate for C is about 4 times that of Au under these conditions. A similar dose-rate dependence for 300-keV ions has also been demonstrated by Linnros and Holmen [45] and Heera et al. [44]. However, Kinomura et al. [43] subsequently varied the ion flux for similar-mass ions over a wide range and found that cascade-density and ion-flux changes do not give identical changes to IBIEC rates. These results are shown in Fig. 12a for Au and Ag ions, where the IBIEC rate seems to vary linearly with defect generation. These data suggest that cascade size and ion flux give rise to separate influences on IBIEC rate, in addition to their common influence on defect-generation rate, as we discuss more fully below. In Fig. 12b we illustrate another case where more extensive data provide further insight into IBIEC processes. These data show that the apparent activation energy of IBIEC extracted from temperature-dependent studies can vary from 0.18 to 0.4 eV, depending on ion mass.

We discuss the significance of these observations in the discussion of IBIEC mechanisms in Sect. 3.4.

3.3 Ion-Cascade Effects on IBIEC: The Role of Atomic Displacements and Mobile Defects

A particularly important question in IBIEC is: if atomic displacements are necessary to induce crystallization, then do such displacements have to be exactly at the amorphous/crystalline interface or can they be induced away from the interface in either the crystalline or amorphous phases? It is clear from a range of early studies [25, 27, 45, 46], where the ion mass and energy were varied to change the magnitude and depth of energy deposition into atomic displacements, that atomic displacements *close* to the interface play the major role, but how close? Irradiation under ion-channeling conditions in the crystalline side of the interface can, in principle, help to clarify where the defects that influence IBIEC are generated, since channeling of ions along crystal lattice rows allows selective reduction in the number of atomic displacements and hence defects produced in the crystalline region. However, in the early measurements using channeling [25, 45, 46], the interpretation of the results (i.e., where the displacements that triggered IBIEC originated from) was not conclusive, mainly because it was difficult to estimate the exact number of point defects generated in the crystalline region

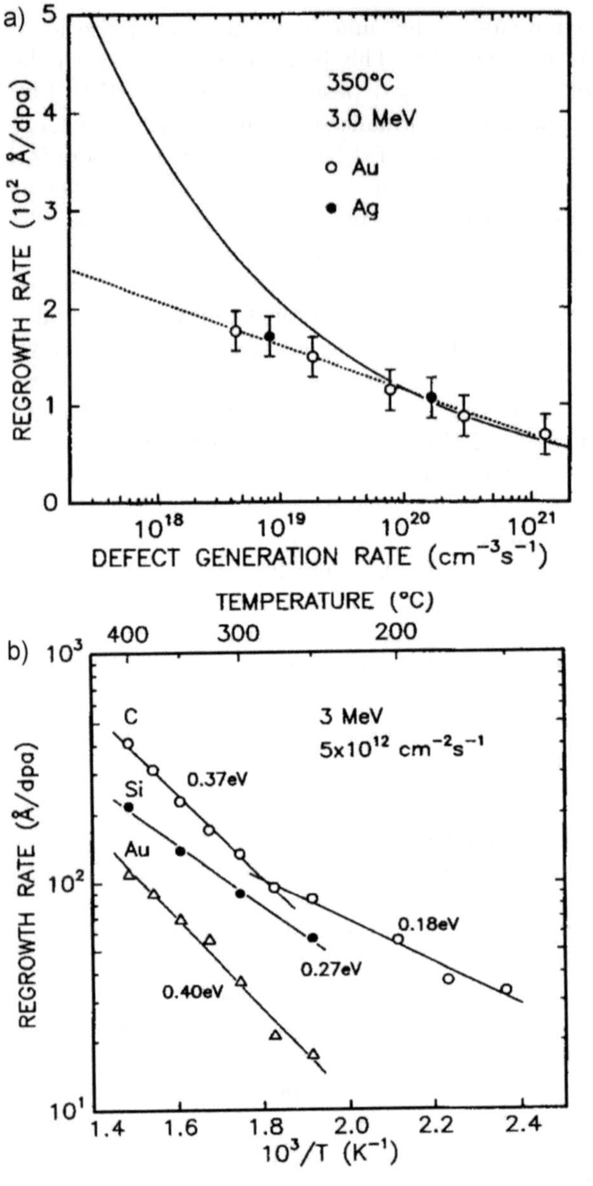

Fig. 12. (**a**) Dose-rate dependence of IBIEC for 3.0-MeV Au and Ag compared with the fitting curve of Fig. 11b (*solid curve*). (**b**) Temperature dependence of IBIEC regrowth rates normalized to the number of displacements for 3.0-MeV Si, Ge and Au with a dose-rate of $2 \times 10^{12}\,cm^{-2}\,s^{-1}$. After [43]

after an ion beam had traversed an amorphous layer before entering the crystal. Ion-channeling irradiations using a buried amorphous layer in which to induce IBIEC were more conclusive [25, 46], since the ion beam can then be channeled in the top crystalline layer before the amorphous layer is entered, thus reducing the number of displacements in the crystal by more than 90%. A large reduction in IBIEC growth rate was observed [25, 26] for the

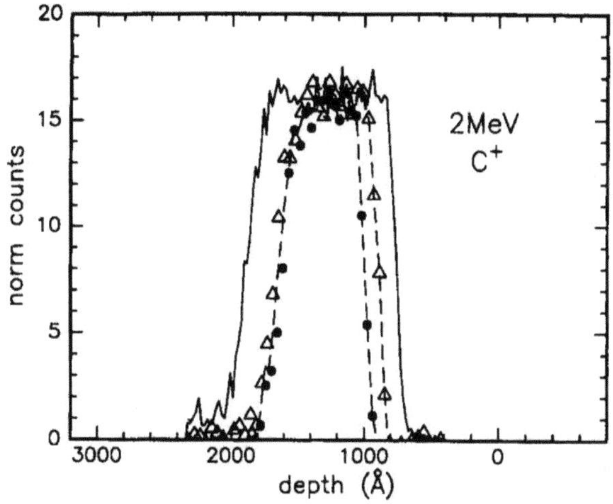

Fig. 13. Random and channeled IBIEC regrowth extracted from RBS/channeling spectra (1.8-MeV He ions) for a 1000 Å amorphous silicon layer buried about 1000 Å below the surface. The data has been corrected for channeled He energy-loss effects. The 2-MeV C ion dose was $1.2 \times 10^{16}\,\mathrm{cm}^{-2}$ in both random (*filled circles*) and channeled (*open circles*) cases. After [47]

near-surface interface of the buried layer (40–100%), compared with a case where the ion beam was randomly oriented in the top crystal. An example of the channeling effect on IBIEC for a buried amorphous layer is shown in Fig. 13 [47]. Here, the regrowth differences are compared for channeled and random irradiations with 2-MeV C ions at 320°C in the top crystalline layer before entering a buried amorphous layer initially 1000 Å thick. Clearly, the front amorphous/crystalline interface under channeling grows only 60% of that under random alignment, whereas the rear interface appears to show no differences between the two irradiations, noting that only roughly 50% of the ion beam will be channeled in the deeper crystal region after transport through the amorphous layer. Overall, this result appears to suggest, in contradiction to earlier reports [25, 45, 46], that there is some role for mobile defects from the crystalline side of the interface rather than displacements exactly at the interface, but there remains a need for accurate simulations of displacement cascades (depth distributions of displacements) under channeling conditions before definitive conclusions can be drawn as to the precise origin of the 'defects' responsible for IBIEC, as we illustrate below.

Prior to reviewing cascade simulations to help interpret experimental IBIEC rates, we note a further difficulty with the IBIEC measurements under channeling conditions that were reviewed above. These measurements obtained the extent of regrowth from ex-situ RBS analysis (which has limited depth resolution) after successive irradiations. More recent studies [48] have used in-situ time-resolved reflectivity (TRR) to more accurately monitor IBIEC growth during irradiation under channeling and random alignment conditions in the silicon crystal that either overlays or underlies the amorphous layer. Results are shown in Figs. 14 and 15 for the cases of surface and buried amorphous layers, respectively. In Fig. 14, for a surface amorphous layer irradiated with 7-MeV Au ions under both random and channeling

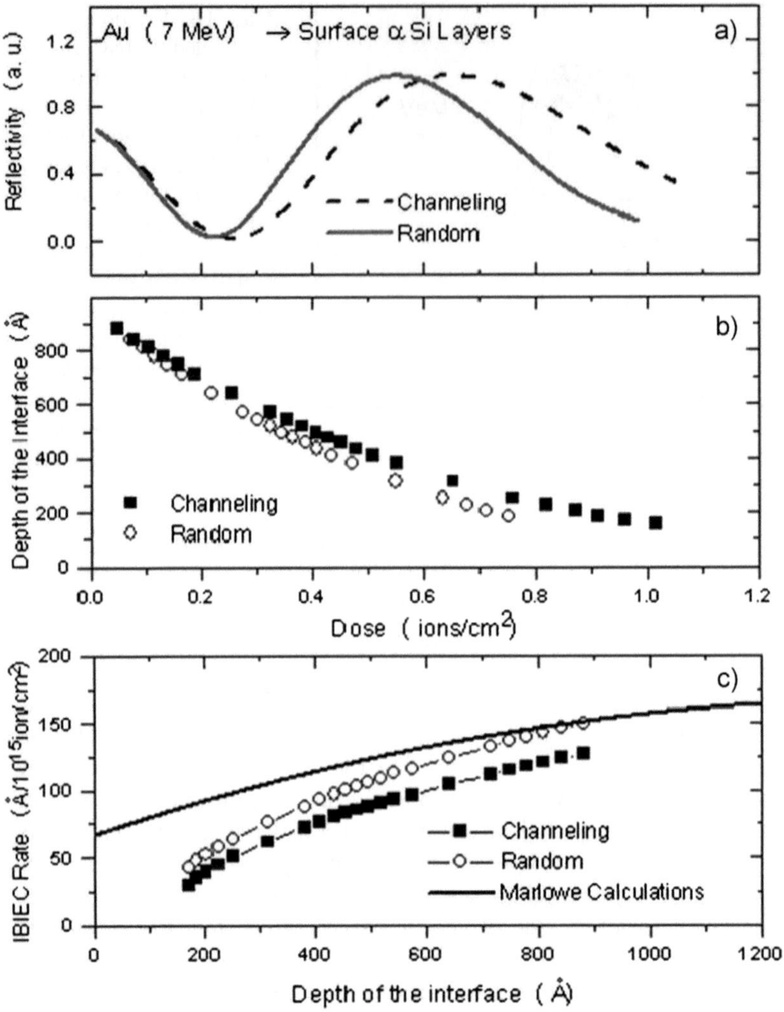

Fig. 14. (**a**) Experimental reflectivity traces, as a function of dose, for 7-MeV Au ions irradiating a surface amorphous layer in silicon. The *solid* and *dashed lines* correspond to the random and channeling irradiations, respectively. (**b**) Depth of the interface as a function of the ion dose. (**c**) IBIEC rates for channeling (*solid symbols*) and random (*open symbols*) cases as a function of the interface depth. The *solid line* corresponds to MARLOWE calculations for the number of vacancies produced per ion per Å at the interface. After [48]

conditions, the experimental TRR traces for random (dashed line) and channeling (solid line) cases are plotted in panel (a). Note that for TRR from silicon using a 6328 Å laser, every 330 Å of growth (interface movement) corresponds to a complete oscillation between a maximum and a minimum of

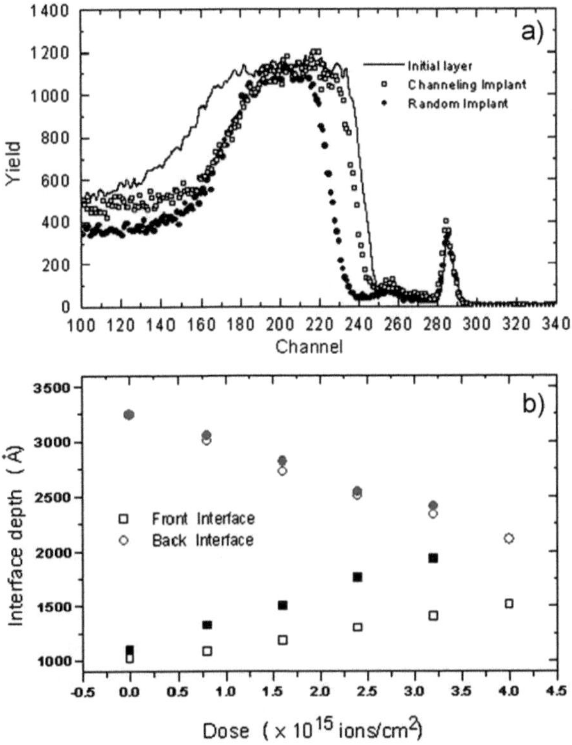

Fig. 15. (**a**) RBS spectra for a buried amorphous layer irradiated with 1.6×10^{15} Au cm^{-2}. The *solid line* is the spectrum for the buried layer before the irradiation. Symbols (*squares* and *circles*) correspond to the random and channeling irradiations, respectively. (**b**) Position of the interfaces as a function of dose. *Open* and *solid symbols* correspond to channeling and random irradiations, respectively. After [48]

the reflectivity. The comparisons between the IBIEC growth under channeling and random alignments of the Au beam are shown in panels (b) and (c) of Fig. 14. The results indicate that there is an effect of channeling in the underlying crystal but it is quite small. For example, the maximum difference in the interface depths between channeling and random alignment cases is of the order of 80 Å and the IBIEC rate for channeling implants is only 20% smaller than the rate observed for random implants.

In Fig. 15, the IBIEC results for buried amorphous layers in silicon are shown [48], again for 7-MeV Au irradiation. Panel (a) displays RBS spectra for the buried layer before irradiation (solid line) and after 3×10^{15} Au cm^{-2} random and channeling bombardments (squares and circles, respectively). Again, a clear channeling effect is observed for the front interface between channels 220–250. This difference is better quantified by an inspection of panel (b), where the position of the amorphous/crystal interfaces is plotted as a function of the ion dose. It is apparent in this figure that the deeper interface (circles) advances at the same rate (281 ± 10 Å and 274 ± 14 Å per 10^{15} ions cm^{-2}) for channeling and random irradiations, respectively. However, the shallower interface (squares) advances much faster in random irradi-

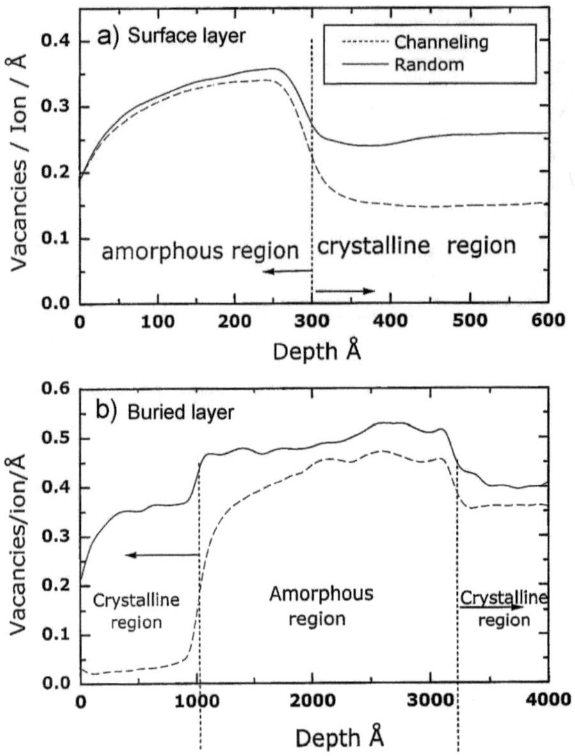

Fig. 16. Point-defect profiles calculated with MARLOWE for channeling (*dashed lines*) and random irradiations (*solid lines*) in surface and buried amorphous layers shown in *upper* and *lower panels*, respectively. After [48]

ations than in channeling cases (262 ± 10 Å and 124 ± 5 Å per 10^{15} ions cm^{-2}, respectively).

The IBIEC growth data presented above (Figs. 14 and 15) are now compared with the results of computer simulations of collision cascades (atomic displacements). For the simulations, all displacements (point defects), both in random and channeled alignments, were calculated with the aid of the MARLOWE code [49, 50]. MARLOWE has been specifically developed for the simulation of atomic displacements in both amorphous and crystalline solids. The code is based on the binary collision approximation (BCA) [51] to construct particle trajectories. The atomic scattering is governed by screened potentials, such as ZBL [32] and Molière [52]. Thermal vibrations are simulated by a random Gaussian distribution of the lattice atoms around their equilibrium positions, with amplitude given by the Debye–Waller [53] model. Figure 16a displays the result of MARLOWE calculations for a 300-Å surface layer [48]. A reduction in the number of vacancies produced per ion per Å (η) in the crystalline region is clearly apparent, even for random bombardments. This feature can be explained as follows. Even though the nuclear-energy dissipation occurs mainly in cascades initiated by high-energy collisions between the incident Au ions and Si target atoms, the average energy transferred to a silicon atom by 7-MeV Au ions is of the order of 0.5 keV only. Such low-energy primary Si knock-ons have a large critical angle for channeling in the crystal (of the order of several degrees) and hence the number of displacement collisions with further Si target atoms that they initiate in crystalline silicon is less than in amorphous silicon because of the high channeling probability in crystalline silicon. This explains the reduction of η in the crystalline region, even for a random orientation of the beam. Furthermore, when the Au beam is aligned with the $\langle 100 \rangle$ channeling direction in the underlying crystalline silicon, the number of vacancies generated at the interface and within the crystalline region is lower than for the random case. It is also interesting to note that, under channeling conditions, η is slightly reduced in the amorphous region, in comparison with random implants. This latter observation implies that cascades initiated in the crystalline region can produce displacements in the amorphous region, even though it is closer to the surface. A comparison of the experimentally observed $\sim$20% lower IBIEC rate for channeling-beam alignment (Fig. 14) with the simulation data in Fig. 16a, indicates that the scale of difference between channeled and random IBIEC rates is more consistent with vacancies produced precisely at the interface than with vacancies produced in the amorphous or crystalline regions.

Figure 16b displays the results of simulations for a buried layer in silicon irradiated with 7-MeV Au ions. It is apparent that, for channeling implants, η is strongly reduced in the crystalline region near to the surface, as one might expect. However, only a small reduction of η is observed after the deeper interface. These features can be explained by the same arguments utilized above to explain the results for surface layers. Therefore, the most important feature displayed in Fig. 16b is that, assuming IBIEC is controlled by point defects generated at the amorphous/crystalline interface, MARLOWE predicts a large channeling effect at the front interface and a very small effect at the back interface, consistent with the experimental data for a buried amorphous layer (Fig. 15). Furthermore, the scale of the experimental IBIEC reduction rate for the front interface under channeling conditions ($\sim$50%) appears to best correlate with the relative number of vacancies produced at the front interface (Fig. 16b), rather than displacements within the amorphous or crystalline regions, as discussed below.

The solid line in Fig. 14c corresponds to the predictions of MARLOWE for vacancies (η) produced at the amorphous/crystalline interface. As can be readily observed, η drops quickly as the interface approaches the surface. This feature is a result of the reduction in the cascade density for shallow depths and the experimental IBIEC rates (γ) display a similar trend. However, γ is clearly steeper than η when the thickness of the amorphous layer is smaller than about 500 Å. We suggest that effects other than defect diffusion within the near-surface region could be responsible for this behavior and for the discrepancies with MARLOWE predictions. For example, it has been demonstrated previously that the IBIEC rate is affected by defect interac-

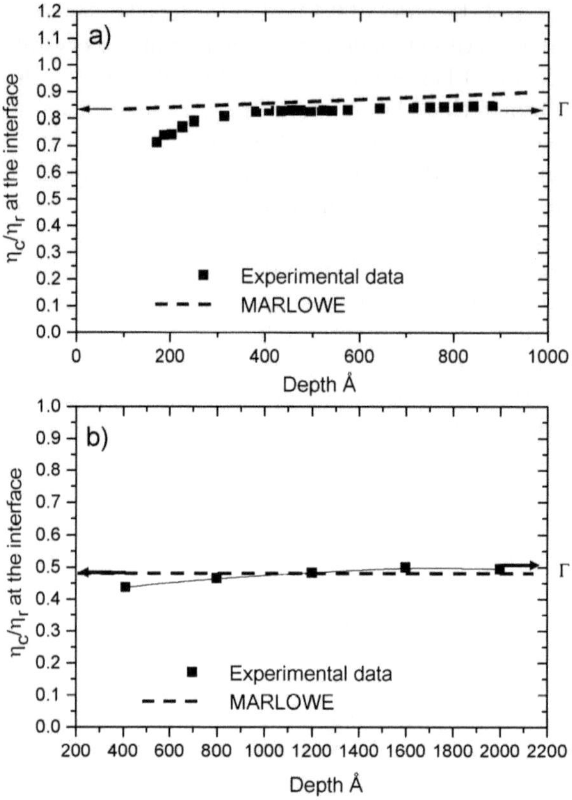

Fig. 17. Ratio of η_c/η_r between the experimental growth rates and the calculated displacements at the interfaces. *Upper* and *lower panels* depict results for surface and buried layers, respectively. Adapted from [48]

tions within individual cascades (i.e. the cascade density) as well as by defect interactions between cascades [43]. This suggests that the observed thickness dependence of γ could be related to a distortion of the point defect profiles at the interface when the interface is close to the surface, due to cascade-density differences and cascade interactions, rather than being related to point defect diffusion. Furthermore, Kinomura et al. [54] have demonstrated that oxygen impurity atoms recoiling from the surface native oxide contribute partially to a decrease in the IBIEC rates close to the surface. Therefore, the comparison of MARLOWE predictions with the experimental results for shallow surface amorphous layers is not straightforward.

In order to more precisely determine the origin of the defects that control IBIEC, the ratio between the channeling and random IBIEC rates ($\Gamma = \gamma_c/\gamma_r$) is compared to the ratio between the corresponding simulated defect profiles. This method removes the influence of shallow surface layers, chemical contamination and cascade interaction effects that are canceled out. In Fig. 17, the ratio Γ between the experimentally determined IBIEC rates under channeling and random conditions is compared to the ratio between the corresponding calculated defect levels (η_c/η_r) at the amorphous/crystalline interface [48]. As can be observed, the magnitude of the experimentally determined Γ is in good agreement with MARLOWE calculations for the ratio η_c/η_r. Furthermore, from the results in Figs. 14 and 15, a dominant role of defects coming from the amorphous region can be eliminated as a possibility since the simulations show that η is reduced by only ~5% (surface amorphous layer) or 10% (buried amorphous layer) under channeling conditions, while the observed channeling effect on IBIEC is of the order of 20% and 50% for surface and buried layers, respectively. On the other hand, the simulations for a buried layer indicate that defects produced in the crystalline region are not likely to be participating in IBIEC, since the simulations predict a 90% reduction of η in the crystalline region close to the surface, while the observed channeling effect is of the order of 50%. Therefore, combining all experimental and simulation comparisons, we conclude that defects produced at or very near the amorphous/crystalline interfaces are most likely to control IBIEC. Although the precise interface defect controlling IBIEC is not revealed by these results, the data is consistent with any crystallization-enabling defect, such as a kink, produced at the interface by the ion beam.

3.4 IBIEC Models

Priolo and Rimini [30] gave an overview of various models to explain IBIEC observations up to about 1990. An early proposal suggested that annealing processes, which occur in the quenching of thermal spikes that overlap the amorphous/crystalline interface, were responsible for IBIEC [55]. Minimum free-energy arguments and differences in free energy of amorphous and crystalline silicon have also been invoked to explain the temperature dependence of ion-induced amorphization and crystallization [56]. However, such proposals do not address many of the observations and also fail to suggest which 'defects' may stimulate IBIEC. Vacancies were suggested by several authors [42, 57, 58] as the prime defect involved. First, the similarity of the initial activation energy of IBIEC (around 0.3 eV) to that of vacancy migration led Linnros et al. [42] to propose that migrating vacancies, produced athermally by the ion beam, mediated IBIEC, whereas, if the temperature was lowered, then the increased stability of divacancies, with a dissociation energy of 1.2 eV, may cause amorphization at the interface. This two-defect model qualitatively explains both the growth of an amorphous layer and the IBIEC process but presupposes the migration of such defects in crystalline silicon to the interface. Other defects proposed to mediate IBIEC are (charged) kinks [25, 30] and dangling bonds [59] that are formed athermally by the ion beam directly at the interface. A difficulty with a single-defect model is the fact that the apparent activation energy of IBIEC has been shown to vary from about 0.18 to 0.4 eV (see Fig. 12b). This led Kinomura et al. [43] to suggest that the rate-limiting effect in IBIEC may involve several different defect-mediated processes, depending on the cascade density at the interface and the temperature. This does not necessarily preclude kinks or other specific interface defects as the final step in the IBIEC process, but rather suggests that more complex defect processes may be involved in the annealing of dense cascades before discrete kinks are formed. A particular concern of vacancy models is that there is now considerable weight, as was indicated through Figs. 14–17, to arguments suggesting that defects produced right at the interface dominate IBIEC.

Another explanation for both the extension of amorphous layers by ion irradiation and IBIEC is due to Jackson [20], who developed an intracascade model in which each ion penetrating through the interface creates a disordered zone. Subsequent local interaction between defects in this zone can either lead to amorphization or crystallization. The onset of either amorphization or crystallization is controlled by a rate equation in which the net rate of interface movement, R, is given by the difference between a crystallization term, R_x, and an amorphization term, R_α, according to:

$$R = \mathrm{d}x/\mathrm{d}\varphi = R_x - R_\alpha, \tag{1}$$

where x is the distance of interface motion and φ is the ion-beam dose. The amorphization term can be written as $R_\alpha = V_\alpha \varphi$, where V_α is the volume of the amorphous zone created by a single ion. Crystallization arises when defects produced by the ion beam annihilate in pairs at the interface. The simplicity of the Jackson model is attractive but it does not adequately account for ion mass and flux effects. Thus, no single existing model appears to adequately explain all observations.

3.5 Interface Evolution

Let us first concentrate on the properties of the interface rather than on the underlying microscopic defect mechanisms leading to the latter. As noted above, Priolo et al. [36] suggested that the kink-and-ledge mechanism [2, 35] devised to explain thermal epitaxial growth (which has since been observed directly via in-situ high-resolution TEM experiments [60]) be extended to analyze ion-beam-induced interfacial growth. In the kink-and-ledge mechanism, the interface is resolved into surfaces of minimum free energy by the formation of terraces with a {111} orientation, separated by [110] ledges so as to maximize the number of bonds with the crystal (Fig. 18). Regrowth involves thermally activated bond breaking and rearrangement at these sites, i.e., depends on the number of (110) ledges formed on (111) terraces – hence on the crystal orientation during growth. The two physical processes to consider are: (1) the probability that a kink (a dangling bond) be created at the interface along the [110] ledges of (111) terraces, where, under thermal equilibrium crystallization conditions, this quantity is essentially zero below a threshold temperature, and grows exponentially above it; (2) the change in growth speed for different orientations is due to the differences in ledge densities, but not to the number of "recrystallized" sites. The latter is constant [61]: each kink "recrystallizes" 200 atoms, the ledge structure remaining

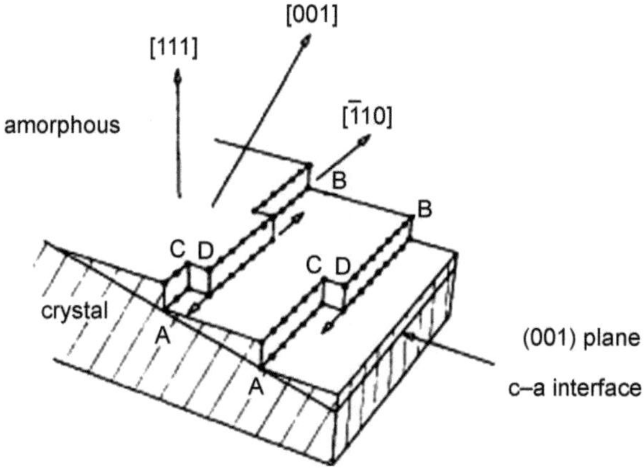

Fig. 18. Kink-and-ledge model of thermal and ion-beam-induced epitaxial crystallization, showing the a/c (001) interface, with kinks (CD) moving along the [110] ledges (AB) on [111] terraces. From [35, 36]

unchanged as growth occurs and the interface roughness is very low in this case.

The extension of this model to IBIEC [62, 63] assumes that growth occurs via a similar kink-and-ledge mechanism as above, but rather than relating the growth's orientation dependence to the ledge density, Monte Carlo simulations were performed assuming that the energy deposited by the ion beam ultimately initiates dangling bonds at *any site on the* {111} *terraces of the* aSi/cSi *interface* (hence enhancing the interface roughness). The probability that a site is efficient in inducing crystallization depends on its surroundings, the most efficient ones being those that have the maximum number of neighbors on the crystal side of the interface. In this picture, the number of recrystallized sites per kink is nearer to unity than to 200, because kink propagation is limited by surface roughening, and the orientation dependence of the crystallization speed is considerably less anisotropic than that found in thermal growth. An excellent fit to the growth-orientation dependence was found by Custer et al. [62] assuming maximum roughening, i.e., no constraint on local configurational energy (Fig. 19). Note that this model – which says nothing of how the kink is created – is compatible with Jackson's model, since the latter does not consider the interface structure.

This result allows us to bridge the gap with the basic physical concepts of surface growth. How does surface roughness change as growth occurs? In most cases [64], the roughness increase with time t follows scaling laws such as

$$\delta(t) \sim t^{\beta}, \tag{2}$$

where β is an exponent that characterizes the growth mechanism. Generally, saturation sets in after a sufficiently long time t_x, the maximum roughness δ_{sat} then being related to the system's size L via

$$\delta_{\text{sat}} \sim L^{\alpha}[t \gg t_x]. \tag{3}$$

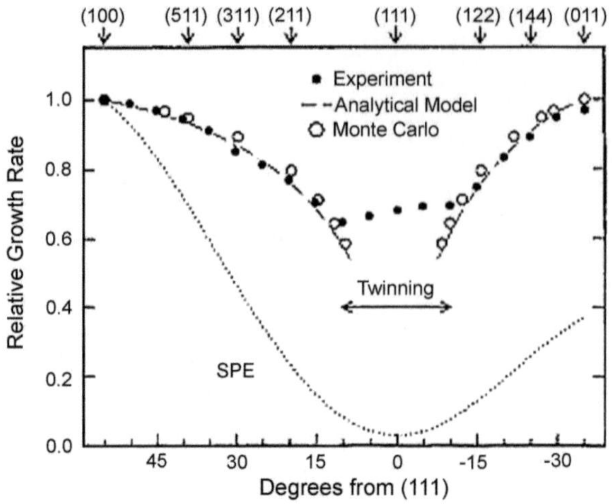

Fig. 19. IBIEC interface growth rate normalized to the value along (100). Comparison of experiments to analytical and Monte Carlo models, assuming the kink-and-ledge growth mechanism and random dangling bond formation. SPE is the thermal solid-phase epitaxy result. From [62]

The time t_x to reach saturation also depends on the system's size according to $t_x \sim L^z$. The exponents α, β, and z characterize the growing system: z is termed the dynamic exponent, while α and β are, respectively, the roughness and the growth exponents. The roughness evolution may be renormalized to the system's size [65] via the scaling law

$$\delta(L,t) \sim L^\alpha f\big(t/L^z\big),$$
$$\text{with } f(u) = u^\beta \text{ if } u \ll 1 \text{ and } f(u) = cst \text{ if } u \gg 1, \tag{4}$$

and the exponents are connected via the scaling law $z = \alpha/\beta$. Different exponent values signal differences in the universality classes of possible surface epitaxial reconstruction mechanisms, essentially as regards the existence of spatial correlations due to surface relaxation during or after adatom deposition. In the absence of such correlations (random columnar deposition), growth is a stochastic process so that $\delta^2 \sim Dt$, where D is an effective diffusion constant characterizing randomness. This leads to an exponent $\beta = 1/2$, whereas α is undefined since the roughness does not level off. Introducing lateral correlations due to relaxation on neighboring lower sites ("random correlated deposition") leads to a linear (Edwards–Wilkinson [66]) equation whose exponents in dimension 1 are $\beta \sim 1/4$ for growth; relaxation-induced lateral correlations lead to roughness saturation in a finite-size system, with $\alpha \sim 1/2$. In the more realistic case where the relaxation mode generates lateral (as well as perpendicular) growth, e.g., when adatoms stick to the nearest occupied site that they find, Kardar–Parisi–Zhang (KPZ) [67] showed that a nonlinear term adds on to the Edwards–Wilkinson equation, and the expo-

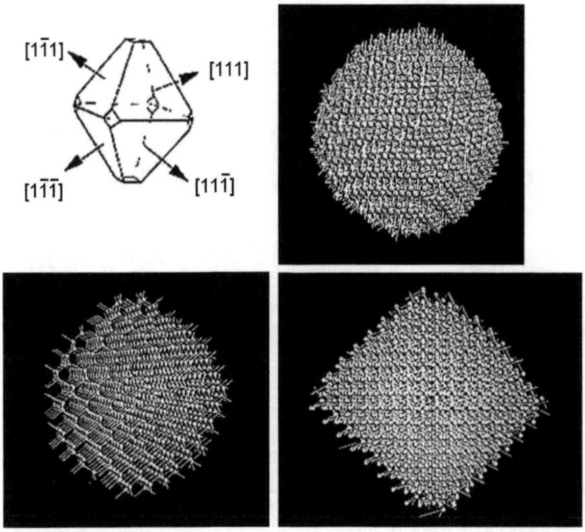

Fig. 20. Monte Carlo simulation of (111)-facet IBIEC on a small precipitate, to emphasize how roughening depends on the number of recrystallized sites per kink. *Lower left, upper right*: two growth stages for $n = 1$; *lower right*: ultimate growth stage for $n = 10$. See text (F. Fortuna, P. Nedellec and H. Bernas, unpublished)

nent values are respectively $\alpha \sim 1/2$ and $\beta \sim 1/3$, with scaling $\alpha + z = 2$. This universality class is particularly important in many areas of growth research, far beyond crystal surfaces. How do these results relate to ion-irradiation-induced interfacial growth? The latter's evolution does not involve any increase in the amount of matter (no adatoms), but the interface motion and roughness are modified during crystallization. This implies that matter has effectively moved along the interface. Can one specify a growth mode in terms of the theories sketched above?

In addition to assuming random initiation of growth sites on the interface, the Monte Carlo simulations of Fortuna et al. ([63] and unpublished work) included local configurational energy minimization: when a kink site was created at random, the neighboring sites – up to 3 near neighbors – were explored to identify whether they belonged (or not) to the crystal. A hierarchy of favorable growth configurations are chosen: first that where 3 neighbors belonged to the crystal, then 2, and 1.

The only free parameter is then the number of sites that the kink may "recrystallize". Simulations were performed for a planar interface, and also for a small, (111)-faceted precipitate in order to emphasize the evolution of the interface roughness. Figure 20 shows the latter case, with (lower left, upper right) two different stages of evolution in the case where each kink only recrystallizes a single site ($n = 1$), and (lower right) a case where each kink recrystallizes up to 10 sites ($n = 10$). The two figures on the RHS correspond to the same number of runs. The effect of roughening is obvious: it is stronger and saturates more quickly when $n = 1$. Note that the rounding of the shape is solely due to kinetic growth – there is no diffusion. The growth speeds for small n agree with experimental IBIEC speed values in the range where the thermal contribution to IBIEC is small, and the roughening amplitude is

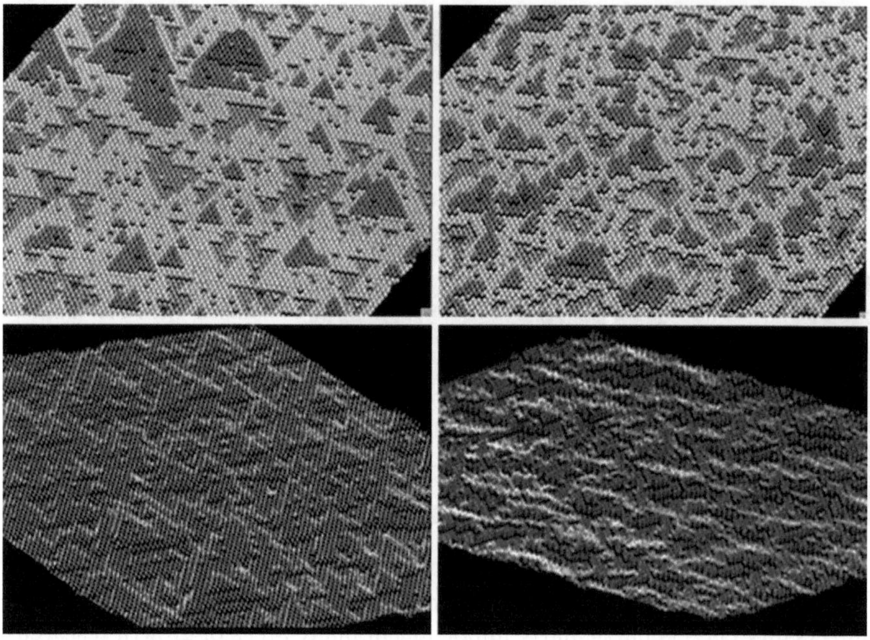

Fig. 21. Early stage of (111)-plane IBIEC simulated by Monte Carlo modeling with kink-and-ledge model as described in text. LHS: $n = 10$, RHS: $n = 1$ (F. Fortuna, P. Nedellec and H. Bernas, unpublished)

quite close to the only experimentally measured value [68]. It is interesting that the simulated values of the growth speed are in reasonable agreement with experiments. The same effect is seen in more detail on the planar interface (Fig. 21), in which the LHS shows the interface when $n = 10$ (note the triangular mounds familiar from STM studies of Si surface growth), whereas the RHS shows two stages of the roughened growth landscape obtained when $n = 1$. A logarithmic plot of the interface roughness δ as a function of the average crystallized thickness $\langle H \rangle$ provides the growth exponent β, shown for $n = 1$ and $n = 10$ in Fig. 22. As indicated above, the roughness exponent α depends on the system size (denoted here by L, number of atoms in a row). The lower part of the same figure shows how α is obtained in the two configurations. Unsurprisingly, the $n = 1$ case (no kink propagation) corresponds to the Edwards–Wilkinson universality class (random correlated deposition), whereas the $n = 10$ case, implying significant lateral growth component, fits the KPZ exponents rather nicely. The transition between the two growth modes takes place for very small values of n (2, 3). Thus, we conclude that the role of the interface roughness in IBIEC is very significant, and perhaps a major one in determining the growth speed in pure silicon. When a sufficient concentration of solute atoms is involved in the IBIEC process (see

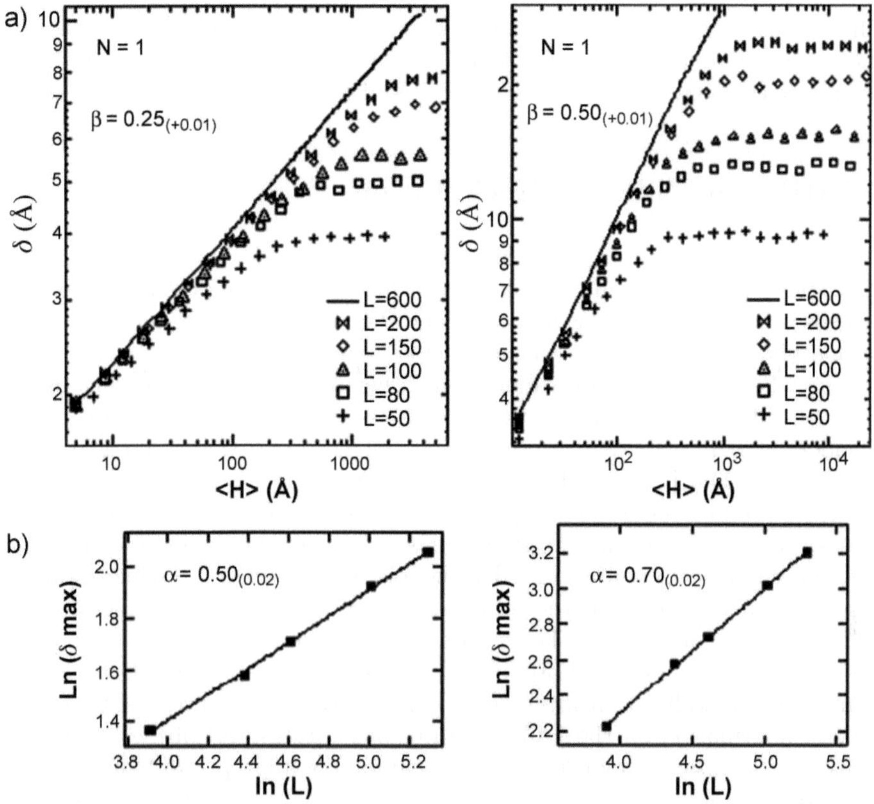

Fig. 22. (**a**) Evolution of interface roughness δ as a function of average crystallized height $\langle H \rangle$ from Monte Carlo simulations as described in the text. Note logarithmic scale. The growth mechanism exponent β is deduced from the slope. LHS: case where $n = 1$, RHS: case where $n = 10$. Different curves correspond to simulations performed for different interface sizes (L = number of sites). This allows (**b**) the roughness exponent α to be deduced from the size dependence of δ (F. Fortuna, P. Nedellec and H. Bernas, unpublished)

below), the interface roughness determines the precipitate density as well as the precipitation process (it is the source of Volmer–Weber growth).

As mentioned previously, these results do not bear upon the microscopic origin of IBIEC or ion-beam-induced amorphization. This was studied in detail via molecular dynamics (MD), combined in some cases with kinetic Monte Carlo simulations [69–71]. A specific, previously known structural bond defect – identified as an interstitial–vacancy (IV) pair when formed by irradiation – was able to account for many features of the amorphization process, including the latter's temperature dependence via the IV recombination probability.

This defect is compatible with the kink structure discussed above. However, these MD simulations are still comparatively "local", and do not yet show the interface geometry over a length scale sufficient to evaluate roughness.

4 IBIEC and Silicide Precipitation

We have seen that the crystallization mechanism for IBIEC is basically the kink-and-ledge mechanism, and that both kinetic and thermodynamic growth processes are involved. We now consider the relation between IBIEC and second-phase precipitation, which provides interesting results for interface physics where two- and three-dimensional phenomena interact strongly. Suppose we diffuse or implant metal species (such as those that easily form silicides) at per cent-range concentrations into the a-Si side of an a-Si/c-Si bilayer, and then perform IBIEC. As the interface moves through the a-Si, it crosses a solute metal "flux": precipitation, and various phase transformations occur on the interface itself. There is a striking analogy between IBIEC and molecular beam epitaxy (MBE): the a-Si/c-Si interface, moving towards the static metal atoms in the a-Si phase, mirrors an incoming metal flux falling on the c-Si surface. The very existence of the interface motion allows us to study some dynamical properties of these transformations. Because the elementary crystallization mechanism is the same in both processes, rather general information on the building up of precipitates and phases at interfaces may be obtained by using the ion beam in the appropriate temperature range to control atomic motion at the interface. Also, such precipitates may be useful for various applications if small enough and if their structures can be controlled. Typically [72, 73], (1) Cross-sectional high-resolution electron microscopy (HREM) pictures taken at differing stages of interfacial growth showed that precipitation occurs on the crystallization front as it progresses; (2) Precipitate sizes depend on the impurity concentration – a concentration profile leads to a size distribution; (3) A detailed study [74] of $FeSi_2$ precipitation in Si showed that the crystallites' structure and epitaxial relation to the c-Si host depends on their size rather than on the equilibrium phase diagram. What is the driving force for precipitation? What determines the precipitate density? What determines the phase structure? In the following, we show that interface roughness determines precipitation, hence the importance of the interfacial energy and of the strain energy in determining the phase and structure of the precipitates. These considerations are directly related to the wealth of experiments and theoretical analyses of surface-based phenomena. The consequences are interesting for precipitate size and structure engineering.

4.1 Precipitate Distribution

In MBE, surface diffusion of deposited atoms leads to island formation via atom-pair formation and growth at island ledges. In IBIEC, trapping will likely occur at "growth sites", which in the IBIEC interface are precisely those that correspond to terrace or ledge roughening. Hence, nucleation and subsequent cluster growth should occur at the "slopes" (rather than the "peaks") in the roughened interface. The average precipitate distance would replicate the average distance between these configurations. It may be deduced by analogy with the classical estimate of islanding density due to diffusion-limited growth by trapping on clusters, (in our case it is reasonable to assume volume, rather than two-dimensional, diffusion). The precipitate density N is [75]

$$N \sim (D/F)^{-\gamma}, \tag{5}$$

where D is the diffusion coefficient and F the number of atoms crossing the moving interface per unit surface and time. $N^{-1} \sim \langle l^2 \rangle$, where l is the distance between precipitates. Experimental values of F and typical diffusion coefficient values lead to typical distances ranging from 40 to 80 nm, in quite good agreement with experiments and with the simulations shown above. Note that for concentrations in the 1–10 per cent range, this leads to Volmer–Weber-type growth and provides a form of "self-organization".

4.2 Phase Composition, Structure and Orientation

Just as in surface growth [76], the interfacial energy and strain energy terms play a crucial role in the Gibbs free energy (FE) relation as long as the surface-to-volume ratio is large. The phase compositions can be deduced from standard clustering thermodynamics (chemical-potential differences, Gibbs–Thomson growth). The formation FE of a nucleus such as that formed by roughness-induced Volmer–Weber-type growth is typically

$$\Delta G = -V\Delta G_{\mathrm{a}} + A\gamma_{\mathrm{i}}, \tag{6}$$

where ΔG_{a}, is the FE difference per atom, V the volume, A the surface and γ_{i} the interfacial energy. V and A depend on the precipitate crystal's orientation versus the substrate, and the latter in turn depends on γ_{i}. On surfaces, the resulting precipitate orientations are determined by the ratio $\gamma_{\mathrm{S}}S/A\gamma_{\mathrm{i}}$ (γ_{i} differs for different orientations) where γ_{S} is the surface energy. At an interface, the equilibrium orientation only depends on the interfacial energy γ_{i}. After the a-Si/c-Si interface's passage, the precipitate orientation can no longer change and pseudomorphic transformations are kinetically blocked as long as the volume term above (i.e., the precipitate radius) is small enough. IBIEC thus produces "phase trapping" of structures with simple epitaxial relations to the host. Increasing the concentration, size (and surface-to-volume ratio) changes modify ΔG and the balance between the terms in the formation FE.

This effect is enhanced by the lattice strain. The total energy E of the epitaxial precipitate is a sum of the bulk total energy under hydrostatic pressure E_{bh} and of a strain-dependent term:

$$E_{\mathrm{e}} = E_{\mathrm{bh}} + q(c_{ij})\Gamma\Delta a^2, \tag{7}$$

where q is a function of the crystal's (orientation-dependent) elastic constants. Γ depends on bulk properties and Δa is the difference in lattice constants. Both terms have a parabolic dependence on the lattice constants' change under pressure, so that an increase in the lattice strain (which adds a negative term to the formation enthalpy of the epitaxial precipitate) may drastically change the FE sequence in IBIEC-induced phase formation [74].

5 Conclusion

Qualitatively, ion-induced disorder and amorphization processes in silicon are reasonably well understood. However, the temperature dependence of defect accumulation and amorphization is quite complex, with a multitude of defect-mediated processes playing major roles depending on the irradiation temperature, ion mass, dose rate and nuclear-energy deposition along the ion track. As a result there is no overall quantitative model (with predictability) that can treat defect accumulation, defect evolution and amorphization over all temperature ranges and irradiation conditions. Available quantitative models (e.g., kinetic Monte Carlo and MD simulations) are reasonably successful at describing observations at either low temperatures (or irradiation conditions) where amorphization is favored, or high temperatures, where defect accumulation and evolution into extended defects occurs, but are only partly successful at best under conditions where both substantial dynamic defect annealing and amorphization processes are occurring together during irradiation. Similarly, there are currently different views as to the importance of specific irradiation-induced defects in the amorphization process, particularly the "growth" of amorphous layers and interface roughness, under elevated-temperature irradiation. Indeed, defect gettering to and trapping at other defects and interfaces can often control disorder accumulation and amorphization behavior but few data and models exist to describe such processes. Finally, a major unknown involves how cascade-energy density determines defect generation and residual disorder. For example, amorphization is not scalable with ion mass and flux and appears to depend in a complex manner on cascade density as well as instantaneous and average defect-generation rates.

In terms of ion-beam-induced epitaxial crystallization, there are several features of the phenomenon that are known and work well. For example, there is now strong evidence that the process is driven by atomic displacements at the amorphous/crystalline interface. The Marlowe simulation code that calculates atomic displacements for random and channeled ion irradiations can successfully predict the effect of channeling on IBIEC growth (i.e., linear scaling of growth rate with atomic displacements at the interface) for individual ion species. The excellent agreement of simulations with experiment, suggests that individual values used in the simulations are accurate, such as nuclear-energy deposition, atomic-displacement distributions for random and aligned irradiations, as well as multiple scattering through amorphous layers and associated angular spreads.

One IBIEC observation that is not understood very well at present is the effect of cascade density on IBIEC growth rates. For example, the dependence of IBIEC growth on ion-mass has no understandable scaling and the trends are the exact opposite to those for the ion mass dependence observed for amorphization. However, the modeling of the near-atomistic interfacial processes involved in IBIEC, associated for example with the sequence of events from initial interfacial atomic displacements, through broken bond and kink formation to "diffusional" and cooperative crystallization processes along the interface are mostly successful in explaining IBIEC observations. Finally, there are clearly a number of areas of ion-induced amorphization and IBIEC covering both observation and modeling that remain to be investigated before a complete understanding of irradiation-induced, defect-mediated processes in silicon is forthcoming.

References

1. A. Zangwill, *Physics at Surfaces* (Cambridge University Press, Cambridge, 1988) 73
2. F. Spaepen, D. Turnbull, in *Laser Annealing of Semiconductors*, ed. by J.M. Poate, J.W. Mayer (Academic Press, New York, 1982), p. 15 74, 87, 98
3. B. Strickland, C. Roland, Phys. Rev. B **51**, 5061 (1995) 74
4. H. Hensel, H.M. Urbassek, Phys. Rev. B **58**, 2050 (1998) 74
5. R.S. Averback, T. Diaz de la Rubia, in *Solid State Physics*, vol. 51, ed. by H. Ehrenreich, F. Spaepen (Academic Press, New York, 1998), p. 282 75
6. L.M. Howe, M.H. Rainville, Nucl. Instrum. Methods Phys. Res. B **19/20**, 61 (1987) 76
7. F.F. Morehead Jr., B.L. Crowder, Radiat. Eff. **6**, 27 (1970) 76
8. J.R. Dennis, E.B. Hale, J. Appl. Phys. **49**, 1119 (1978) 76
9. M.L. Swanson, J.R. Parsons, C.W. Hoelke, Radiat. Eff. **9**, 249 (1971) 76
10. F.L. Vook, H.J. Stein, Radiat. Eff. **2**, 23 (1969) 76
11. L.A. Christel, J.F. Gibbons, T.W. Sigmon, J. Appl. Phys. **52**, 7143 (1981) 76
12. J.S. Williams, Unpublished, 1998 77
13. J.S. Williams, MRS Bull. **17**, 47 (1992) 77, 80
14. S. Takeda, M. Kohyama, A. Ibe, Philos. Mag. A **70**, 287 (1994) 77

15. R.D. Goldberg, J.S. Williams, R.G. Elliman, Phys. Rev. Lett. **82**, 771 (1999) 78, 79, 80
16. R.D. Goldberg, J.S. Williams, R.G. Elliman, Nucl. Instrum. Methods Phys. Res. B **106**, 242 (1995) 78, 79
17. J.S. Williams, H.H. Tan, R.D. Goldberg, R.A. Brown, C. Jagadish, Mater. Res. Soc. Symp. Proc. **316**, 15 (1994) 78
18. J.S. Williams, R.D. Goldberg, M. Petravic, Z. Rao, Nucl. Instrum. Methods Phys. Res. B **84**, 199 (1994) 78
19. R.G. Elliman, J. Linnros, W.L. Brown, Mater. Res. Soc. Symp. Proc. **100**, 363 (1988) 80, 81, 82
20. K.A. Jackson, J. Mater. Res. **3**, 1218 (1988) 80, 98
21. J.S. Williams, Trans. Mater. Res. Soc. Jpn. **17**, 417 (1994) 80
22. R.D. Goldberg, R.G. Elliman, J.S. Williams, Nucl. Instrum. Methods Phys. Res. B **80/81**, 596 (1993) 82
23. J. Linnros, R.G. Elliman, W.L. Brown, J. Mater. Res. **3**, 1208 (1988) 82, 84, 85, 87
24. R.G. Elliman, J.S. Williams, W.L. Brown, A. Leiberich, D.A. Maher, R.V. Knoell, Nucl. Instrum. Methods Phys. Res. B **19/20**, 435 (1987) 82, 83, 84, 85
25. J.S. Williams, R.G. Elliman, W.L. Brown, T.E. Seidel, Phys. Rev. Lett. **55**, 1482 (1985) 84, 85, 89, 90, 91, 97
26. G.L. Olson, R.A. Roth, Mater. Sci. Rep. **3**, 1 (1988) 84, 87, 90
27. J. Linnros, G. Holmén, B. Svensson, Phys. Rev. B **32**, 2770 (1985) 84, 85, 89
28. F. Priolo, C. Spinella, A. La Ferla, E. Rimini, G. La Ferla, Appl. Surf. Sci. **43**, 178 (1989) 84, 85, 87
29. G. Lulli, P.G. Merli, M. Vittori Antisari, Phys. Rev. B **36**, 8038 (1987) 85
30. F. Priolo, E. Rimini, Mater. Sci. Rep. **5**, 319 (1990) 85, 87, 97
31. J.S. Williams, R.G. Elliman, W.L. Brown, T.E. Seidel, Mater. Res. Soc. Symp. Proc. **37**, 127 (1985) 85, 86
32. J.F. Ziegler, J.P. Biersack, U. Littmark, *The Stopping and Range of Ions in Solids* (Pergamon, New York, 1985) 85, 89, 94
33. S. Cannavo, A. La Ferla, S.U. Campisano, E. Rimini, G. La Ferla, L. Gandolfi, J. Liu, M. Servidori, Mater. Res. Soc. Symp. Proc. **51**, 329 (1986) 87
34. D.M. Maher, R.G. Elliman, J. Linnros, J.S. Williams, R.V. Knoell, W.L. Brown, Mater. Res. Soc. Symp. Proc. **93**, 87 (1987) 87
35. J.S. Williams, R.G. Elliman, Phys. Rev. Lett. **51**, 1069 (1983) 87, 98, 99
36. F. Priolo, C. Spinella, E. Rimini, Phys. Rev. B **41**, 5235 (1990) 87, 98, 99
37. E.F. Kennedy, L. Csepregi, J.W. Mayer, T.W. Sigmon, J. Appl. Phys. **48**, 4241 (1977) 87

38. J.M. Poate, D.C. Jacobson, J.S. Williams, R.G. Elliman, D.O. Boerma, Nucl. Instrum. Methods B **19/20**, 480 (1987) 87
39. F. Priolo, C. Spinella, A. La Ferla, A. Battaglia, E. Rimini, G. La Ferla, A. Carnera, A. Gasparotto, Mater. Res. Soc. Symp. Proc. **128**, 563 (1989) 87
40. F. Spaepen, E. Nygren, A.V. Wagner, in *Crucial Issues in Semiconductor Materials & Processing Technologies* (Kluwer Academic, Boston, 1992), p. 483 87
41. J.M. Poate, J. Linnros, F. Priolo, D.C. Jacobson, J.L. Batstone, M.O. Thompson, Phys. Rev. Lett. **60**, 1322 (1988) 87
42. J. Linnros, W.L. Brown, R.G. Elliman, Mater. Res. Soc. Symp. Proc. **100**, 369 (1988) 87, 97
43. A. Kinomura, J.S. Williams, K. Fuji, Phys. Rev. B **59**, 15214 (1999) 87, 88, 89, 90, 96, 97
44. V. Heera, T. Henkel, R. Kögler, W. Skorupa, Phys. Rev. B **52**, 15776 (1999) 87, 89
45. J. Linnros, G. Hólmen, J. Appl. Phys. **59**, 1513 (1986) 89, 91
46. R.G. Elliman, J.S. Williams, D.M. Maher, W.L. Brown, Mater. Res. Soc. Symp. Proc. **51**, 319 (1986) 89, 90, 91
47. J.S. Williams, I.M. Young, M.J. Conway, Nucl. Instrum. Methods Phys. Res. B **161–163**, 505 (2000) 91
48. G. de M. Azevedo, J.S. Williams, I.M. Young, M.J. Conway, A. Kinomura, Nucl. Instrum. Methods B **190**, 772 (2002) 91, 92, 93, 94, 96, 97
49. M.T. Robinson, I.M. Torrens, Phys. Rev. B **9**, 5008 (1974) 94
50. M.T. Robinson, Nucl. Instrum. Methods B **48**, 408 (1990) 94
51. G. de M. Azevedo, J.C. Martini, M. Behar, P.L. Grande, Nucl. Instrum. Methods B **149**, 301 (1999) 94
52. G. Molière, Z. Naturforschung: Sect. A-A J. Phys. Sci. **2a**, 133 (1947) 94
53. W. Eckstein, *Computer Simulation of Ion-Solid Interactions* (Springer, Berlin, 1991). And references therein 94
54. A. Kinomura, A. Chayahara, N. Tsubouchi, C. Heck, Y. Horino, Y. Miyagawa, Nucl. Instrum. Methods B **175–177**, 319 (2001) 96
55. G.A. Kachurin, Sov. Phys. Semicond. **14**, 461 (1980) 97
56. H.A. Atwater, C.V. Thompson, H.I. Smith, Phys. Rev. Lett. **60**, 112 (1988) 97
57. G. Lulli, P.G. Merli, M. Vittori Antisari, Mater. Res. Soc. Symp. Proc. **100**, 375 (1988) 97
58. J. Nakata, M. Takahashi, K. Kajiyama, Jpn. J. Appl. Phys. **20**, 2211 (1981) 97
59. L.E. Mosley, M.A. Paesler, Appl. Phys. Lett. **45**, 86 (1984) 97
60. J.P. Guillemet, B. de Mauduit, R. Sinclair, T.J. Konno, in *Int. Conf. Electr. Micros. (ICEM-13)*, Paris, France, 1994 98
61. G.Q. Lu, E. Nygren, M.J. Aziz, J. Appl. Phys. **70**, 5323 (1991) 98

62. J.S. Custer, A. Battaglia, M. Saggio, F. Priolo, Phys. Rev. Lett. **69**, 780 (1992) 99, 100
63. F. Fortuna, P. Nedellec, M.O. Ruault, H. Bernas, X.W. Lin, P. Boucaud, Nucl. Instrum. Methods B **100**, 206 (1995) 99, 101
64. A.L. Barabási, H.E. Stanley, *Fractal Concepts in Surface Growth* (Cambridge University Press, Cambridge, 1995) 99
65. F. Family, T. Vicsek, J. Phys. A **18**, L75 (1985) 100
66. S.F. Edwards, D.R. Wilkinson, Proc. R. Soc. Lond. A **381**, 17 (1982) 100
67. M. Kardar, G. Parisi, Y.-C. Zhang, Phys. Rev. Lett. **56**, 889 (1986) 100
68. M. Lohmeier, S. de Vries, J. Custer, F. Vlieg, M.S. Finney, F. Priolo, A. Battaglia, Appl. Phys. Lett. **64**, 1803 (1994) 102
69. M.-J. Caturla, T. Diaz de la Rubia, L.A. Marques, G.H. Gilmer, Phys. Rev. B **54**(16), 683 (1996) 103
70. D.M. Stock, B. Weber, K. Gärtner, Phys. Rev. B **61**, 8150 (2000) 103
71. L. Pelaz, L. Marquès, J. Barbolla, J. Appl. Phys. **96**, 5947 (2004) 103
72. J. Desimoni, M. Behar, H. Bernas, Z. Liliental-Weber, J. Washburn, Appl. Phys. Lett. **62**, 306 (1993) 104
73. X.W. Lin, M. Behar, J. Desimoni, H. Bernas, Z. Liliental-Weber, J. Washburn, Appl. Phys. Lett. **63**, 105 (1993) 104
74. X.W. Lin, Z. Liliental-Weber, J. Washburn, H. Bernas, J. Desimoni, J. Appl. Phys. **75**, 4686 (1994). And refs. therein 104, 106
75. J.A. Venables, G.D. Spiller, M. Hanbücken, Rep. Prog. Phys. **47**, 399 (1984) 105
76. M. Zinke-Allmang, L.C. Feldman, M.H. Grabow, Surf. Sci. Rep. **16**, 377 (1992) 105

Index

Voids and Nanocavities in Silicon

J.S. Williams and J. Wong-Leung

Department of Electronic Materials Engineering, RSPE, Australian National University, Canberra, 0200, Australia, e-mail: jsw109@rsphy1.anu.edu.au

Abstract. In silicon, defects that are normally observed following ion implantation and annealing are interstitial based, that is they arise from the agglomeration of interstitials that are produced during ion irradiation. Vacancies that are produced in equal numbers to interstitials during irradiation only agglomerate into larger open-volume defects (almost exclusively voids) under special implantation and annealing conditions as a result of the instability of many vacancy-based defects. Hence, the observation of open-volume defects (voids and nanocavities) requires careful control of implantation and annealing conditions. Nevertheless, they have significant scientific and technological consequences and have been under active study recently. This chapter reviews open-volume defects, or nanocavities, in silicon beginning with the two main methods for producing them by ion bombardment: namely, by high-dose hydrogen or helium irradiation to first produce gas bubbles and then annealing to expel the gas and leave cavities, and during sufficiently high-dose irradiation under implantation conditions that do not amorphize the silicon to give rise to small vacancy clusters and voids at depths within the first half of the projected ion range. Such voids and nanocavities once produced have a number of interesting properties. They are very attractive trapping sites for a number of interstitial diffusers in silicon, particularly metal atoms and silicon interstitials themselves. Some intriguing nonequilibrium precipitation phenomena can be observed to occur at cavities and there are a number of ways in which cavities can be induced to shrink and disappear under subsequent irradiation and/or annealing. These aspects are especially reviewed. From the technological point of view, open-volume defects can be detrimental in terms of electronic or optoelectronic device performance but there are also beneficial applications such as the so-called "smart-cut" process, whereby a thin silicon layer can exfoliate from the host wafer under specific hydrogen implantation and annealing conditions, and also metal impurities can be removed from active device regions by strategically placing a band of voids to strongly trap them during thermal processing.

1 Introduction

Defects in silicon, how they form and their thermal stability, are of considerable basic interest. The fundamental point-defects are vacancies and interstitials. When a silicon atom is removed from a lattice site in a "perfect" single-crystalline lattice, a vacancy–interstitial pair can be produced. This can be induced thermally, and as the temperature increases the equilibrium

H. Bernas (Ed.): Materials Science with Ion Beams,
Topics Appl. Physics **116**, 113–146 (2010)
DOI: 10.1007/978-3-540-88789-8_5,

number of such defect pairs increases. However, since the lowest energy state is the perfect lattice, there is a driving force for interstitials to annihilate with vacancies. Thus, at any temperature there is a balance between defect creation and annihilation that determines the equilibrium defect concentration. The real situation is more complex since mobile point-defects can become trapped or annihilate at surfaces, impurities or at other (existing) defects in the silicon. In addition, there is a possibility that point-defects can also agglomerate into more complex multivacancy or multi-interstitial defects, where defect formation, migration and dissociation energies control the equilibrium concentrations of all defects at a given temperature. Such processes depend on the perfection of the starting material and the nature and concentration of existing defects. In silicon there are a range of extended defects that can form during crystal growth such as dislocations and stacking faults, as well as impurities such as oxygen and carbon that can become trapping sites for migrating point-defects. It is interesting that existing (extended) defects in silicon tend to be interstitial based. That is, dislocations and defect clusters are observed to contain additional atoms. The reason for this is related to the thermal stability of small interstitial clusters (up to a few atoms) whereas vacancy clusters (di- and trivacancies) and even vacancy dislocation loops are thermally unstable [1, 2]. Recently, however, voids (clusters of 50 or more vacancies) have been found in silicon after growth from the melt [3] and this has raised the possibility that if very high vacancy concentrations can be produced in silicon, they could evolve into voids or larger cavities. Disorder from ion bombardment can also clearly produce large concentrations of vacancy–interstitial pairs but mostly these have been observed to evolve into extended defects of interstitial character on annealing, as previous chapters have illustrated. However, as we illustrate in this chapter, voids can now be produced under a range of implantation conditions. From a technical point of view, such voids, if they exist, are important. First, like interstitial-based defects discussed in previous chapters, they may be detrimental to the electrical properties of silicon devices. Secondly, voids or open-volume defects may also be useful technologically since, as we further illustrate in this chapter, they can trap migrating impurities in silicon that may otherwise have been detrimental to device performance. Indeed, the process of impurity trapping at voids, the decoration of cavity walls with metals and the ultimate precipitation of nanoparticles in voids are interesting processes in their own right.

Open-volume defects can be formed in silicon by ion irradiation in two ways. The first method begins with the implantation of a species that forms gas bubbles, such as hydrogen or helium. Gas bubbles form either during implantation or during the early stages of annealing. Annealing at sufficiently high temperatures can drive the gas from the bubbles, leaving nanocavities. This is the most straightforward method of producing observable open-volume defects. So-formed nanocavities are typically large (of diameter up to

50 nm) and are more stable than small vacancy clusters resulting from direct implantation by a second method as below.

The second method involves ion irradiation under conditions that facilitate and maintain a vacancy-excess in crystalline silicon within the first half of the projected ion range. Whereas extended defects of interstitial character are easily observed by electron microscopy, as shown in a previous chapter, open-volume defects require more stringent irradiation, annealing and characterization conditions to uncover them. A major reason for this behavior is that there are a number of metastable forms of interstitial-based defects in silicon, particularly line defects such as {311} defects and interstitial-based dislocation loops, as illustrated previously, whereas similar vacancy-based defects are quite unstable. We might expect that the spatial separation of vacancies and interstitials within a collision cascade would lead to a vacancy-rich region close to the surface and an interstitial excess near the end of range of the implanted ions. However, the interstitials invariably evolve into readily observable defects on annealing, whereas the vacancies are more likely to annihilate. Nevertheless, vacancy-excesses can survive annealing to result in vacancy clusters and voids under certain conditions. We examine those conditions in this chapter but note the difficulty in observing vacancy-based defects and some controversy as to their positive identification. For example, for some considerable time unidentified defects in the region where vacancy-excesses might be expected were simply termed $R_\mathrm{p}/2$ defects, that is, evidence for disorder at about half the projected ion range following annealing. Such defects have been variously interpreted as either interstitial or vacancy related. Now, it is universally accepted that so-called $R_\mathrm{p}/2$ defects are predominantly of vacancy character and consist of vacancy clusters and voids.

This chapter examines the formation, properties and stability of nanocavities or open-volume defects formed by both methods. Nanocavities also exhibit a range of interesting (often nonequilibrium) properties, such as efficient trapping sites for fast-diffusing metals and interstitial-based defects, precipitation of second phases within the open volume, and preferential amorphization and shrinkage of nanocavities during subsequent irradiation. These processes are also illustrated and discussed in this chapter.

2 Formation of Nanocavities and Voids by Ion Irradiation

We first treat the more straightforward case of the formation of nanocavities by the high-dose implantation of H and He into silicon. Such nanocavity formation has been the subject of considerable recent attention, partly as a result of the interesting properties that cavities exhibit for the gettering of metal impurities [4–9], and partly because they are an important precursor to the cleaving of a thin layer of silicon from a host wafer, which takes place in the so-called smart-cut process [10].

2.1 Nanocavity Formation by H and He Irradiation

During high-dose implantation of silicon with He or H ions, the precursor for cavity formation is the agglomeration of the implanted species into small gas bubbles. It is important to understand how such bubbles might form and where they form in relation to the implantation damage profile. For example, Fig. 1 illustrates the separation between the ion distribution and the generated displacement (vacancy) distribution for 100-keV He-implanted silicon, obtained by simulations using the TRIM code [11]. As expected, there is a reasonable separation in depth between the peak of the damage and the peak of the He distribution. Some studies (see, for example, [12]) have proposed that bubbles (and subsequently cavities) form during annealing in a depth interval that closely corresponds to the vacancy distribution, rather than the ion distribution. Such behavior is illustrated by the hatched region in Fig. 1, which represents the width of a band of cavities observed with transmission electron microscopy (TEM) by Raineri et al. [12] for a He dose of $3 \times 10^{16}\,\text{cm}^{-2}$, after annealing at 950°C. The following model has been suggested to explain this behavior. During implantation or in the early stages of annealing, He atoms migrate to and are trapped at vacancies, thus suppressing annihilation with interstitials. Bubbles then grow via agglomeration of He-filled vacancy clusters. Initially, the bubbles are elongated in the (100) planes parallel to the surface but develop into spherical shapes on annealing via Ostwald ripening. On further annealing to around 700°C, the He gas is expelled, leaving a band of nanocavities located at depths that mirror the original damage or vacancy distribution [5, 12]. There are alternate views [13] that the He bubbles and hence cavities can form close to the He ion range. This difference may lie with the different implantation and annealing conditions used and also the correlation of as-implanted He distributions, rather than final He distributions, with the depth of the cavity band.

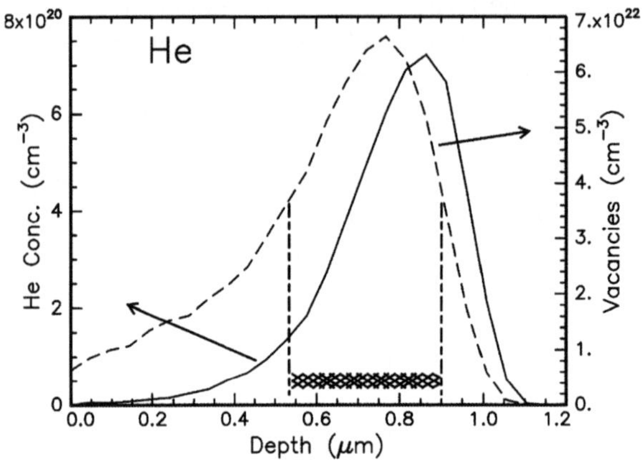

Fig. 1. TRIM95 simulations [11] of 100-keV He ion range and vacancy profiles in silicon. The *hatched area* in (**a**) corresponds to the cavity-band region as obtained in [12] for a He dose of $3 \times 10^{16}\,\text{cm}^{-2}$

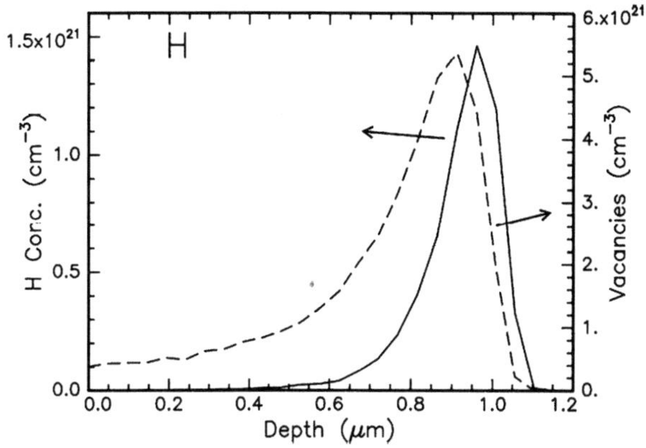

Fig. 2. TRIM95 simulations [11] of 100-keV H ion range and vacancy profiles in silicon

The TRIM simulations in Fig. 2 illustrate that, for H-implanted silicon, the ion and vacancy distributions are narrower than the corresponding distributions for He. Figure 3a shows secondary ion mass spectrometry (SIMS) profiles of 100-keV H, to a dose of $3 \times 10^{16}\,\mathrm{cm}^{-2}$, as a function of annealing temperature [14]. The initial ion distribution matches that given by TRIM (Fig. 2) quite well and, furthermore, the H is completely removed by 750°C. The cross-sectional TEM (XTEM) images in Figs. 3b and c show the evolving cavity band, for the case illustrated in Fig. 3a, following annealing at 500°C and 750°C. At 500°C many of the partly filled cavities are still elongated parallel to the surface and there is considerable disorder surrounding the cavity-band region. However, following 750°C annealing, when all the H is expelled, the cavities are well formed with few other defects (e.g. dislocation loops) in their vicinity. For the case of 100-keV H implantation to a dose of $3 \times 10^{16}\,\mathrm{cm}^{-2}$, the integrated surface area of the so-formed cavities (per cm^2) is about 1/10-th of that of the sample surface. Furthermore, the depth and width of the cavity-band corresponds quite closely with the H-ion distribution, but the closeness of the damage and H distributions makes it difficult to say whether the H is in fact, like the He case, actually decorating the implantation damage.

When the H-ion dose is increased beyond that illustrated in Fig. 3, the silicon surface can blister or delaminate on annealing, which is the basis of the ion-cut process mentioned earlier. Because of the strong technological interest in this process, there have been several recent studies directed at understanding the mechanism of H-induced exfoliation of silicon. Such studies in turn have provided insight into the early stages of cavity formation and it is therefore useful to review them here. For example, Weldon et al. [15] have shown that substantial chemical bonding (trapping) of hydrogen occurs during implantation at disorder within the H profile. This chemical bonding consists of hydrogenated point-defect complexes, trapping at vacancy clusters and formation of platelets near the peak of the implantation profile. Anneals

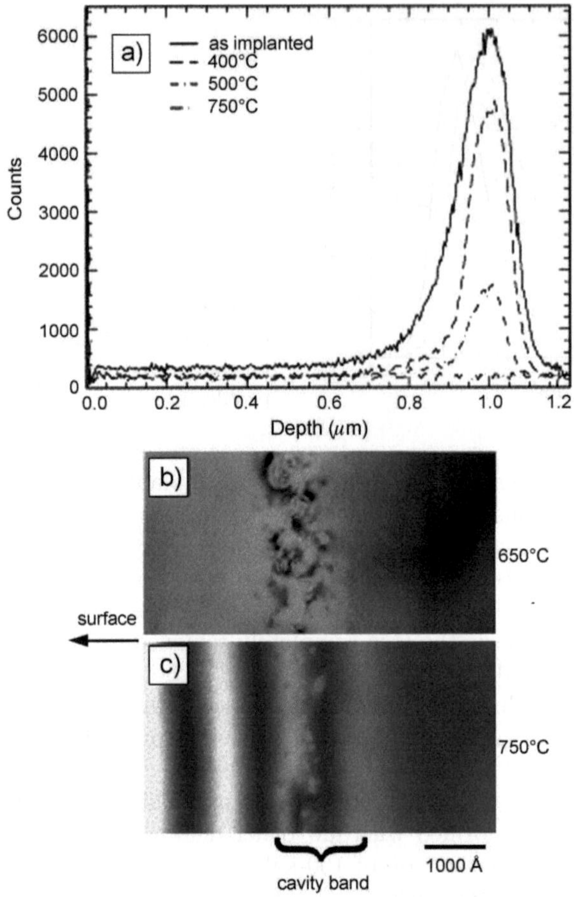

Fig. 3. (**a**) SIMS profiles of 100-keV, $3 \times 10^{16}\,\mathrm{H\,cm^{-2}}$ implanted into silicon following annealing for 1 h at various temperatures [14]. (**b**) and (**c**) Cross-sectional TEM micrographs of H-implanted silicon as in (**a**) following 650°C (**b**) and 750°C (**c**) annealing (from [14])

to temperatures up to 400°C result in the collapse of the defect structure, the formation of H_2 gas bubbles and the agglomeration of the bound H into vacancy–defect complexes. At higher temperatures, the defect structure reorganizes into H-terminated {100} and {111} surfaces and the trapping of H_2 in the microvoids between these surfaces. As the pressure builds up, these microvoids can develop into bubbles, macroscopic cracks and then blisters in unconstrained silicon, and to complete exfoliation in cases where the surface is constrained. For the case shown in Fig. 3b, where the H dose is slightly lower than that required for blister formation, the structure of bubbles and microvoids, which are elongated along the (100) plane after annealing to 500°C, is consistent with this picture. In terms of the ion-cut process, Höchbauer et al. [16] have recently correlated the depth at which the surface exfoliates with both the implantation damage and the H profile. Figure 4 shows data from this work that compares the depth profiles of damage (from ion channeling) and H (from heavy-ion elastic recoil) for 175-keV H-implantation into silicon,

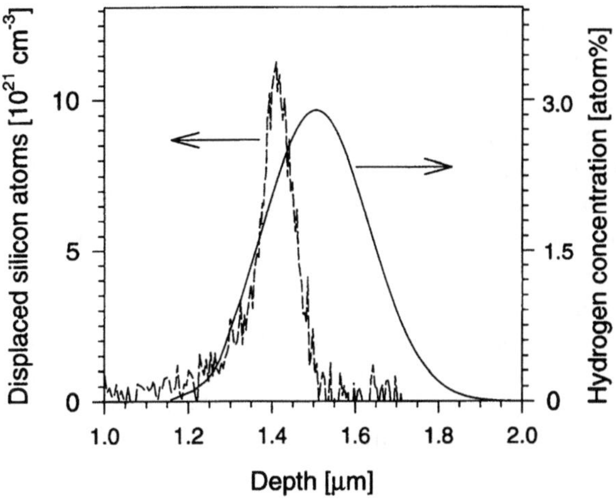

Fig. 4. In-depth distribution of 175-keV H implanted into silicon to a dose of $5 \times 10^{16}\,\mathrm{cm}^{-2}$ as measured by ERDA and damage distribution as measured by RBS/channeling (after [16])

where the expected separation of the damage and H profiles is clearly shown. XTEM was used to measure the thickness of the exfoliated layer and the nature of the disorder below each of the newly cleaved surfaces. Results show that the layer separates at the peak of the damage distribution (1.42 μm) rather than at the H peak. Furthermore, the cleavage was shown to occur at regions where H-decorated (100) platelets had formed and it was assumed that this took place at the peak of the damage distribution as a result of the stress field within H-implanted silicon. Consistent with previous studies [15], Höchbauer et al. [16] proposed that hydrogen gas accumulated at such (100) platelets on subsequent annealing, bubbles grow by Ostwald ripening [15] and ultimately lead to crack initiation and propagation.

An interesting issue in bubble and nanocavity formation is to account for the silicon expelled in generating the open volume. Although it could be argued that the region around the cavities could expand elastically to create the open volume, the lack of stress fields around cavities following annealing to 750°C (see Fig. 3c) suggests that silicon is more likely to be transported away during annealing. Indeed, Raineri et al. [12] have shown that excess silicon interstitials, in the case of He implantation, migrate to the surface and annihilate during annealing. There is also evidence [9] that the same behavior (out diffusion of silicon interstitials) may also occur during H-induced cavity formation. The influence of silicon interstitials on other cavity properties will be discussed in Sects. 3 and 5.

2.2 Irradiation-Induced Vacancy Excess and Void Formation

During implantation into silicon, the ballistic processes within collision cascades produce a net flux of displaced silicon atoms in the forward direction. This leads directly to a local excess of silicon atoms (or an interstitial excess) at depths near the end-of-ion-range and a net deficiency of silicon (or a vacancy-excess) at shallower depths [17]. There is currently considerable interest in semiconductor technology as to the stability of such "defects" and whether they predominantly annihilate during annealing or can evolve into larger vacancy-rich or interstitial-rich defect clusters. As indicated in a previous chapter, interstitial-based residual defects such as {311} rod-like defects and dislocation loops are clearly stable to annealing up to 600–700°C and such defects are readily observed by TEM following implantation and annealing [18–20]. However, since vacancy-based defects are usually not observed after annealing, it is interesting to examine the stability of vacancy-based defects in silicon. First, like isolated silicon interstitials, single vacancies are mobile at low temperatures and divacancies dissociate by 200°C. Thus, neither of these defects is expected to be observed above their equilibrium concentrations in silicon, following annealing above about 200°C [1]. It is further interesting to note that vacancy-based dislocation loops have never been observed in silicon but small vacancy clusters and voids have recently been identified in as-grown Czochralski (Cz) silicon [3]. Furthermore, there have been attempts to model the nucleation and growth of vacancy-based defects, such as the work by Plekhanov et al. [2], who showed that vacancy-based dislocation loops were thermodynamically unstable whereas vacancy clusters above a critical size were stable to temperatures above 1000°C, consistent with the observations of voids in as-grown silicon. Thus, voids would seem to be the only thermodynamically stable vacancy-based defects observable following implantation and annealing.

Voids have been observed in ion-implanted and annealed silicon but only for high-dose implants in cases where care was taken to avoid amorphization, such as by undertaking elevated temperature implantation or using MeV implants. For example, in MeV implantation, where the spatial separation of the vacancy and interstitial excesses is large, Zhou et al. [21] and Ellingboe et al. [22] observed voids in TEM following elevated temperature implants of O and Si to doses exceeding $10^{17}\,cm^{-2}$. Also, Holland et al. [23] showed contrast in TEM images consistent with clusters of voids, for lower-energy P- and As-implanted silicon to doses above $10^{16}\,cm^{-2}$ at 200°C. In this latter case the authors indicated that the chemical nature of the implanted species may have helped to stabilize the voids. However, despite these few reports, the direct observation of voids by TEM following implantation of silicon is not common and almost certainly a combination of a narrow range of implant and anneal conditions is necessary to produce stable voids. As a result of this difficulty of observing voids, indirect methods have been used to determine vacancy-excesses by techniques such as positron annihilation [24] and also the labeling of vacancy clusters by fast-diffusing metals [25–27]. Results from such measurements, along with more recent TEM studies of voids in ion-implanted silicon, will be treated in Sect. 4. However, before concluding this section, we illustrate the typical implantation defects that result from the separation of the vacancy and interstitial excess regions following annealing of

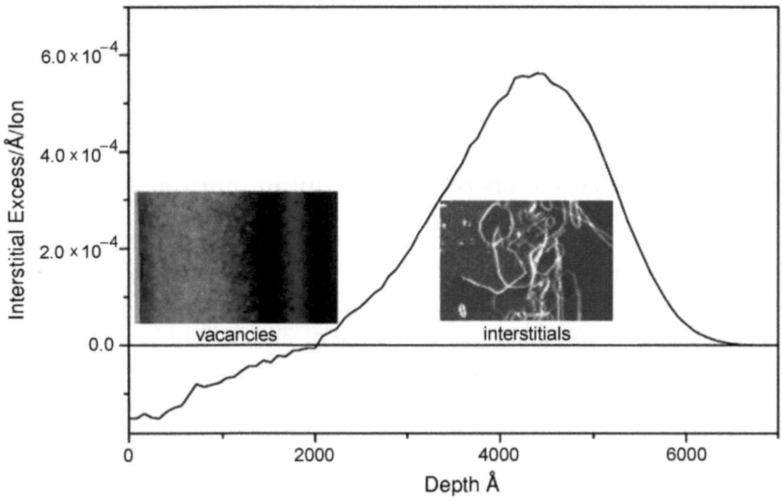

Fig. 5. TRIM simulation indicating the net vacancy and interstitial excess in silicon as a function of depth following 245-keV Si implantation. The insets show TEM micrographs of typical residual defects in the vacancy-excess region (note the well-defined voids) and at the end-of-range region (note the interstitial-based loops) following annealing at 850°C for 1 h (data from [28])

silicon implanted under conditions that favor the formation of voids. Figure 5 shows a TRIM simulation indicating the net vacancy and interstitial excess in silicon as a function of depth following 245-keV Si implantation. Note that the projected ion range, R_p, for 245-keV Si ions in silicon is 3800 Å. The insets in Fig. 5 show TEM micrographs of typical residual defects in the vacancy-excess region (note the well-defined voids) and at the end-of-range region (note the interstitial-based loops) following annealing at 850°C for 1 h [28]. The implant conditions used to give rise to the TEM micrographs were 245-keV Si implanted to a dose of $1.4 \times 10^{16}\,cm^{-2}$ at 100°C. Further details are given in Sect. 4

3 Interaction of Impurities with Nanocavities

The use of hydrogen implantation in silicon for smart-cut processes has generated much research interest in understanding defect evolution and annealing of implantation damage. As mentioned in the previous subsection, the implantation disorder evolves on annealing firstly into a band of H-filled bubbles and then into nanocavities as hydrogen is released. Furthermore, the strain and defects in silicon surrounding the band of nanocavities is greatly reduced by high-temperature annealing. A remarkable feature of such nanocavities within otherwise defect-free silicon is their ability to act as strong sinks for metallic impurities. This property has also stimulated much research into the use of nanocavities in silicon as strong sites for the gettering and hence removal of metal impurities from active device regions. We review such gettering studies below.

3.1 Interactions at Low Levels of Metal Contamination

Figure 6a shows Au profiles from Rutherford backscattering (RBS) analysis in a silicon sample with a band of preformed nanocavities before and after a 1-h anneal at 850°C [7]. The band of nanocavities was created at a depth of 1 μm by a 100-keV H implantation to a dose of $3 \times 10^{16}\,\mathrm{cm}^{-2}$ and a 1-h anneal at 850°C. Clearly, Fig. 6a illustrates that the Au originally implanted in the near surface at an energy of 95 keV to a dose of $5 \times 10^{13}\,\mathrm{cm}^{-2}$ is effectively relocated to the depth of the cavity-band after a 1-h anneal at 850°C. TEM analysis of this sample revealed the presence of a band of nanocavities, some of which were faceted. A typical micrograph of this sample is shown in Fig. 6b [7]. The nanocavities exhibit dark contrast in TEM, arising from the presence of Au on their internal walls. We note that the band of nanocavities in Fig. 6b offers an internal surface areal density of about $1 \times 10^{14}\,\mathrm{cm}^{-2}$. Hence, a dose of $5 \times 10^{13}\,\mathrm{cm}^{-2}$ Au corresponds to about half a monolayer of Au on the nanocavity internal surfaces after gettering. Such efficient gettering to cavity walls has been observed for a number of other implanted metals including Cu [6], Ni [29], Fe [30], Co [31], Pt and Ag [32], as long as the metal concentration is insufficient to saturate the cavity walls. In most of these cases the cavities appear to be very strong sinks for diffusing metal impurity atoms and can even remove metals from solid solution in silicon if the binding to the cavity walls is sufficiently strong [33], as we explore more fully below.

In addition to efficiently trapping moderately low concentrations of metals that have been introduced into silicon by ion implantation, cavities also appear to be very efficient for trapping extremely low levels of metal contamination, introduced by thermal processing. For example, Fig. 7 shows SIMS profiles of the two Cu isotopes at cavities in silicon following annealing in a contaminated furnace tube [34]. The total areal density of Cu at cavities is $4 \times 10^{12}\,\mathrm{cm}^{-2}$. In this case, it was of interest to find the level of Cu remaining in the bulk (in solid solution) to quantify the efficiency of trapping. Figure 8 shows the results [34] of neutron activation analysis (NAA) on contaminated samples, both with and without cavities. Before annealing or contamination, the wafers had a Cu areal density of around $3 \times 10^{10}\,\mathrm{cm}^{-2}$. After annealing at 850°C in a contaminated furnace, this increased to around $10^{12}\,\mathrm{cm}^{-2}$ and did not decrease after a light surface etch to remove any surface Cu. However, after etching away a depth greater than the cavity-band, the sample containing the cavities did not reveal any Cu in solid solution (i.e., Cu was below the detection limit of $2 \times 10^{10}\,\mathrm{cm}^{-2}$). In the control sample without the cavity-band, Cu remained after the deep etch, indicating that Cu was

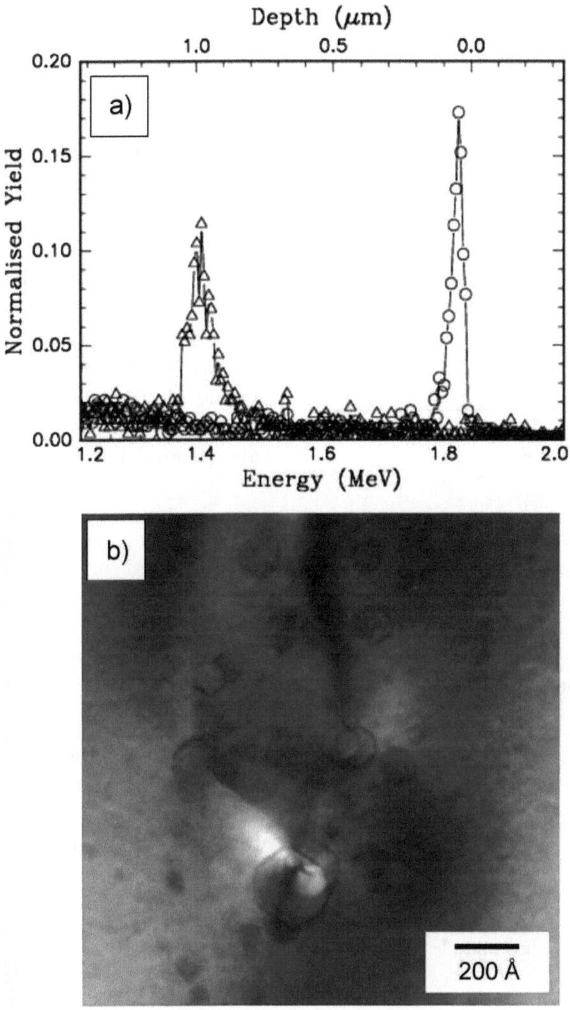

Fig. 6. (**a**) RBS profiles of implanted Au (50 keV, $5 \times 10^{13}\,\mathrm{cm}^{-2}$) after implantation (*filled circles*) and following annealing at 850°C (*open triangles*) for a silicon sample containing a cavity-band at a depth of 1 μm. (**b**) XTEM micrograph showing the cavity-band region of (**a**) after annealing (adapted from [14])

distributed throughout the bulk in solid solution in this case. This behavior illustrates the fact that Cu can be removed from solid solution to trapping sites at the walls of cavities, clearly as a result of very strong bonding of Cu atoms to cavity walls. This is consistent with similar observations by Myers et al. [33].

The very strong trapping of metal atoms at cavity walls has been modeled by Myers et al. [31, 33] and is thought to arise from a chemisorption-like reaction at cavity walls with a high binding free energy that is slightly temperature dependent. Myers et al. [31, 33] have determined a binding energy of 2.2 eV for Cu at cavities (at 700°C), 2.3 eV for Au (800°C) and 1.4–1.5 eV for Co and Fe (800°C). Such binding is significantly greater than that for these

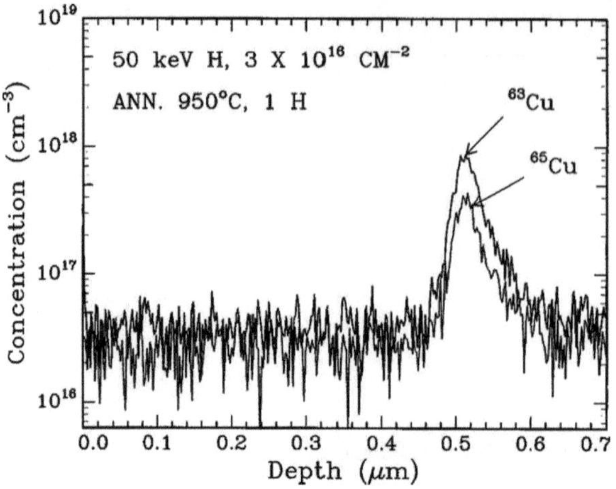

Fig. 7. SIMS profiles of Cu in a silicon sample (containing cavities at 0.5 μm), following annealing at 950°C in a contaminated furnace tube (from [34])

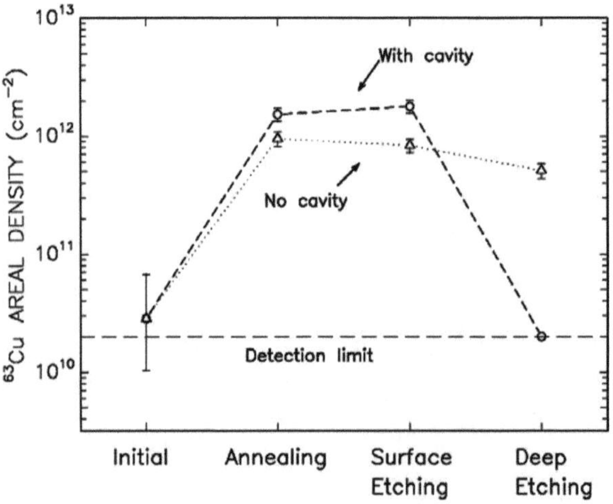

Fig. 8. NAA data indicating the amount of Cu contained in a silicon wafer after annealing in a contaminated furnace and also following etching to different depths (after [34])

metal atoms in solid solution. For example, for Co and Cu at 800°C (binding free energy 1.4 eV) the fractional occupation of cavity walls is around 10^6 times the occupation of solution sites [33]. In addition, the stronger trapping of the monovalent Cu and Au compared with multivalent Fe and Co is thought to arise from the difficulty of accommodating high bonding coordination on the cavity surfaces in the latter cases [31]. Thus, the bonding configuration(s) of the metal atoms on the internal surfaces of cavities will determine the average binding free energy and hence the efficiency of gettering by this trapping mechanism. The situation of cavity-wall decoration by metals should be analogous to that for metal decoration of flat external surfaces, where ultra-high-vacuum surface studies have revealed multiple binding energies, different metal configurations on different surface orientations and ordered island surface states [35]. Indeed, strongly preferred {111} faceting of cavities decorated with Au in Fig. 6b is almost certainly a manifestation of higher binding energy of Au to silicon {111} surfaces than to {110} [35].

The thermal stability of the above gettering process is of interest. For example, how stable are cavities themselves in silicon and how stable are the metal atoms at cavity walls under high-temperature annealing? In terms of thermal stability of cavities, annealing at temperatures greater than 1100°C is needed to cause a significant reduction in their size [5, 13]. Indeed, large cavities appear to grow slightly at the expense of small cavities during high-temperature annealing. It has also been observed that extended annealing at high temperatures can cause some metals to desorb from cavities walls. For example, annealing at 950°C for 24 h caused the amount of Au trapped at cavities in Cz silicon to be greatly reduced [36]. This was attributed to migration of oxygen (O) to the cavities and the replacement of Au by O. Such a process did not occur in float-zone silicon wafers where the O concentration in the bulk of the wafer is much lower.

3.2 Interactions at High Metal Concentration Levels

Interestingly, similarly efficient gettering behavior has also been observed for high-doses of Au and Cu in silicon, where the metal-atom areal density substantially exceeds a monolayer coverage on the cavity walls. Figure 9a illustrates the case of a sample with a Au dose of $8 \times 10^{14}\,\mathrm{cm}^{-2}$ [9]. After 1-h anneal at 850°C, most of the implanted Au is relocated to the cavity band. This amount of Au is well above a monolayer coverage of the cavity walls and TEM analysis of this sample [9] showed well-faceted nanocavities, where the cavity walls are decorated with Au (see Fig. 9b). However, there is clearly Au precipitation in some of the cavities, as shown by the dark round features that are not faceted. The driving force for Au to precipitate in cavities in this high dose case is believed to be a result of the non-equilibrium aspect of ion implantation in which it is possible to implant Au at concentrations well above the solid solubility limit. During annealing, the implanted Au can diffuse and the open volume of the cavities is a preferred precipitation site. The details of the diffusion and precipitation mechanisms are discussed in the following section, but the order of implantation and annealing is important to observe cavity precipitation. Furthermore, the Au precipitates within cavities are believed to be in the form of a Au silicide [9].

To examine the thermal stability of Au precipitates at cavities, long-time high-temperature annealing can be carried out. Figure 10 shows a typical result for the sample in Fig. 9, annealed at 950°C for 48 h. As shown in Fig. 10, a decrease in the amount of Au at the cavity-band was observed after this anneal. Furthermore, the reduction in the amount of Au at the cavity band, if distributed throughout the sample, corresponded to a concentration of Au close to the solubility level of Au at 950°C. This decrease in Au [9] was further investigated by NAA to ascertain the distribution of Au throughout the wafer

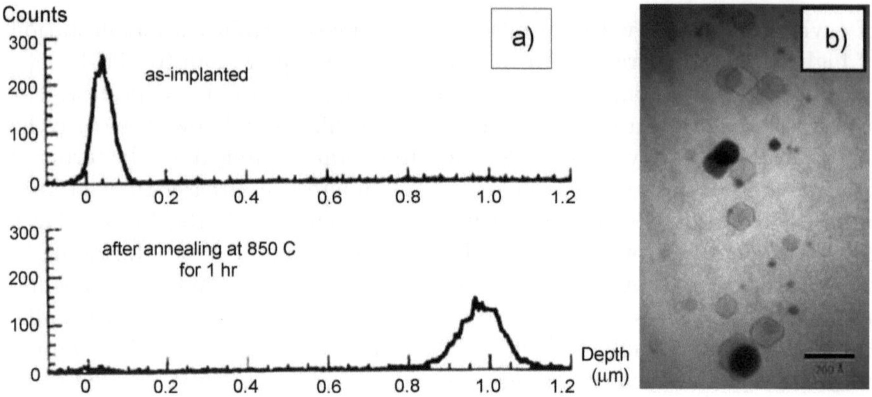

Fig. 9. (**a**) RBS profiles showing the relocation of implanted Au (70 keV, $8 \times 10^{14}\,\mathrm{cm}^{-2}$) to cavities at a depth of 1 μm following annealing at 850°C. (**b**) XTEM micrograph showing Au decoration of cavity walls as well as Au precipitation in cavities (adapted from [9])

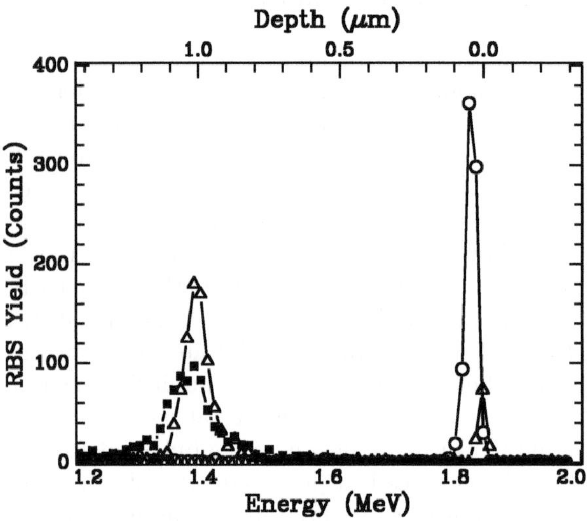

Fig. 10. RBS spectra showing Au profiles in a silicon sample with a cavity band at 1 μm following Au implantation to a dose of $10^{15}\,\mathrm{cm}^{-2}$ (*open circles*) and after annealing at 850°C for 1 h (*open triangles*) and 950°C for 48 h (*filled squares*) (after [9])

and the possible role of evaporation of Au from the sample. NAA studies confirmed that the amount of Au in the wafer remained constant after the long high-temperature anneal. Moreover, by controlled etching of the wafer surface followed by NAA measurements, the Au was clearly shown to be in solution throughout the wafer. TEM analysis of these samples showed a difference in the microstructure of these Au samples in the cavity-band region after the long annealing sequence at 950°C, whereby well-faceted cavities could not be observed and small Au precipitates had undergone an Ostwald ripening process, resulting in much larger and fewer Au precipitates with defects pinned at the precipitates. The results in Figs. 9 and 10 illustrate interesting

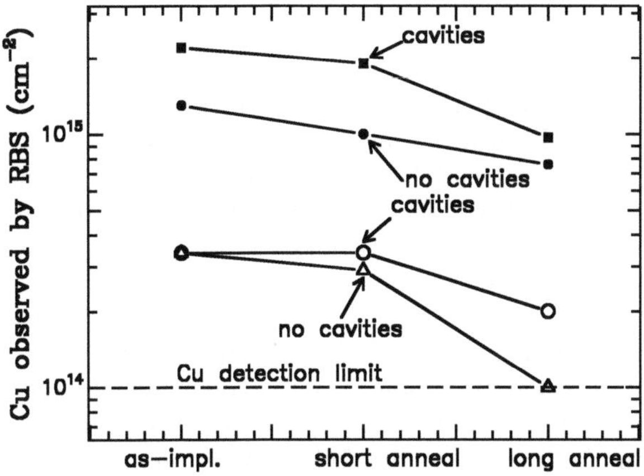

Fig. 11. Graph showing the amount of Cu detected by RBS in samples with and without cavities after different annealing treatments: at the surface for samples without cavities, and at cavities and the surface for samples with cavities. The measured as-implanted Cu doses are not the same for samples with and without cavities for the nominal $2 \times 10^{15}\,\mathrm{cm}^{-2}$ implanted dose (from [9])

nonequilibrium behavior in high-dose ion-implanted silicon and the pathways to equilibrium during annealing. Initially, the implanted Au is introduced in the near-surface region of silicon at concentrations well above the equilibrium solubility limit but at room temperature it cannot form well-defined precipitates. During the initial stages of annealing the cavities constitute a favorable precipitation site close to the supersaturated Au distribution and this local precipitation occurs before the entire system has reached equilibrium. Further annealing causes some of the precipitated Au to dissolve to achieve the equilibrium solubility limit throughout the wafer at 950°C.

Similar cavity precipitation results were also observed for high-dose Cu-implanted samples. Again implanted Cu can precipitate at cavities during the initial stages of annealing when the dose is above that required for a monolayer coverage of cavity walls [6, 14]. Furthermore, Wong-Leung et al. [9] have examined changes in the Cu distribution and subsequent dissolution of Cu from the cavity band after a long anneal at 780°C. Figure 11 summarizes typical results (obtained from RBS analyses) for two doses of Cu, in samples both with and without cavities [9]. The two nominal doses were $3.4 \times 10^{14}\,\mathrm{cm}^{-2}$, which, if distributed throughout the Si wafer, is equivalent to a concentration of Cu below the known solubility level of Cu at the annealing temperature of 780°C, and $2 \times 10^{15}\,\mathrm{cm}^{-2}$, which is above the bulk solubility level. After implantation, RBS detects the Cu within the near surface region (<100 nm) consistent with the implant energy and much shallower than the cavity band at a depth of around 1 µm. After a short anneal at 780°C (1 h) for the lower-dose Cu (open symbols in Fig. 11), essentially all of the Cu is observed by RBS for both cavity and no cavity samples (less than 10% not detected) but the main location of the Cu is different. In the sample without cavities, the Cu remains within the near-surface region but in the sample with cavities, it has almost entirely relocated to the cavity band where some precipitation (Cu silicide) is observed by TEM [9]. TEM also indicated that Cu precipitation had occurred in the sample without cavities, both at the surface and within the original implant profile. In contrast, after the long annealing sequence (24 h), the total amount of Cu observed by RBS in the sample without cavities was reduced to a level below the detection limit of Cu in Si, while the sample with cavities showed only around 30% reduction in the total amount of Cu detected at cavities. Detailed NAA analysis confirmed that the reduction of Cu was a result of Cu dissolution and distribution throughout the wafer and not a consequence of evaporation [9]. For the sample without cavities, the Cu that had initially precipitated locally in the near-surface region dissolved and was redistributed in solution as the system attained equilibrium. However, equilibration in the sample containing cavities was a little different since the Cu attached to the cavity walls (between 1–2 $\times$ $10^{14}\,cm^{-2}$) remained strongly bound, whereas the precipitated Cu dissolved and was redistributed in solution.

The high-dose situation in Fig. 11 is a little different after a long anneal to achieve equilibrium since this amount of Cu, when distributed throughout the wafer, exceeds the solubility limit at 780°C. Hence, in both samples, the reduction in Cu observed by RBS reflects the Cu solubility limit. The remaining Cu in the near-surface region for the sample without cavities is in the form of precipitates (as observed by TEM [9]) and Cu at cavities is both attached to cavity walls and in precipitates. These results indicate two important features of implantation and cavities: nonequilibrium processes can lead to transient phenomena such as local precipitation at cavities, whereas equilibration allows the system to reach its lowest free-energy configuration and a balance between impurities in solution, strong trapping at cavities and local precipitation at favorable sites. These features are discussed further in the following section.

3.3 Mechanisms for Metal Trapping and Precipitation at Cavities

As we have illustrated in the previous section, the overall gettering behavior of ion-implanted metals to cavities illustrates several nonequilibrium aspects of ion implantation. Indeed, implantation can introduce defects and impurities into silicon in concentrations that substantially exceed thermal equilibrium values. Thus, the material system can achieve some intriguing metastable local (free-energy-minimum) states during subsequent annealing. For example, there can be kinetic barriers to achieving the final thermal equilibrium state, such as formation energy for defects and second phases and the activation energy for silicide dissolution. We can illustrate such local minima using the observed Cu precipitation behavior at cavities outlined above [9]. When a high concentration of Cu is implanted into silicon, there is a strong driving force for it to diffuse and precipitate (as a silicide) during annealing since the equilibrium solubility of Cu in silicon is relatively low. To achieve thermal equilibrium, one would expect that Cu would diffuse through the wafer to first saturate at the solubility limit at the annealing temperature and then the excess Cu would precipitate at preferential sites. However, the enormous supersaturation of Cu and the high defect concentrations in the implanted region in close proximity to cavities give rise to unexpected behavior. Initially, there is a great driving force for Cu precipitation created by the supersaturation of Cu in the near-surface region. The cavities are strong sinks for interstitial Cu diffusing away from the supersaturated region. As we have shown, Cu can preferentially relocate to first saturate the internal walls of the cavities then to precipitate within cavities. This nucleation of bulk phase (a Cu_3Si silicide phase in the case of Cu) at the cavities is a metastable situation achieved with very little Cu in solution in the wafer and hence there is a further driving force for the local dissolution of these precipitates to achieve the equilibrium solubility of Cu with respect to the silicide phase formed. However, once Cu_3Si has locally precipitated to fill cavities, there is a kinetic barrier to dissolution. For example, in order to conserve volume and minimize the local free energy, dissolution of Cu_3Si requires the availability of silicon interstitials to initiate dissolution. This constitutes a kinetic barrier or bottleneck to dissolution and hence long annealing times and high temperatures are required for the system to achieve thermal equilibrium [9]. Figure 12 schematically illustrates this situation. The implanted metals are initially highly supersaturated within a disordered region of silicon (solid curve). During annealing, the supersaturated metals diffuse and are trapped or precipitate at strong sinks such as cavities, with little soluble metal in the

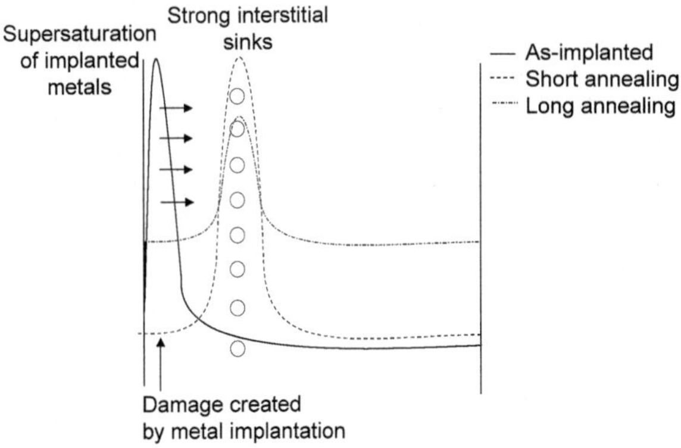

Fig. 12. Schematic illustrating slow equilibration of implanted metals in silicon containing cavities during annealing (from [9])

bulk (dashed curve). We suggest that this transient first stage constitutes a pseudoequilibrium, or metastable state, due to a defect-mediated kinetic barrier to dissolution. Thermal equilibrium can only be reached (dotted curve) when such a barrier is overcome by supplying thermal energy or the requisite defects (e.g., silicon interstitials). The order in which cavities are formed, implantation of metals occurs and annealing steps are performed, can control the observed diffusion, solubility and precipitation behavior. For example, Myers et al. [4, 8, 13, 33] do not observe bulk-phase formation (precipitation) of Cu and other metals at a He-induced cavity-band. Under their experimental conditions, the Cu_3Si and other metal silicides were initially formed by high-dose implantation and annealing with the equilibration of metals throughout the wafer up to the solubility level with respect to the silicide phase. The cavities were then introduced and provided strong sinks for metal impurities in solution to decorate the cavity walls but there is no free-energy benefit in relocating bulk phase Cu to the cavity-band in this case. Thus, the apparent inconsistency between the work of Myers and coworkers and that of Wong-Leung and collaborators (as illustrated in Figs. 9–11) can be explained in terms of different experimental steps and hence different pathways to final equilibration.

A further illustration of the importance of understanding the pathways to equilibrium in interpreting experimental data is given by the following example. Wong-Leung et al. [37] have shown a variation in the amount of Cu gettered to a cavity band between samples where the cavity band was preformed before metal implantation and samples where the metal implantation occurred prior to cavity formation. When the Cu dose was higher than that required to saturate the cavity walls, the amount of Cu gettered to the cavity band was significantly higher for the case where the cavities were not preformed. This effect can be explained as follows. Cavity formation from H implantation and annealing necessitates the release of Si interstitials, which diffuse away from the cavity band and can interact with processes occurring in the region of implanted Cu. If the cavities are forming while the Cu begins to diffuse and precipitate within the implant layer, the flux of silicon interstitials from the evolving cavities can suppress Cu_3Si formation in the implant region since this metal-rich silicide requires volume expansion. Conversely, the cavities are very attractive sites for precipitation of diffusing Cu since the open volume facilitates Cu_3Si formation. In the case where cavities are already preformed, Cu precipitation at the cavity band is lower since more Cu is locked up in silicide precipitates within the implant layer. Thus, competition between various diffusion, defect-trapping and precipitation processes clearly determines the extent of cavity gettering during annealing in cases where both the cavities and metal impurities are introduced by implantation.

It is important to note that the behavior illustrated above for Cu diffusion and precipitation at cavities is not universal for all diffusing metals. The observed behavior is very dependent on the particular diffusion mechanisms, solid solubility and the nature and structure of the precipitating phase. Indeed, metal transport mechanisms in silicon are different (see [33]) and the magnitude of interstitial to substitutional solid solubilities can play a role in transport and in diffusion mechanisms. For example, most metals are fast diffusers since they predominantly diffuse by metal interstitial motion through the silicon lattice. Cu is one of the fastest interstitial diffusing metals, where the interstitial solubility of Cu in silicon is orders of magnitude higher than the substitutional solubility (at 900°C the interstitial solubility is around 10^{17} atoms cm^{-3} or about 10^{-3} atom per cent of Si atoms). On the other hand, Au does not diffuse in silicon via a pure interstitial mechanism, its effective diffusivity is more than 2 orders lower than Cu, and the substitutional solubility of Au is several orders of magnitude higher than the interstitial solubility (Au solid solubility at 900°C is around $3 \times 10^{15}\,cm^{-3}$). Indeed, Au is well known to diffuse by the kick-out mechanism [38] that requires the intervention of Si interstitials to kick out the metal atoms occupying substitutional sites for the diffusion process to take place.

Fast-diffusing metals also display very different properties in terms of their ability to form silicides. For example, Au and Ag do not readily form silicides in the solid phase (Au requires a eutectic melt to facilitate formation) whereas Ni and Cu readily react with silicon at temperatures well below 500°C to form silicides and there are a number of stable silicides (of varying metal content) shown in the phase diagram for each of these metals. Indeed, the particular silicide phase, its structure and silicon content determine its density relative to the silicon host and whether its formation leads to a contraction or expansion of the lattice. As we illustrated for the formation of Cu_3Si, metal-rich silicides cause an expansion of the lattice and ejection of silicon interstitials in order to accommodate the growing silicides phase. On the other hand, the formation of silicon-rich silicides such as $FeSi_2$ and $NiSi_2$ is usually accompanied by a contraction in the lattice and formation results in the consumption of Si interstitials. These differences affect the ability of metals to precipitate in silicon and at cavities in particular. The influence of particular silicide-forming metals on gettering to cavities is further illustrated by the gettering of implanted Ni to cavities [29]. Efficient gettering of implanted Ni is observed when the Ni amount is below a monolayer coverage of cavity walls. However, higher implant concentrations above a monolayer coverage usually cause precipitation of $NiSi_2$ in the near-surface region rather than at cavities. This occurs because precipitation of $NiSi_2$, which has a lower number of Si atoms per unit volume compared to pure Si, prefers to occur within the silicon lattice rather than within the open-volume of cavities. Similar behavior to that of Ni has been observed for Fe [30] and Pt [32], which both form silicon-rich silicides preferentially within the implanted region. Ag, a nonsilicide former that precipitates as metallic Ag, behaves somewhat like Au in that it readily precipitates within cavities at high concentrations [32].

4 Trapping and Precipitation at So-Called $R_p/2$ Defects

As indicated in an earlier section, ion-implantation-induced lattice displacements and the consequential knock-on effect result in a separation between the interstitial and vacancy distributions [17, 39, 40]. As we have already indicated (Fig. 5), this separation can lead to both visible voids closer to the surface than the ion range and extended defects of interstitial character slightly beyond the ion range. The direct evidence for voids comes from TEM for high-dose implants under specific conditions [22, 23, 28], where voids can be observed at a depth equivalent to roughly the $R_p/2$ depth. Recently, electrical measurements [40] have also indirectly shown this displacement in vacancy and interstitial distributions. For MeV implants, the displacement between the vacancy and interstitial distributions is larger and hence easier to resolve spatially. However, several studies [23–25] have shown that the presence of vacancies at the $R_p/2$ range can be detected by positron annihilation spectroscopy (PAS) even when no voids can be seen in TEM. Positron annihilation is sensitive to open-volume defects but the vacancy concentration needs to be high ($\sim 4 \times 10^{16}\,\mathrm{cm}^{-2}$) [41] and there is also a question as to the influence of clusters and voids of different sizes on the sensitivity of this method. It has also been shown [26–28] that Au can be a very sensitive detector for vacancy clusters formed at the $R_p/2$ depth. For example, Fig. 13 shows the RBS Au profiles in silicon implanted with 245-keV Si at 100°C, then reimplanted with Au in the near-surface region and further annealed at 850°C for 1 h [28]. The projected range R_p for 245-keV Si^+ ions in Si is 3800 Å. The results are illustrated for two different doses of Si ions. For the lowest Si dose of $3 \times 10^{15}\,\mathrm{cm}^{-2}$, some of the Au is relocated to a shallow region between half of the projected range ($R_p/2$) and the surface, while most of the Au is still left at the surface close to the original Au implant depth.

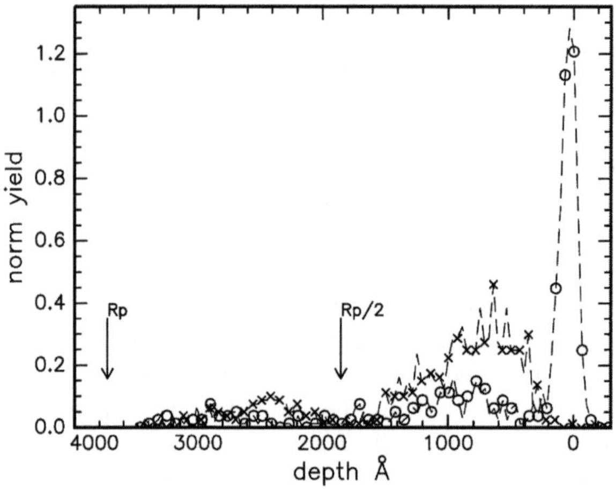

Fig. 13. RBS spectra of Au profiles after annealing at 850°C for 1 h in Si samples previously implanted with 245-keV Si^+ ions at 100°C to doses of $3 \times 10^{15}\,\mathrm{cm}^{-2}$ (*open circles*) and $1.4 \times 10^{16}\,\mathrm{cm}^{-2}$ (*crosses*). After Si implantation a 30-keV Au implant to a dose of $5 \times 10^{13}\,\mathrm{cm}^{-2}$ was carried out at room temperature (from [28])

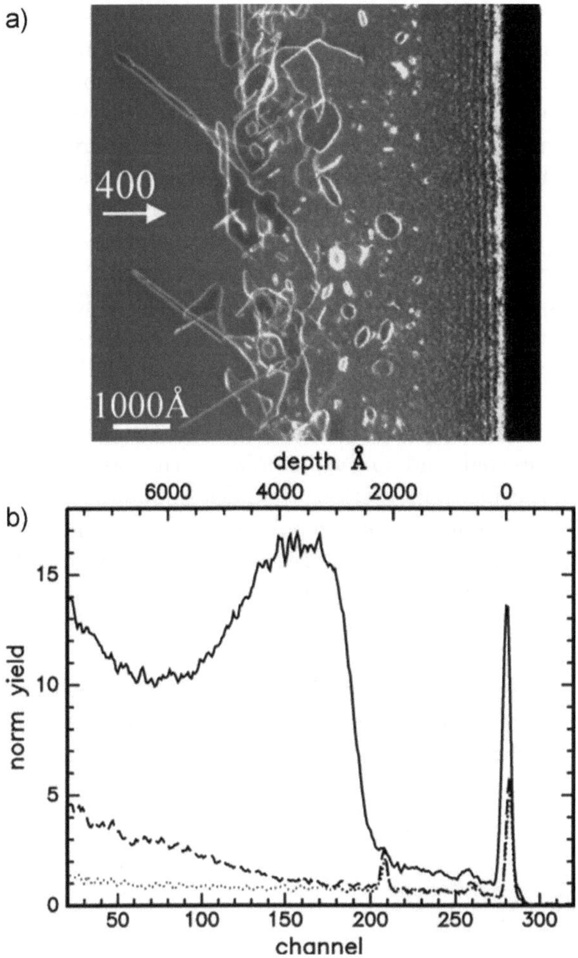

Fig. 14. (**a**) XTEM micrograph of the $1.4 \times 10^{16}\,\mathrm{cm}^{-2}$ sample from Fig. 13. (**b**) RBS/channeling spectra of this same sample before (*solid curve*) and after (*dashed curve*) annealing at 850°C. The *dotted curve* depicts a spectrum from an unimplanted sample (from [28])

In the case of the Si dose of $1.4 \times 10^{16}\,\mathrm{cm}^{-2}$, the amount of Au at that same region was much higher with no Au left at the surface. This dose also exhibited a Au peak at a depth of about 2400 Å that can be explained with reference to Fig. 14. The channeled RBS spectra corresponding to a Si dose of $1.4 \times 10^{16}\,\mathrm{cm}^{-2}$ are shown in Fig. 14b both following implantation and after annealing. This Si dose initially created a buried amorphous layer ~1600 Å thick, with the shallow amorphous/crystalline (a/c) interface located at a depth of 2400 Å (see as-implanted profile (solid line)). The small Au peak for the $1.4 \times 10^{16}\,\mathrm{cm}^{-2}$ dose in Fig. 13 is located just shallower than this original a/c interface after annealing and the amorphous layer is clearly shown to be recrystallized in the corresponding spectrum in Fig. 14b. The defect microstructure in the annealed sample is illustrated by the weak-beam TEM image in Fig. 14a, which indicates a dislocation-free region up to a depth of ∼1500 Å followed by a band of small loops (diameter ≤500 Å) extending to a depth of about 3000 Å. Deeper in the sample, we observe the presence of large dislocation networks, which suggests that the layer was probably not totally amorphous, presumably a consequence of the elevated implant temperature. We cannot ascertain the nature of the defects that trap Au specifically at 2400 Å but they could still be vacancy clusters. However, in the dislocation-free region, TEM analysis has shown some voids with a diameter of ∼50 Å at depths less than $R_p/2$. A typical TEM micrograph of the $1.4 \times 10^{16}\,\mathrm{cm}^{-2}$ dose case at this depth region is shown in Fig. 15. Some of the voids are faceted and show a dark contrast similar to the Au-decorated nanocavities discussed previously. A precipitate phase was also observed as shown by the moiré fringes in the inset in Fig. 15. This phase is presumably Au or gold silicide formed as a result of the Au concentration at these voids exceeding a monolayer coverage of the internal void surface, similar to the case of the larger cavities illustrated in the previous section. Samples implanted to Si doses higher than that in Fig. 15 showed a broader Au profile, with most of the Au observed at depths equivalent to the $R_p/2$ for the Si implant. TEM showed a similar microstructure to that in Fig. 15 except that the defect-free region near the surface extended to greater depths and a higher density of Au precipitates at depths up to $R_p/2$ was observed.

Several groups have now reported the decoration of the vacancy-rich regions resulting from implantation with various metals and other species. For example, Brown et al. [25] reported the gettering of O, Cu, Fe and Ni to the

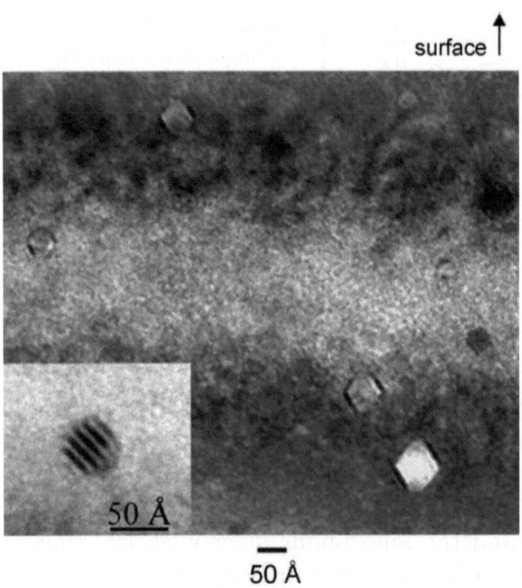

Fig. 15. XTEM micrograph from a depth (460–1100 Å from the surface) corresponding to the region denuded of interstitial-based defects but containing Au for the $1.4 \times 10^{16}\,\mathrm{cm}^{-2}$ sample from Fig. 14. The inset shows a precipitate that is crystalline and presumably Au rich (from [28])

$R_p/2$ depth region of MeV Si-implanted Si after annealing. The use of PAS in this study also showed the presence of open-volume defects in some of the samples but TEM investigation showed no visible defects at this range, presumably as a result of the small size of vacancy clusters or their low density. Venezia et al. [26] in reporting Au decoration of and precipitation in vacancy-rich regions in MeV Si-implanted samples suggested that, under appropriate conditions, Au could fully saturate the open volume and can be used as a measure of open volume and vacancy content. However, care must be taken in such experiments that conditions favor precipitation at voids in contrast to the equilibrium studies of Myers et al. [42], where prior equilibration and precipitation of Au at surfaces before cavities are introduced, prevented Au precipitation at cavities. Kalyanaraman et al. [27] have used the Au labeling technique to quantify the excess vacancy defects created by high-energy ion implantation, clearly a case where Au saturation of vacancy clusters and voids is possible. In this study, vacancies were first introduced by a 2-MeV Si implant and an annealing process. Au was then introduced by implantation into the near-surface region and the samples were further annealed to saturate the vacancy-excess region of the original Si implant. The amount of Au decorating the vacancy-excess region is then a measure of the number of vacancies but the absolute number of vacancies can only be estimated if the number of Au atoms that occupy a single vacancy is known. Si interstitials were injected into the vacancy-excess region by implantation to interact with and annihilate vacancies. As described in the next section, it is assumed that each implanted ion leads to one excess interstitial (the so-called "+1" model [43]) that is available to interact with vacancies. Using such a calibration method, it was found that one Au atom decorated the volume of 1.2 vacancies. The experiments upon which this calibration method is based is outlined in the next section.

5 Stability Under Subsequent Irradiation

In the previous sections, the strong ability of cavities to getter fast-diffusing interstitial metals was demonstrated. Indeed, cavities have also been shown to be very efficient trapping sites for any diffusing interstitials including oxygen [36], boron [44] and even silicon [45]. In the latter case, ion-implanted interstitial-based defects are depleted in the region surrounding cavities following an annealing step. In this section we examine the interaction of irradiation-induced defects with open-volume defects both during irradiation and during subsequent annealing. We treat first the case where amorphization does not occur during irradiation and the open-volume defects can act as a sink for diffusing silicon interstitials, then we examine a process of preferential amorphization around cavities and finally a process where open-volume defects can shrink and disappear during amorphization.

5.1 Interaction of Defects with Voids and Nanocavities

To ensure that amorphization does not occur during irradiation, it is necessary to use low ion doses or elevated implantation temperatures. For examining the interaction of defects with nanocavities, Williams et al. [46] irradiated preformed cavities with Si ions at 300°C. The cavities were initially formed by implanting with 20-keV H ions to a dose of $3 \times 10^{16}\,\mathrm{cm}^{-2}$, followed by annealing to 850°C for 1 h. Samples with and without cavities were then re-irradiated with 245-keV Si ions to various doses and then analyzed by TEM to examine the defect microstructure surrounding nanocavities. This particular Si ion energy was chosen to place the end-of-ion-range (i.e., the region where an interstitial excess is expected) at the depth of the cavity-band. A typical result of this irradiation is illustrated in Fig. 16 for a Si ion dose of 1.4×10^{16} Si cm^{-2}. Both the sample with cavities and that without cavities exhibit considerable disorder, which, for 300°C irradiation, consists of interstitial-based clusters, small loops and possibly also {311} defects. However, for the cavity sample in Fig. 16a, it is clear that there is a zone around

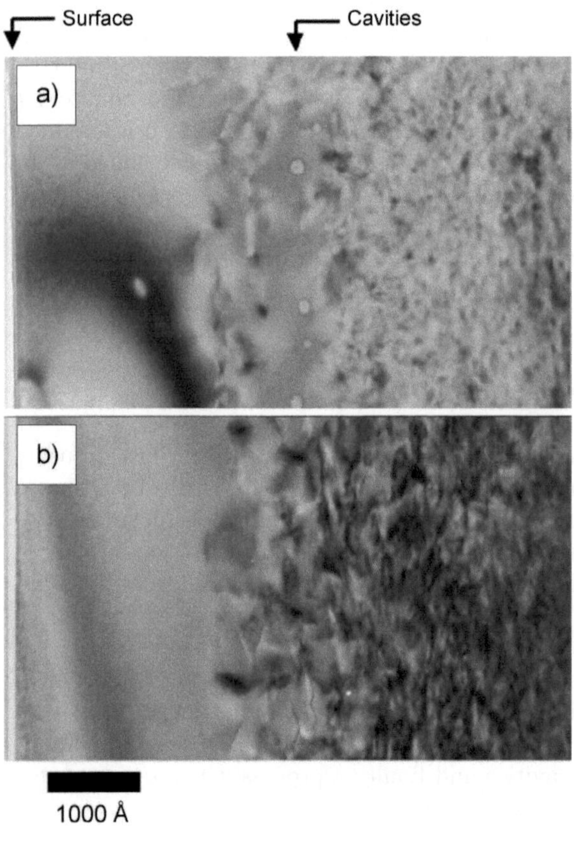

Fig. 16. XTEM micrographs of samples with cavities (**a**) and without cavities (**b**) after irradiation with 245-keV Si^+ ions to a dose of $1.4 \times 10^{16}\,\mathrm{cm}^{-2}$ at 300°C (from [46])

each cavity that is denuded of defects. This indicates that, at 300°C at least, the cavities are strong sinks for mobile Si interstitial defects during irradiation and that interstitial–interstitial interactions that form "stable" interstitial clusters are inhibited within a critical distance (about 400 Å) surrounding a cavity. This critical distance is clearly related to the diffusion length or mean free path of mobile, irradiation-produced defects before they cluster or annihilate at cavities at 300°C. One might expect such a diffusion length to be strongly dependent on irradiation parameters, including temperature and defect-production rate, the latter depending on the cascade density and the ion flux. Such parameters have not yet been investigated in any detail for cavities but it is clear that reducing the implantation temperature reduces the radius of the defect-denuded zone around cavities and the number of interstitials trapped at cavities [47]. Indeed, the diffusion and annihilation of interstitials at cavities appears to be the mechanism for the observed shrinkage of cavities during ion irradiation at elevated temperatures, when no amorphization occurs [47].

It is not surprising that silicon interstitials can also diffuse to smaller vacancy clusters and voids if the implant and annealing conditions are suitable. Indeed, Sect. 4 indicated that such small open-volume defects are favored sites for the accumulation of fast-diffusing interstitial metal atoms. To examine the ability of silicon interstitials to similarly accumulate at vacancy clusters, Kalyanaraman and coworkers [27, 47, 48] carried out the following experiments. First, silicon was irradiated with 2-MeV Si ions at 70–80°C to a dose of $10^{16}\,\mathrm{cm}^{-2}$ to form vacancy-related defects within the $R_\mathrm{p}/2$ region, which were stabilized as vacancy clusters by annealing at 815°C. The existence of vacancy clusters was confirmed by a Au-labeling experiment, similar to those described in the previous section. A further Si implant at 600 keV, termed the +I implant, was then carried out to doses up to $2 \times 10^{14}\,\mathrm{cm}^{-2}$ at room temperature to inject Si interstitials into the $R_\mathrm{p}/2$ region of the previous implant. Complete interaction of the excess interstitials with vacancy clusters was promoted by a further annealing step at 765°C. The rationale here is that the 600-keV Si implant is expected to inject Si interstitials into the vacancy region at a rate of one per ion (the so-called "+1" model [43] as indicated in the previous section) and every injected interstitial is assumed to annihilate one vacancy in this region. Although the +1 model is a reasonably good estimate of excess interstitials following annealing, the assumption that every excess interstitial annihilates with a vacancy in the vacancy-excess region is somewhat contentious. For example, some interstitials particularly for the higher implant doses may be expected to cluster and form {311} rod-like defects before annihilation, whereas others may diffuse out of the vacancy-excess region. Nevertheless, somewhat surprisingly, the "+1" model is found to be a good estimate of the number of excess interstitials that are available for vacancy annihilation. Finally, Au-labeling experiments were carried out to detect the presence of remaining vacancy clusters at $R_\mathrm{p}/2$. Results showed that increasingly less Au accumulated at the $R_p/2$ region with increasing dose of the +I implant, suggesting that interstitials had accumulated and annihilated at vacancy clusters. By a +I dose of $2 \times 10^{14}\,\mathrm{cm}^{-2}$, that is $2 \times 10^{14}\,\mathrm{cm}^{-2}$ excess interstitials, there appeared to be no remaining open volume at $R_p/2$, suggesting that there were $2 \times 10^{14}\,\mathrm{cm}^{-2}$ vacancies in the vacancy-excess region. These results are consistent with the observations of Si interstitial interactions with larger nanocavities (Fig. 16) and indicate that, if the implant and annealing conditions are appropriately chosen, implant-injected interstitials can annihilate at open-volume defects in cases when the Si surrounding such open-volume defects is not amorphized. In the following section, we treat the case of implant conditions that do lead to amorphization.

5.2 Preferential Amorphization

At implant temperatures below about room temperature, amorphous layers readily form and there is little difference between the manner in which damage builds up in samples with and without pre-existing defects, whether they are interstitial-based defects, cavities or voids. However, at slightly elevated temperatures, where some defect annihilation and agglomeration might be expected, there is quite a substantial difference in the way irradiation-induced disorder, and amorphization in particular, occurs. This is illustrated for the case of cavities by the RBS/channeling spectra and TEM micrographs [46] in Fig. 17 for Si ion irradiation of a sample containing pre-existing cavities at 100°C. At a dose of $8 \times 10^{15}\,\mathrm{cm}^{-2}$, the sample with cavities exhibits a sharp peak in the channeling spectrum at the cavity depth (Fig. 17), whereas the control sample does not. This difference can be interpreted following inspection of the corresponding cross-sectional TEM micrographs. Figure 18a, for example, shows that the sample with the cavities contains clear regions of amorphous silicon. Furthermore, at the center of each amorphous zone there appears to be a cavity. In contrast, the control sample shown in Fig. 18b does not appear to contain any amorphous silicon at this dose, but rather, the residual disorder consists of a dense network of defect clusters. Selected-area diffraction patterns confirm the presence of amorphous silicon in the case of the sample with cavities and the absence of any amorphous silicon in the sample without cavities. The nature of the amorphous zone surrounding a particular cavity is further illustrated in the higher magnification image in Fig. 19. Detailed inspection indicates that each cavity is surrounded by a zone of amorphous silicon. As the Si ion dose increases, the thickness of these zones has been observed to increase until they overlap to form a continuous amorphous layer.

The above result clearly indicates that cavities, or the cavity walls, can act as preferential amorphization sites in cases where substantial dynamic annealing takes place during implantation. If cavities are indeed sinks for defects, we might have expected the opposite behavior, that is, the cavities to be surrounded by a region with a lower concentration of residual disorder,

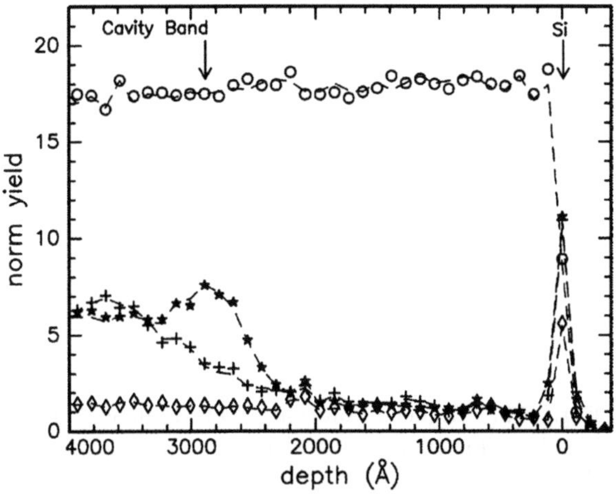

Fig. 17. Normalized RBS/channeling spectra showing the residual damage in a Si sample containing cavities (*filled stars*) and a sample without cavities (+) following 245-keV Si^+ bombardment to a dose of 8×10^{15} cm^2 at 100°C. A random spectrum (*open circles*) and virgin spectrum (*open diamonds*) are shown for comparison (from [46])

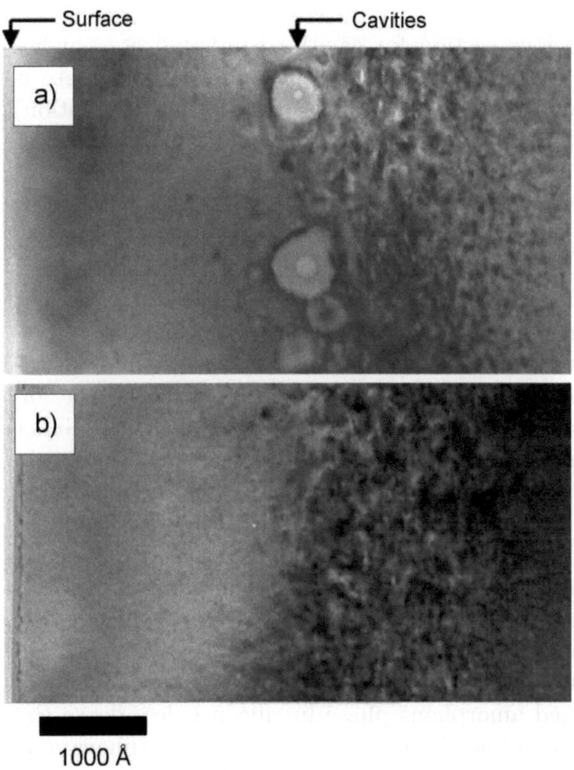

Fig. 18. XTEM micrographs of samples with cavities (**a**) and without cavities (**b**) corresponding to the 8×10^{15} Si cm^{-2} irradiation shown in Fig. 17 (from [46])

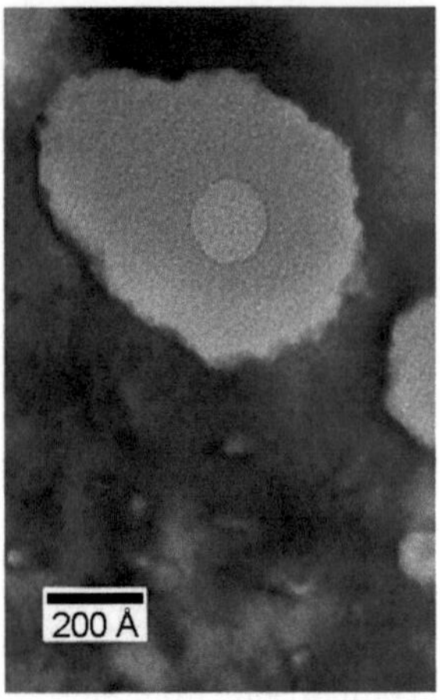

Fig. 19. Higher-resolution XTEM image from Fig. 18a showing a cavity within an amorphous zone (from [46])

as was the case in the previous section for cavities irradiated at a higher temperature. However, at critical irradiation temperatures, where there is a close balance between the rates of defect production (by ion irradiation) and dynamic defect annealing (as a result of defect mobility), the manner in which damage builds up in silicon, often ultimately leading to amorphization, has previously been shown to be quite complex [50, 51]. Amorphous layers can form at high enough doses, when the local free energy of the system (as a result of the accumulation of defect clusters with increasing dose) exceeds that of the amorphous phase. In such cases, it has been shown that amorphization is nucleation limited [51] and is observed to nucleate at preexisting defects, surfaces or interfaces. Figure 18 clearly shows that cavities can also constitute preferential nucleation sites for amorphization. The following explanation has been proposed [46, 52] to explain this behavior. Under the implantation conditions used to obtain the data in Fig. 18, despite substantial dynamic annealing of implantation-produced disorder, defect clusters initially build up in silicon as a result of incomplete defect annihilation and thus raise the local free energy. Reduced defect mobility and increased defect stability at 100°C, compared with that at 300°C for example (see Fig. 16), results in the accumulation of such defect clusters very close to the cavities. As the defect density increases with increasing ion dose, amorphization will be favored when the local free energy exceeds that of the amorphous phase. However, since a fully relaxed amorphous phase in silicon is less dense (by about 1.8% [53]) than the crystalline phase, nucleation of an amorphous zone will normally strain the surrounding crystal, thus raising the local free energy above that of the amorphous phase alone. Thus, at elevated temperatures defect clusters will tend to accumulate in silicon until the local free energy of the defective region substantially exceeds that of amorphous silicon before amorphization occurs, unless a more energetically favorable nucleation site exists. Cavities would appear to be a particularly attractive nucleation site since, if amorphization is first nucleated at cavity walls, the amorphous phase is free to expand into the cavity open-volume without straining the surrounding silicon crystal. Thus, the system free energy is minimized if amorphous zones nucleate at cavities. Furthermore, a sample containing cavities will amorphize at a lower dose than a sample without cavities, as a result of the additional strain energy associated with amorphization in the latter case. Although appropriate experiments have not yet been attempted, it is expected that smaller open-volume defects such as vacancy clusters and voids will similarly act as preferential nucleation sites in silicon under suitable irradiation conditions.

5.3 Shrinkage and Removal of Open-Volume Defects During Amorphization

Once the region surrounding cavities is amorphized, there is another interesting phenomenon that occurs with continued bombardment, namely the gradual shrinkage of cavities, leading ultimately to their complete disappearance [47, 54]. For example, cavities have been observed to disappear within ion-irradiated amorphous silicon and on subsequent recrystallization no evidence for open-volume defects was obtained [54]. Recently, Si irradiation of samples containing cavities in an electron microscope [54] has shown the evolution of cavity shrinkage, as illustrated in Fig. 20, which shows in-situ TEM micrographs of a single nanocavity during Si irradiation at ~21°C. In this case, a cross-sectional foil was made of a sample containing cavities, which was subsequently irradiated with 100-keV Si ions that passed through the foil. As is readily apparent, the mean nanocavity diameter decreases as a function of Si irradiation dose. No shrinkage was observed to take place unless the ion beam was present. Also, preferential amorphization of the Si substrate in close proximity to the nanocavity is evident for a Si ion dose of 2×10^{15} cm^{-2}, consistent with the behavior reported in the previous section. At greater ion doses, the Si surrounding the nanocavity is completely amorphized and thereafter the greatest change in nanocavity diameter occurs. By measuring the rate of shrinkage as a function of dose it was noted [47] that the nanocavity *volume* (proportional to diameter cubed) did not decrease proportional to irradiation dose. Indeed, after amorphization, except for the very final stage of shrinkage, the dependence of cavity diameter on dose appeared to be closer to linear. Such a dependence may suggest that shrinkage of a cavity only occurs when an ion cascade passes through it, thus causing an incremental inward expansion of the surrounding amorphous material within the cascade volume. However, more detailed measurements with a statistically meaningful number of nanocavities would be necessary to accurately determine the irradiation-parameter dependencies.

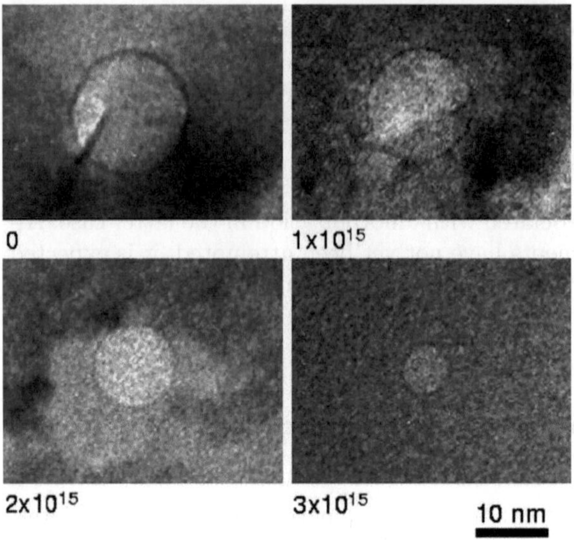

Fig. 20. In-situ TEM micrographs of a single nanocavity for an irradiation temperature of ∼21°C as a function of Si ion dose (cm^2) (from [47])

Possible mechanisms for cavity shrinkage during irradiation have been proposed as follows [47, 54]. As indicated in the previous section, amorphization of the region immediately surrounding cavities can allow a less-dense amorphous phase to expand into the open cavity volume to minimize free energy. This behavior presupposes that amorphous silicon can flow under ion irradiation. Indeed, there is clear evidence that amorphous Si flows during ion irradiation at room temperature from studies that measured the strain and expansion in silicon during amorphization (see, for example, [55]). In these studies, the flow of amorphous silicon resulted in measurable step heights between implanted and unimplanted regions [55] and could also result in mass transport in the direction of the ion beam [56]. Thus, plastic flow of amorphous silicon under ion irradiation could allow the amorphous phase to achieve its equilibrium density and at the same time reduce the open-volume of cavities. Alternatively, the amorphous silicon may flow via the diffusion of vacancies into amorphous silicon from a cavity source, noting that the equilibrium vacancy concentration in the amorphous phase is much higher than in the crystalline phase [57]. This presupposes that such vacancies are mobile in amorphous Si under ion irradiation, a fact that is yet to be established. In addition to expansion and flow of amorphous silicon to minimize the free energy of the amorphous phase, there are additional driving forces that could contribute to cavity shrinkage. For example, cavities raise the local free energy through, in part, the surface-energy contribution [58], and this term might be expected to increase (per unit volume) as the cavities decrease in size to atomic dimensions (quantum size effect). Hence, removal of the open-volume during shrinkage will decrease the excess free energy and hence drive the shrinkage process. Finally, in order to clarify the mechanism of shrinkage, more detailed in-situ measurements of nanocavity shrinkage as functions of irradiation dose and temperature will be required.

6 Conclusions

Nanocavities are not only of technological interest in silicon processing, as efficient gettering sites and for giving rise to "smart-cut" silicon, but also for the opportunities that they offer for studying metal–cavity and defect–cavity interactions, which often involve nonequilibrium behavior. For example, we illustrated that fast-diffusing metals can preferentially bond to cavity walls and precipitate within cavities if the metal concentration is high enough. It was shown that the formation and dissolution of metal silicides at cavities is often controlled by the availability or removal of silicon interstitials. For short annealing times, pseudoequilibrium can be reached, with metals such as Au and Cu, where essentially all the metal is precipitated at cavities and the amount of metal in the bulk is far below the equilibrium level.

When silicon that contains cavities is implanted at slightly elevated temperatures, we demonstrated that cavities can act as preferential nucleation sites for amorphization. This behavior was explained in terms of local free-energy minimization as implantation-induced residual defects build up with increasing ion dose. When the region surrounding cavities is completely amorphized, continued implantation can lead to shrinkage and the eventual disappearance of nanocavities. It was proposed that the plastic flow of the less-dense amorphous silicon phase into the cavities during irradiation is responsible for this behavior.

References

1. G.D. Watkins, J. Phys. Soc. Jpn. (Suppl. II) **18**, 22 (1963) 114, 120
2. P.S. Plekhanov, U.M. Gösele, T.Y. Tan, J. Appl. Phys. **84**, 718 (1998) 114, 120
3. J.-G. Park, H. Kirk, K.-C. Cho, H.-K. Lee, C.-S. Lee, G.A. Rozgonyi, in *Semiconductor Silicon 1994*, ed. by H.R. Huff, W. Bergholz, K. Sumino (Electrochemical Society, Pennington, 1994), p. 370 114, 120
4. S.M. Myers, D.M. Follstaedt, D.M. Bishop, Mater. Sci. Forum **143–147**, 1635 (1994) 115, 130
5. D.M. Follstaedt, Appl. Phys. Lett. **62**, 1116 (1993) 115, 116, 125
6. J. Wong-Leung, C.E. Ascheron, M. Petravic, R.G. Elliman, J.S. Williams, Appl. Phys. Lett. **66**, 1231 (1995) 115, 122, 127
7. J. Wong-Leung, E. Nygren, J.S. Williams, Appl. Phys. Lett. **68**, 416 (1995) 115, 122
8. S.M. Myers, G.A. Petersen, Phys. Rev. B **57**, 7015 (1998) 115, 130

9. J. Wong-Leung, J.S. Williams, A. Kinomura, Y. Nakano, Y. Hayashi, D.J. Eaglesham, Phys. Rev. B **59**, 7990 (1999) 115, 119, 125, 126, 127, 128, 129
10. M. Bruel, Electron. Lett. **31**, 1201 (1995) 115
11. J.F. Ziegler, J.P. Biersack, U. Littmark, *The Stopping and Range of Ions in Solids* (Pergamon, New York, 1985) 116, 117
12. V. Raineri, S. Coffa, E. Szilagyi, J. Gyulai, E. Rimini, Phys. Rev. B **61**, 937 (2000) 116, 119
13. D.M. Follstaedt, S.M. Myers, G.A. Petersen, J.W. Medernach, J. Electron. Mater. **25**, 151 (1996) 116, 125, 130
14. J. Wong-Leung, PhD Thesis, The Australian National University, 1997 117, 118, 123, 127
15. M.K. Weldon, V.E. Marsico, Y.J. Chabal, A. Agarwal, D.J. Eaglesham, J. Sapjeta, W.L. Brown, D.C. Jacobson, Y. Caudano, S.B. Christman, E.E. Chaban, J. Vac. Sci. Technol. B **15**, 1065 (1997) 117, 119
16. T. Höchbauer, A. Misra, R. Verda, C.H. Bauer, A. Misra, R. Verda, Y. Zheng, S.S. Lau, J.W. Mayer, M. Nastasi, Nucl. Instrum. Methods B **175–177**, 169 (2001) 118, 119
17. A.M. Mazzone, Phys.-Status Solidi A **95**, 149 (1986) 120, 132
18. D.J. Eaglesham, P.A. Stolk, H.-J. Gossmann, J.M. Poate, Appl. Phys. Lett. **65**, 2305 (1994) 120
19. P.A. Stolk, J.L. Benton, D.J. Eaglesham, D.C. Jacobson, J.-Y. Cheng, J.M. Poate, S.M. Myers, T.E. Haynes, Appl. Phys. Lett. **68**, 51 (1996) 120
20. J.L. Benton, S. Libertino, P. Kringhøj, D.J. Eaglesham, J.M. Poate, S. Coffa, J. Appl. Phys. **82**, 120 (1997) 120
21. D.S. Zhou, O.W. Holland, J.D. Budai, Appl. Phys. Lett. **63**, 3580 (1993) 120
22. S.L. Ellingboe, M.C. Ridgway, Nucl. Instrum. Methods B **127**, 90 (1997) 120, 132
23. W.O. Holland, L. Xie, B. Nielsen, D.S. Zhou, J. Electron. Mater. **25**, 99 (1996) 120, 132
24. C.S. Szeles, B. Nielsen, O. Asoka-Kumar, K.G. Lynn, A. Anderle, T.P. Ma, G.W. Rubloff, J. Appl. Phys. **76**, 3403 (1994) 120, 132
25. R.A. Brown, O. Kononchuk, G.A. Rozgonyi, S. Koveshnikov, A.P. Knights, P. Simpson, F. Gonzales, J. Appl. Phys. **84**, 2459 (1998) 120, 132, 134
26. V.C. Venezia, D.J. Eaglesham, T.E. Haynes, A. Agarwal, D.C. Jacobson, H.-J. Gossmann, F.H. Baumann, Appl. Phys. Lett. **73**, 2980 (1998) 120, 132, 135
27. R. Kalyanaraman, T.E. Haynes, V.C. Venezia, D.C. Jacobson, H.-J. Gossmann, C.S. Rafferty, Appl. Phys. Lett. **76**, 3379 (2000) 120, 132, 135, 137
28. J.S. Williams, M.J. Conway, B.C. Williams, J. Wong-Leung, Appl. Phys. Lett. **78**, 2867 (2001) 121, 132, 133, 134

29. B. Mohadjeri, J.S. Williams, J. Wong-Leung, Appl. Phys. Lett. **66**, 1889 (1995) 122, 131
30. B. Mohadjeri, J.S. Williams, J. Wong-Leung, Unpublished, 1996 122, 131
31. S.M. Myers, G.A. Petersen, C.H. Seager, J. Appl. Phys. **80**, 3717 (1996) 122, 123, 124
32. A. Kinomura, J.S. Williams, J. Wong-Leung, M. Petravic, Appl. Phys. Lett. **72**, 2713 (1998) 122, 131
33. S.M. Myers, M. Seibt, W. Schroter, J. Appl. Phys. **88**, 3795 (2000) 122, 123, 124, 130, 131
34. A. Kinomura, J.S. Williams, J. Wong-Leung, M. Petravic, Y. Nakano, Y. Hayashi, Appl. Phys. Lett. **73**, 2639 (1998) 122, 124
35. R. Plass, L.D. Marks, Surf. Sci. **380**, 497 (1997) 124, 125
36. J.S. Williams, M.J. Conway, J. Wong-Leung, P.N.K. Deenapanray, M. Petravic, R.A. Brown, D.J. Eaglesham, D.C. Jacobson, Appl. Phys. Lett. **75**, 2424 (1999) 125, 135
37. J. Wong-Leung, J.S. Williams, M. Petravic, Mater. Res. Soc. Symp. Proc. **469**, 457 (1997) 130
38. U. Gosële, W. Frank, A. Seeger, Appl. Phys. **23**, 361 (1980) 131
39. M. Tamura, Mater. Sci. Rep. **6**, 141 (1991) 132
40. P. Pellegrino, P. Lévêque, J. Wong-Leung, C. Jagadish, B.G. Svensson, Appl. Phys. Lett. **78**, 3442 (2001) 132
41. B. Nielsen, O.W. Holland, T.C. Leung, K.G. Lynn, J. Appl. Phys. **74**, 1636 (1993) 132
42. S.M. Myers, G.A. Petersen, T.J. Headley, J.R. Michael, T.L. Aselage, C.H. Seager, Nucl. Instrum. Methods B **127–128**, 291 (1997) 135
43. M.D. Giles, J. Electrochem. Soc. **138**, 1160 (1991) 135, 137
44. J. Wong-Leung, J.S. Williams, M. Petravic, Appl. Phys. Lett. **72**, 2418 (1998) 135
45. V. Raineri, U. Campisano, Nucl. Instrum. Methods B **120**, 56 (1996) 135
46. J.S. Williams, X.F. Zhu, M.C. Ridgway, M.J. Conway, B.C. Williams, F. Fortuna, M.-O. Ruault, H. Bernas, Appl. Phys. Lett. **77**, 4280 (2000) 136, 138, 139, 140
47. X.F. Zhu, J.S. Williams, M.J. Conway, M.C. Ridgway, F. Fortuna, M.-O. Ruault, H. Bernas, Appl. Phys. Lett. **79**, 3416 (2001) 137, 141, 142
48. R. Kalyanaraman, T.E. Haynes, V.C. Venezia, D.C. Jacobson, H.-J. Gossmann, C.S. Rafferty, Mater. Res. Soc. Symp. Proc. **610**, B9.2.1 (2001) 137
49. R. Kalyanaraman, T.E. Haynes, D.C. Jacobson, H.-J. Gossmann, C.S. Rafferty, Mater. Res. Soc. Symp. Proc. **610**, B9.4.1 (2001)
50. J.S. Williams, H.H. Tan, R.D. Goldberg, R.A. Barown, C. Jagadish, Mater. Res. Soc. Symp. Proc. **316**, 15 (1994) 140
51. R.D. Goldberg, J.S. Williams, R.G. Elliman, Phys. Rev. Lett. **82**, 771 (1999) 140

52. J.S. Williams, M.C. Ridgway, M.J. Conway, J. Wong-Leung, X.F. Zhu, M. Petravic, F. Fortuna, M.-O. Ruault, H. Bernas, A. Kinomura, Y. Nakano, Y. Hayashi, Nucl. Instrum. Methods B **178**, 33 (2001) 140
53. J.S. Custer, M.O. Thompson, D.C. Jacobson, J.M. Poate, S. Roorda, W.C. Sinke, F. Spaepen, Mater. Res. Soc. Symp. Proc. **157**, 689 (1990) 140
54. X.F. Zhu, J.S. Williams, D.J. Llewellyn, J.C. McCallum, Appl. Phys. Lett. **74**, 2313 (1999) 141, 142
55. C.A. Volkert, J. Appl. Phys. **70**, 3521 (1991) 142
56. L. Cliche, S. Roorda, M. Chicoine, R.A. Masut, Phys. Rev. Lett. **75**, 2348 (1995) 142
57. D.C. Jacobson, R.G. Elliman, J.M. Gibson, G.L. Olson, J.M. Poate, J.S. Williams, Mater. Res. Soc. Symp. Proc. **74**, 327 (1987) 142
58. F. Schiettekatte, C. Wintgens, S. Roorda, Appl. Phys. Lett. **74**, 1857 (1999) 142

Index

Damage Formation and Evolution in Ion-Implanted Crystalline Si

Sebania Libertino and Antonino La Magna

CNR-IMM, Stradale Primosole 50, 95121, Catania, Italy,
e-mail: sebania.libertino@imm.cnr.it

Abstract. Damage formation during ion implantation in crystalline Si and its evolution as a function of annealing have been widely investigated both theoretically and experimentally in the last few decades. The increasing knowledge of the damage features helps scientists in the understanding and modeling of many phenomena such as transient enhanced diffusion of dopants and extended-defect evolution. Nevertheless, many questions on defect agglomeration and evolution upon annealing are still unsolved. Despite the large efforts devoted in recent years, it is not clear how point-like defects agglomerate, forming more stable and complex structures such as defect clusters, and how they evolve into extended defects. Such knowledge would be the basic foundation to implement simulation programs of device processing. In this chapter we review the main results on defect agglomeration, from the elementary defects generated during ion implantation, interstitials and vacancies, to extended defects and their implication on our knowledge of the clustering mechanism. The results are used to clarify some of the unsolved puzzles on the mechanisms of implantation-damage formation and evolution in crystalline Si.

1 Introduction

In a solid at thermodynamical equilibrium, a finite concentration of point defects is present. Many of their properties are determined by the change in free energy required for their formation and migration in the lattice. The equilibrium defect concentration at each temperature is exponentially dependent on the defect-formation energy. On the other hand, the defect-migration energy controls phenomena such as defect recombination and agglomeration.

In the crystalline Si lattice, schematically shown in Fig. 1, two elementary defects can be defined: the vacancy (V) and the Si self-interstitial (I). When an energy as high as ~5 eV is transferred to a Si atom, it is displaced from its lattice position creating a vacant site (V), while it will occupy an interstitial position in the lattice (I) [1, 2]. The defect structures and properties have been the target of several investigations, both experimental and theorical, but a comprehensive understanding of their properties is far from being reached.

The similar formation energy value for Is and Vs results in their simultaneous presence in equilibrium conditions. Moreover, their concentration is very low: only ~10^{13} defects/cm^3 are present in Si at thermodynamical equilibrium even at ~1200°C [3]. They are, however, highly mobile in the lattice:

H. Bernas (Ed.): Materials Science with Ion Beams,
Topics Appl. Physics **116**, 147–212 (2010)
DOI: 10.1007/978-3-540-88789-8_6,

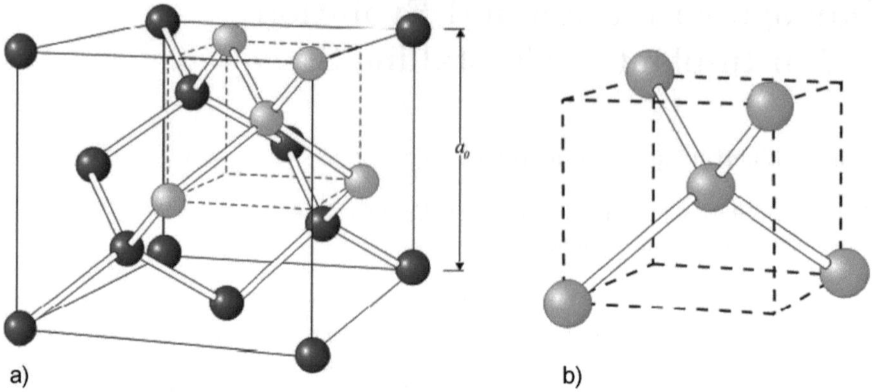

Fig. 1. Schematic of the Si lattice: (**a**) elementary cell. The *gray atoms within the dashed cube* represent the unit cell, magnified in (**b**)

in fact the activation energy for migration was estimated to be very low, ~0.3 eV and ~1 eV for V and I, respectively [2, 4]. Since the equilibrium defect concentration in Si is very low, the analysis of their properties requires the use of nonequilibrium processes, able to produce a high defect concentration. Among the different approaches, irradiation with high-energy electrons was, traditionally, the most used technique to generate a large number of point defects in the crystalline structure and to study their properties [5], while in recent studies ion implantation is used to generate defects in Si [6]. The use of these nonequilibrium processes produced new phenomena and opened new interesting research areas, also driven by the increasing importance of ion implantation in device processing. In fact, in VLSI (very large scale integration) technology dopant incorporation is achieved by the means of ion implantation [7], thanks to a precise control of the final dopant concentration in the material. The main drawback of this process is the damage formation in the crystalline host, since the defects presence produces significant modifications of the Si structural and electrical properties. Moreover, defects are known to mediate the impurities' diffusion mechanisms. For these reasons, the research on defect formation and evolution in ion-implanted Si has attracted an increasing attention.

As previously mentioned, the defect migration energy in Si is quite low, even at room temperature. Defects migrate until they either recombine, according to relation $I + V = \varnothing$, or are stored in stable defect complexes. As will be discussed in Sect. 2, point defects (I and V) in Si strongly interact among themselves, with impurities (e.g., C and O) and dopant atoms (e.g., B, P) of the substrate, forming point-like defects that are stable at room temperature (RT). These interactions significantly affect the defect's migration and agglomeration properties. Moreover, the point-like defect presence strongly modifies the Si electrical properties. A detailed study on the point-like defect introduction rates and properties was performed over the last forty years [1–8], suggesting that the fundamental properties of this class of defects is well established. As will be shown, this is not fully correct, in fact the state-of-the-art device performances require Si substrates with O and C contents below $1 \times 10^{15}\,\mathrm{cm}^{-3}$ (epitaxial Si) and with metal impurity concentration in the lattice below $1 \times 10^{10}\,\mathrm{cm}^{-3}$. The impurity concentration is much lower than that available when most of the studies on point-like defects were performed. Since the defects strongly interact with impurities, impurity concentration affects their evolution. The reduction in the impurity content is due to the different Si wafer fabrication methods. In Czochralski (CZ) Si, high O and C concentrations are incorporated in the crystal, since they are equal to the C and O solid solubility in Si at the seed/bath interface temperature. Typical values are: $\sim 10^{18}\,\mathrm{O/cm}^3$ and $\sim 10^{17}\,\mathrm{C/cm}^3$. The most used growth method, today, is the epitaxial growth of silicon layers through chemical vapor deposition (CVD) procedures. It is very clean and both O and C concentrations are below $10^{15}\,\mathrm{cm}^{-3}$. The reduction of more than three order of magnitudes in the O concentration plays a major role in the damage storage and evolution during the device fabrication steps.

Point defects in Si exist in different charge states, hence both their migration and agglomeration properties are affected by the charge state presence, further complicating the analysis. For example, the defect's equilibrium concentration depends on the Fermi-level position in the sample. A change in the charge state of a defect can also result in a change of its equilibrium atomic structure. Both V and I can be embodied into different lattice configurations having slightly different formation energies. This effect has dramatic consequences on their migration properties. Detailed studies have shown that the barrier for I migration is strongly dependent on the charge state and even athermal migration paths exist. As a consequence Is can migrate athermally even at 4 K in p-type crystalline Si [9]. The defects formed at room temperature dissociate upon annealing releasing the I and V they stored. It is interesting to note that dopant migration is assisted by the presence of V or I that act as diffusion vehicles mediating the transport of matters in Si [10]. Upon annealing, these nonequilibrium defects will interact among themselves and with the dopant atoms. Both the final dopant distribution and the defect's configuration strongly depend on the thermal budget provided to the sample. Moreover, the thermal budget necessary to fully recover the lattice is strongly dependent on the damage structure. While simple defects are mainly recovered after annealing at 500°C, temperatures as high as 1200°C are needed to anneal out extended defects.

The rearrangement in the defect structure of ion-implanted crystalline Si upon thermal annealing is the target of several investigations [6, 11, 12] driven by both the scientific challenge and technological needs. The evolution of point defects into extended defects and the full recover of the Si crystallinity upon thermal annealing, as well as the role played by the impurity content of the substrate on defect migration and evolution are problems of particular importance. Their solution is made more critical by the stringent requirements in the miniaturization of Si devices, which require a complete control of the processing steps. The high level of control on the doping level and localization in Si reached using ion implantation allowed engineers to reduce the spatial dimension of Si devices, but the device shrinkage requires a reduction of the thermal budget in order to minimize the dopant diffusion. The dominant role played by defects on dopant migration and final distribution is at the base of the renewed interest in defect formation and evolution.

For a light ion, such as Si, implanted in crystalline Si (c-Si), the residual damage structure depends on the ion-implantation dose and annealing temperature. To better understand the field of investigation a schematic, summarizing our current knowledge of residual damage structure after a MeV Si ion implantation in Si, is plotted in Fig. 2. The implantation dose–annealing temperature graph is divided into regions, and each of them is labeled according to the dominant defect type in that domain. Of course, the positions of the domain borders are only tentative. Before describing Fig. 2, it should be noted that this schematic is only valid for a light ion. During implantation of a heavy ion, clustering occurs within the single collision cascade and the formation of amorphous pockets can be detected regardless of the ion-implantation dose.

For low-dose Si implants, $\leq 1\times10^{10}$ cm^2, and low annealing temperatures, up to ~300°C, the damage is mainly formed by "*point-like defects*". Their structural and electrical properties have been widely studied with deep level transient spectroscopy (DLTS) [8, 13–15], electron paramagnetic resonance (EPR) [5, 9] and photoluminescence (PL) [16, 17] (see Sect. 2).

When the annealing temperature is increased a region labeled as "*defects in trap*" in the figure is encountered. They are obtained after annealing at temperatures in the range 300–550°C for implantation doses of $1\times10^9 - 1\times10^{11}$ Si/cm^2. The dominant defects in this region are second-order

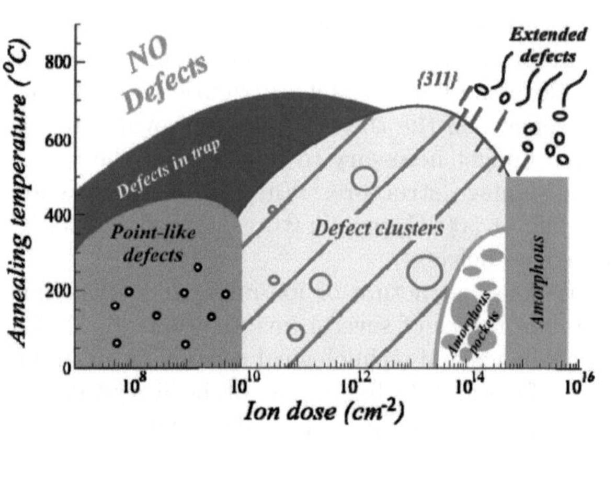

Fig. 2. Schematic showing the principal damage features as a function of the ion dose and annealing temperature for MeV Si self-implantation

point defects formed by annealing and/or migration of the known point-like defects [8, 18] (see Sect. 3). Upon a further increase in temperature they anneal out and the lattice crystallinity is fully recovered. The borders of this region strongly depend on the impurity concentration, since the annealing temperature of these defects depends on the sample impurity content.

For high implantation doses, $\geq 1 \times 10^{13}$ Si/cm^2, amorphous pockets and, eventually, continuous amorphous layers are formed in the as-implanted state. Annealing at temperatures $\geq 600°$C causes the formation of extended defects, such as {113} defects [19], dislocation loops, or dislocations. A critical dose for the formation of amorphous layers can be defined, which increases decreasing the implanted ion mass and increasing the target implantation temperature [20]. Thermal processes are performed after implantation to "activate", electrically and/or optically, the implanted species. Just after implantation the dopant atoms do not occupy a substitutional position. Therefore, thermal processes are necessary to allow them to move in the closest lattice site. Finally, using thermal processes it is possible to provide enough energy to the dopant atom to diffuse in the lattice. One of the main drawbacks of this process is the formation of extended defects. The extended-defects study shows that the Vs or Is forming the defect are accommodated in the lattice in order to reduce the strain and the number of dangling bonds. Such studies also showed that the growth of extended defects is governed by an Ostwald ripening (OR) mechanism [21]. This process consists in the dissolution of smaller clusters and the capture of the migrating elemental defects (Vs and Is) by bigger clusters, more stable in temperature. However, the only experimental evidence of OR from the point-like to extended defects regime is the existence of extended defects and the studies on their size distribution. These defect structures have been widely studied with techniques for structural characterization like TEM [22] or Rutherford backscattering spectrometry (RBS) [23], hence very different from those used to characterize point defects. By using TEM analysis it was possible to understand that a substantial rearrangement of the lattice occurs in order to accommodate defects with the minimum stress. The lattice reconstruction around them makes such defects very stable with temperature and they represent, therefore, a big problem in device fabrication.

Among the extended defects, particular attention has been devoted in the last few years to the {113} extended defects. They are rod-like defects elongated along the $\langle 110 \rangle$, which consist of interstitials precipitating on {113} planes as a single monolayer of hexagonal Si. They are known to form during oxidation [24], electron irradiation [25], and ion implantation [26]. They consist of interstitials forming 5- and 7-membered rings [27]. This atomic arrangement allows one to insert planes of self-interstitials without dangling bonds, they are very stable in temperature. In Fig. 3 the high-resolution TEM image of a {113} defect is shown. The dark shadow along the plane orthogonal to the [113] direction shows the lattice strain. In the same figure

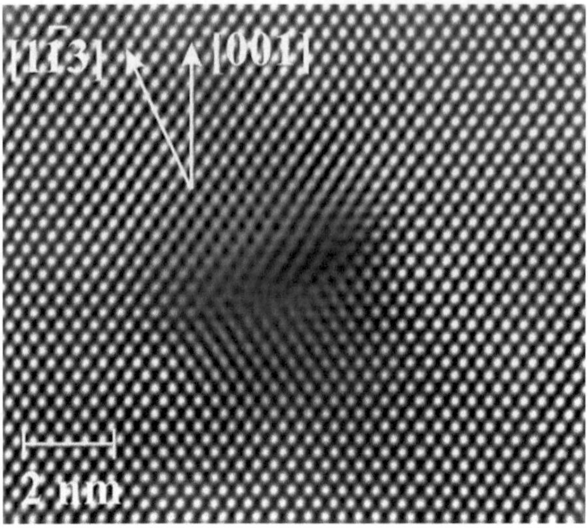

Fig. 3. High-resolution TEM of a {113} defect. The [001] and the [113] directions are also indicated

the extra plans, caused by the I presence are visible. Experimental evidences of OR were provided by TEM observations of the {113} evolution as a function of the time [26]. While the defect density decreases as a function of the annealing time, the average defect size increases, providing strong evidence of the OR for extended defects.

Although extremely powerful, the techniques normally used to monitor and study extended defects fail to monitor small defect clusters. On the other hand, the high implantation doses (typically higher than $10^{13}\,\mathrm{cm}^{-2}$) used in device fabrication are outside the domain usually explored by DLTS and PL. Therefore, it is extremely difficult to link the results of the different techniques into a coherent picture.

Due to the lack of experimental techniques to detect such defects, there is a wide region in Fig. 2 labeled as the "*defect clusters*" region, where only recently experimental data and simulations have been provided. They are formed after implantation to doses $\geq 1 \times 10^{12}\,\mathrm{Si/cm}^2$ and annealing temperatures $\geq 550°\mathrm{C}$. Most of the information today available on this region come from simulations [27–32]. They show that I and V generated during high-dose ion implantation tend to interact, forming defect clusters either during the implantation process or in the first stages of annealing. Such defect clusters evolve, migrating during annealing and rearranging their structure in the lattice. Experimental data in this region are fundamental for the comprehension of the damage evolution paths from point to extended defects. Some interesting experiments carried out by Cowern and coworkers [28], simulations [29–33] and the interesting results of Goss et al. [31] will be discussed in Sect. 3. Such comprehension would also be of significant technological interest since it is crucial for the understanding and modeling of defect-related phenomena such as transient-enhanced diffusion (TED) of dopants in Si, and extended defect formation. Both occur during the thermal process necessary to activate the dopant and their importance is becoming more and more pronounced as the thermal budget and the overall dopant diffusion reduce. According to the Fick diffusion law [2], and using the equilibrium thermal diffusivity of boron, D^*_{B}, the diffusion length after annealing should be much shorter than experimentally observed [26]. Also, the depth profiles of the electrically active B, obtained by spreading resistance profiling (SRP) show a diffused tail longer than expected. Boron diffuses through an interstitialcy or kick-out mechanism [2], according to the relation $\mathrm{B_s} + \mathrm{I} = \mathrm{B_i}$ [34, 35]. A substitutional B atom will be displaced from its lattice position by a Si interstitial, through a ballistic process. Once B occupys an interstitial position, it can migrate through the lattice, from one interstitial "site" to another, as the activation energy for this process is very low (~0.3 eV). In order to return to a substitutional site, it has to overpass a barrier of ~0.6 eV [36]. If a supersaturation of interstitials I/I^* is present in the sample, an enhancement of B diffusivity (D_{enh}) occurs, according to the equation:

$$D_{\mathrm{enh}} = D^*_{\mathrm{B}} \frac{I}{I^*}. \qquad (1)$$

The I supersaturation during annealing can be caused by the dissolution of I-related defects such as extended defects (dislocation loops or rod-like defects), or I clusters storing the ion-beam residual damage. The enhancement in the B diffusivity is transient since the I excess will eventually anneal to the surface [37–39] or to other sinks. Moreover, the enhancement in the B diffusivity is nearly constant for the duration of the transient, before dropping abruptly to its equilibrium value [40]. Furthermore, the TED duration after Si self-implantation depends on the implant condition. This behavior has been explained [26, 28] assuming that TED arises from the equilibrium between I-clusters and free interstitials. An interstitial in a cluster may spontaneously escape at a rate given by the interstitial hopping frequency and the binding energy to clusters, while the cluster grows, whenever diffusing interstitials are trapped by the cluster itself. It is obvious that, in order to understand and model TED and related phenomena, it is necessary to understand how Is are stored by and released from defects.

It has been proposed that the Is generated by the extra implanted ion determine the excess in the early stages of annealing. This assumption provides the physical basis for the "+1" phenomenological model [41]. It assumes that when almost all the defects have recombined, an imbalance between V and I concentrations due to the extra implanted ion is left in the bulk. The extra interstitials will cluster and eventually form extended defects. Finally, when the extended defects dissolve, the released interstitials drive the TED phenomenon. Stolk and coworkers [26] confirmed this model in the extended-defect region. They monitored the interstitial density in {113} defects, by TEM cross sections and plan views in the first stages of the annealing. Earlier studies on the formation and growth of dislocation loops also show a good agreement with the plus one model in the first stages of annealing [34]. No evidences of V-type defects in concentration $\geq 5 \times 10^{15}\,\mathrm{cm}^{-3}$ have been detected by positron annihilation spectroscopy [42] in such a dose and annealing temperature range.

Experimental results, as well as simulation have greatly improved our knowledge of defect agglomeration, mainly in the extended defect region, where the comparison is possible. Today, a valid explanation of TED in the temperature–time range where extended defect dissolution is detected has been provided. However, the TED occurs even without the formation and dissolution of extended defects, e.g., for low-dose implants or during the early stages of the annealing process, before the formation of extended defects [43, 44].

In this chapter, the main results on the defect formation and evolution in ion-implanted Si are reviewed. The two elemental defects (I and V) in Si, their main properties, as obtained in the last forty years of investigation and the simple complexes they form with impurities and dopant atoms are discussed in Sect. 2. The evolution of point-like defects, the effect of the sample impurity content on their annealing and the evolution to secondary defects is the main topic of Sect. 3. The experimental results obtained on the formation and evolution of defect clusters and the extended defects are discussed in Sect. 4. Finally, the main results reviewed in this chapter will be summarized.

2 Point-Like Defects Formation and Evolution

Quantitative information on room-temperature formation and migration properties of V and I, and on their interaction with dopants and impurities atoms of the substrate, are critical for the comprehension of defect accumulation in crystalline Si. In spite of the large scientific and technological interest in ion-implanted Si, most of our current knowledge of defect structures and their annealing behavior still relies on studies performed forty years ago on electron-irradiated samples [8, 9, 45–49]. These studies were performed using DLTS [8, 45, 46], PL [16, 17], or EPR [5, 9] measurements. They allowed scientists to fully characterize the defect production and migration to dopant or impurity atoms. According to those investigations, electron irradiation, or equally, light-ion implantation to low dose [6] in Si generates Frenkel pairs. They consist of V and Si self-interstitials, which will migrate through the lattice until they either recombine with each other, according to the relation $\mathrm{I} + \mathrm{V} = \varnothing$, or are stored in room-temperature-stable defect complexes. A schematic summarizing the principal room-temperature-stable defect complexes is shown in Fig. 4. A V migrates until it either finds another V and forms a divacancy (VV), or it pairs with impurities to create point defects such as oxygen–vacancy, OV, phosphorous–vacancy, PV [8, 45, 46], if P is in the lattice or BV in the presence of B [50].

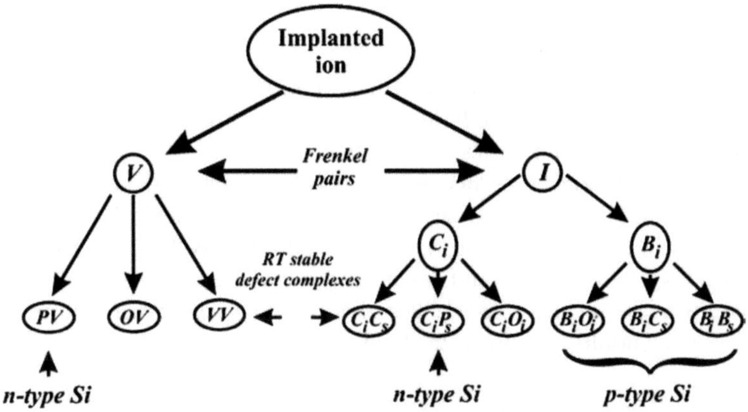

Fig. 4. Room-temperature-stable defect complexes generated during ion implantation (light ion) or electron irradiation in Si

The other branch of the Frenkel pair reaction in Fig. 4 shows I-based interactions. A Si self-interstitial creates, by the Watkins' replacement mechanism [51], boron interstitial, B_i, and carbon interstitial, C_i. The favored branch of the reaction depends on the ratio between the impurity concentrations (B vs. C). These interstitial species diffuse at RT until they are trapped and form I-impurity pairs. If C_i is generated, then carbon substitutional–carbon interstitial, C_sC_i, carbon interstitial–oxygen interstitial, C_iO_i, carbon interstitial–phosphorous, C_iP_s, are formed [45, 52]. When B_i is generated in a suitable concentration, the formation of boron-related defect complexes such as boron interstitial–oxygen interstitial, B_iO_i, boron interstitial–carbon substitutional, B_iC_s and boron interstitial–boron substitutional, B_iB_s, [53–55] is favored. Recent studies indicate that Si I can pair together forming the I_2 complex [56].

Most of these simple point-defect pairs have been widely characterized and a brief review of their main properties is provided hereafter. All of them anneal at temperatures below 450°C. The wealth of published information on the identity, introduction rates, annealing behavior, structural (by EPR), electrical (by DLTS) and optical (by PL) signatures of V- and I-related point defects in electron-irradiated Si provides a firm foundation for a comprehensive study of the introduction and evolution of implantation-induced defects in Si. In the rest of this section, the main properties of Vs and Is, the current knowledge on their formation and the formation of point-like defects by Is and Vs pairing with impurities is investigated. Similarities and differences of ion implantation and electron irradiation are also provided. Finally, the annealing of RT-stable point defects, the unquestionable base for every investigation of defect evolution, is discussed in detail.

2.1 Point Defect Properties

The properties of "intrinsic point defects" in Si, i.e., the simplest structural defects that can be formed in Si, are here reviewed. They are defined in terms of lattice positions and no impurities are involved.

In order to understand the structure of a point defect and the lattice rearrangements around it, the Si lattice structure has to be briefly described. As previously mentioned the Si lattice is a diamond cubic cell (Fig. 1) with a lattice constant $a_0 = 5.43\,\text{Å}$. Is and Vs in Si are the elemental part of all other more complex structural defects. For this reason their structural and electrical properties should be known in detail.

Vacancy and Vacancy-Type Defects

The *isolated vacancy* (V) has been observed after electron irradiation at 4.2 K [57, 58]. Its electrical and diffusional properties, as well as its interactions with other defects and impurities in the Si lattice, have been well characterized. Since an isolated V is essentially given by a missing atom in the tetrahedral Si lattice, the unsaturated covalent bonds around the defect tend to rearrange in order to minimize the number of dangling bonds, hence the total energy. This rearrangement causes a local distortion of the lattice around the point defect. The V has been observed in five different charge states: $V^{=}$, V^{-}, V^{0}, V^{+} and V^{++}. EPR measurements and theoretical calculations [5, 59] allowed the different structures to be defined. The main difference among the various charge states consists in the way in which the broken-bond orbitals of the four Si atoms surrounding the vacant site are saturated. EPR and DLTS measurements allowed scientists to identify the two donor levels introduced by the vacancy in the bandgap: $E_V + 0.13\,\text{eV}$ $(++/+)$ [8, 60, 61] and $E_V + 0.05\,\text{eV}$ $(+/0)$ [62]. Unfortunately, the deep levels of the two vacancy acceptor charge states have not been determined. The failure of DLTS measurements suggests that such states are deeper than $E_C - 0.17\,\text{eV}$ [63]. In fact, for temperatures above 90 K the single V becomes mobile and the formation of V-related defects (see Fig. 4) is achieved. On the other hand, temperatures above 90 K are necessary to detect trap levels deeper than $\sim E_C - 0.17\,\text{eV}$ by DLTS.

The activation energy for V migration depends on its charge state, hence on the material doping. In low-resistivity n-type Si, where the V is prevalently in its double negative charge state, an activation energy for migration of $0.18 \pm 0.02\,\text{eV}$ was estimated [64]. On the other hand, in low-resistivity p-type Si the V^{++} has an activation energy for migration of $0.32 \pm 0.02\,\text{eV}$ [49], while DLTS studies applying a reverse bias voltage showed an activation energy for migration of $0.45 \pm 0.04\,\text{eV}$, attributed to V^{0} [2]. Since the V can be considered as a microscopic cavity, its migration involves the motion of one of the atoms that migrates around its macroscopic surface [65]. Moreover, athermal V diffusion at 4.2 K has been observed under optical excitation or electrical injection [58].

V interaction with interstitial O causes the formation of a defect consisting of an almost substitutional oxygen bonded with two Si atoms [58], the oxygen vacancy (OV) complex. It exhibits a single acceptor level in the Si bandgap, at $E_C - 0.17$ eV. Its atomic configuration is plotted in Fig. 5a where the O atom occupies the empty subsitutional position of the V [8, 47]. A schematic of the Si bandgap and the OV deep level is plotted in the same figure. During its motion a V can also reach a dopant atom (i.e., P or B) and pair with it to form the phosphorous vacancy (PV) or the boron vacancy (BV) complexes, shown in Figs. 5b and c, respectively [58]. The P atom sits close to the vacant site and the PV-related deep level has an activation energy of $E_C - 0.46$ eV [8], shown in the Si bandgap schematic. The B atom of the BV complex sits on the second neighbor of the V and this defect introduces a deep level at $E_V + 0.19$ eV. It should be mentioned that the BV is not stable at RT and its characterization was obtained after low-temperature electron irradiation [66].

If two Vs pair, a divacancy, whose atomic structure is shown in Fig. 5d, is formed (VV) [8, 67]. The missing atoms are represented as dashed circles. The divacancy introduces three different levels in the Si bandgap, related to its various charges states: two acceptor levels at $E_C - 0.22$ eV $(= /-)$, and $E_C - 0.40$ eV $(-/0)$ and a donor level at $E_V + 0.20$ eV $(0/+)$ [8, 68, 69], as shown in the schematic. VVs can form directly during implantation [68].

Finally, due to V migration, or as a consequence of a heavy-ion implantation, higher-order V-complexes can form. Examples are the trivacancy (V_3), the four-vacancy complex (V_4), that can cluster in a planar and in a nonplanar configuration, etc. [71, 72]. The vacancies in the clusters will accommodate in order to minimize the number of dangling bonds created at the cluster boundary [73]. Recent simulation results [29], performed using a tight-binding molecular-dynamics (TBMD) approach in a 1000-atom cell, show that the V_6

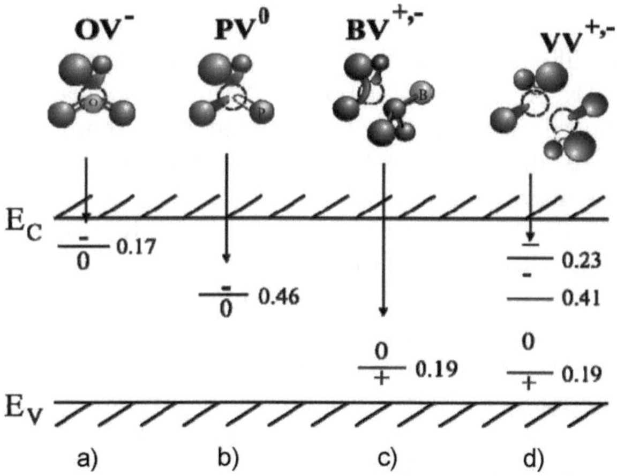

Fig. 5. Structural configuration and deep-level schematics of: (**a**) oxygen–vacancy (OV), (**b**) phosphorous–vacancy (PV) and (**c**) boron–vacancy (BV), (**d**) divacancy (VV) in all its charge states; as obtained by EPR and DLTS measurements

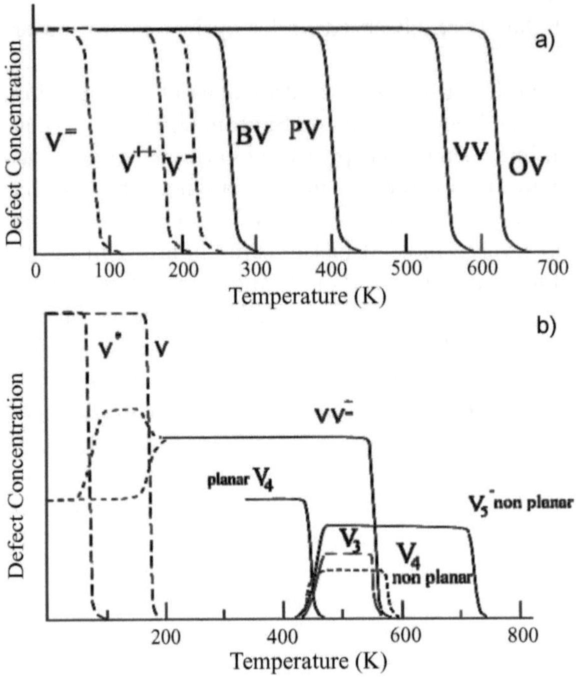

Fig. 6. Annealing behavior of: (**a**) V-type defects; (**b**) small defect clusters, such as V_3, V_4 and V_5. The curves are compared with the V^0, V^- and VV annealing curves (*dashed lines*)

is the most stable configuration. In particular, they removed atoms from the 6-membered bond rings present in the c-Si structure and the result is the formation of hexagonal ring clusters. The V-clusters arrangement is such as to minimize the number of dangling bonds.

The thermal stability of each V complex reflects the binding energy between the V and the impurity, or between the single V forming the cluster. The experimental annealing curves of some of these structures, obtained for isochronal (15 min) annealing at temperatures ranging from 4 K to 700 K are shown in Fig. 6 [8], where the number of defects as a function of the annealing temperature is plotted. The different single V charge states anneal well before room temperature, ~230 K. Also, the BV is not stable, and its annealing is achieved at ~270 K. The other V-complexes are stable at room temperature. The PV complex anneals at ~150°C (~420 K), while the VV anneals at ~250°C (~520 K) [67], or migrates to form V_2O [18]. Finally, the most stable vacancy complex is the OV that anneals at ~350°C (~620 K) [74].

The measured thermal stability of higher-order vacancy complexes is shown in Fig. 6b, where the defect concentration as a function of the annealing temperature is shown. It is interesting to observe that all the higher-order structures, anneal before the VV, with the exception of the nonplanar five-vacancy that anneals at ~450°C (~700 K) [67]. Similar results have been obtained by Alippi et al. [29]. Their simulations show that the V_2 is more stable than V_3 and V_4, while a higher stability is achieved for V_5 and V_6.

Detailed studies on the V-introduction rate as a function of the substrate impurity content [75] showed that it is strongly dependent on the doping level of the Si sample used for V characterization. In fact, it has been found than the V production is quite efficient in p-type Si and much lower in n-type Si. This result was explained by suggesting that both Is and impurities can effectively modify the Vs introduction rate. Also, for this reason it is necessary to fully understand the impurity role in the defect formation and evolution. In the next sections, evidences in this direction are provided.

Interstitial and Interstitial-Type Defects

A self-interstitial is a host atom displaced from its lattice position. Electron irradiation performed on Si samples at 4.2 K did not allow the identification of the Si self-interstitial. In fact, even at such low temperatures, Si Is are free to migrate in the lattice and, in p-type Si, only evidences of trapped I at dopant atoms were found [69, 75]. It was proposed that, in p-type Si, self-interstitials can follow athermal migration paths even at 4.2 K, by changing their charge state, until they find a B atom in a substitutional position. The Si I displaces the B through the Watkins' replacement mechanism [51] creating a B_i. The B atom prefers to nestle between two nearest Si neighbors and the Si substitutional-B_i atom bond can be oriented in the various crystallographic orientations ($\langle 100\rangle$ or $\langle 111\rangle$ directions). The different charge states have different atomic structure: the B_i^- is oriented along the $\langle 111\rangle$, while B_i^+ is along the $\langle 100\rangle$ direction. B_i introduces an acceptor level at $E_C - 0.37\,\text{eV}$ $(-/0)$ and a donor level at $E_C - 0.13\,\text{eV}$ $(0/+)$ [67, 76]. After electron irradiation at 4.2 K and annealing at ~100 K evidences of I trapped at C atoms, C_i, were found. A C atom shares a single substitutional site with a Si one (dumbell configuration), introducing a level in the Si bandgap. In particular, it shows an acceptor state at $E_C - 0.10\,\text{eV}$ $(-/0)$ and a donor state at $E_V + 0.27\,\text{eV}$ $(0/+)$ [77].

The interstitial evolution is less clear in n-type Si, where the dopant (P) is not an efficient I trap. EPR studies after irradiation of n-type Si at 4.2 K showed that Vs created by irradiation are perturbed by some other defect nearby. At temperatures of ~100 K I trapped at C atoms start to be detected and a concomitant annealing of V-type defects was observed [78]. On the other hand, in n-type Si partially compensated with B evidences of long-range I migration and B_i atoms were found, even if this last process is less efficient than in p-type Si. It was supposed that, in n-type Si, I are frozen into the lattice nearby the V right after their production, probably forming I–V complexes. I–V recombination occurs as soon as the temperature allows the complex to break and the interstitial to migrate. According to tight binding calculations [79, 80] a barrier to the I–V recombination and to the I migration exists. When Is and Vs are close to each other, they do not recombine immediately, but form a complex (I–V). The recombination process occurs only if the temperature is sufficient to break two of the interstitial bonds to

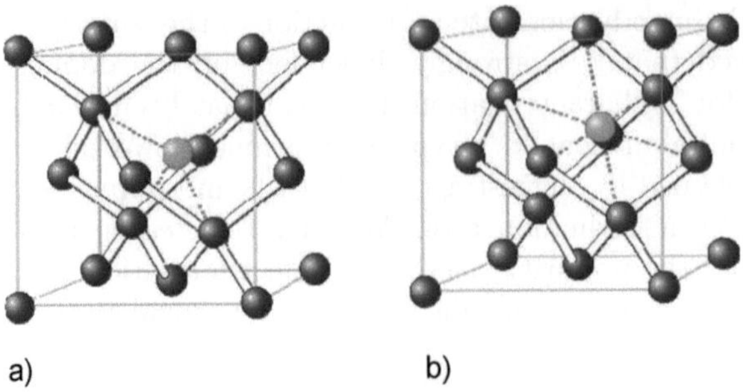

Fig. 7. Schematic of two possible structural configurations for the Si self-interstitial: (**a**) tetrahedral and (**b**) hexagonal position. The Si interstitial is represented with a *green ball*, while the symmetry is evidenced by the *dashed lines*

the lattice to allow the rotation and the reconstruction of the perfect Si lattice [80]. The simulations were performed assuming a starting structure for the I. Since the Si self-interstitial structure was not detected experimentally, its atomic configuration has been inferred by first-principles calculations [82], and studying the atomic configuration of the I trapped at impurities. Two possible I configurations are plotted in Fig. 7: the tetrahedral (7a) and the hexagonal (7b) [2].

Finally, according to calculations, a Si I can occupy a different configuration, called a *silicon split interstitial.* It consists of two Si atoms sharing the same substitutional site and the direction of the complex is different along the $\langle 111 \rangle$, $\langle 100 \rangle$ and $\langle 110 \rangle$ directions. Since both the tetrahedral (hexagonal) and the split interstitial accommodate an extra Si atom in the lattice, they will be both referred as I in the following.

All the possible I configurations listed above exhibit, according to simulations, different activation energies for migration and the variety of possible configuration suggested that I migration can be achieved athermally through changes in its charge state [83] by successive captures of electrons and holes. In this way, the migration in the Si lattice can occur without overpassing potential barriers [4, 5, 67]. The predicted stable configurations are different for the three charge states, hence, a change in the charge state allows saddle points to be avoided and athermal migration occurs. Two deep levels with activation energies at about $E_V + 0.05\,\text{eV}$ and at $E_C - 0.40\,\text{eV}$ have also been predicted. Unfortunately, no experimental data are available to confirm the calculations.

As previously mentioned, at $\sim$100 K C_i form and C-related defects, such as carbon interstitial–carbon substitutional (C_iC_s), carbon interstitial–oxygen (C_iO_i) and phosphorous–carbon interstitial (P_sC_i) complexes, are detected. Their atomic configurations are plotted in Figs. 8a, b and c, respectively. Also, these complexes introduce deep levels in the gap, as shown in the bandgap schematic in the same figure. The C_iC_s, formed by a C_i having a substitutional C atom in the second-neighbor position, introduces an acceptor level at $E_C - 0.10\,\text{eV}$. Moreover, it is visible in PL since it introduces an optical band at 969 meV [16, 17]. This center can be used as a sensitive tool to

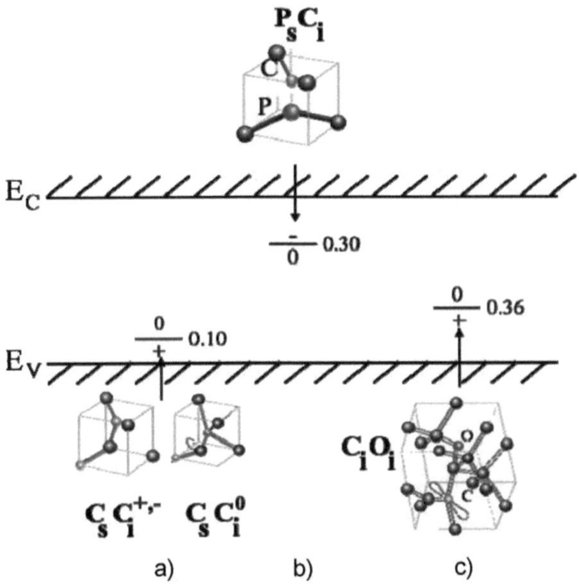

Fig. 8. Structural configuration and deep levels generated by: (**a**) carbon substitutional–carbon interstitial (C_sC_i), (**b**) phosphorous substitutional–carbon interstitial (P_sC_i) and (**c**) carbon interstitial–oxygen interstitial complexes (C_iO_i)

measure the sample C content. The maximum concentration of this center depends on the sample temperature and on the presence of other impurities. The C_iO_i is obtained when a C_i atom has an O atom as second neighbor, and introduces a donor level at $E_V + 0.36\,\text{eV}$ [57, 64]. Also, this level introduces an optical band, located at 789 meV [16, 17]. The P_sC_i is formed by a C_i atom sitting close to a substitutional P atom, sharing the same lattice position. This complex exhibits at least four metastable acceptor levels at $E_C - 0.30\,\text{eV}$, $E_C - 0.29\,\text{eV}$, $E_C - 0.23\,\text{eV}$ and $E_C - 0.21\,\text{eV}$ [57]. Finally, C acts as a nucleation center for Is, hence other C_i-related defects form at C atoms [77].

Besides C, a Si self-I can B_i whose migration will cause the formation of B-related defects, such as a boron interstitial–boron substitutional complex (B_iB_s), with a donor level at $E_V + 0.29\,\text{eV}$, a boron interstitial-carbon substitutional complex (B_iC_s), with a donor level at $E_V + 0.26\,\text{eV}$ and a boron interstitial–oxygen complex (B_iO_i), with an acceptor level at $E_C - 0.26\,\text{eV}$ [53–55]. Unfortunately, since no EPR data are available on B-related defects their atomic structure is not known in detail.

The annealing kinetics of the I-related defects, as obtained by DLTS studies [53–55] is plotted in Fig. 9. A schematic of the reactions producing the different I-type structures as a function of the annealing temperature is shown in Fig. 9a. At low temperature (∼−100°C), a Si interstitial creates C_i or B_i and they will pair with O. It is not a very efficient trap for Is, hence below room temperature (∼−50°C) the pair will break and the released Is form extra B_i or C_i. B_i will pair, below room temperature (∼0°C) with impurities forming B_iB_s, B_iO_i or B_iC_s. When the temperature is increased the C_i

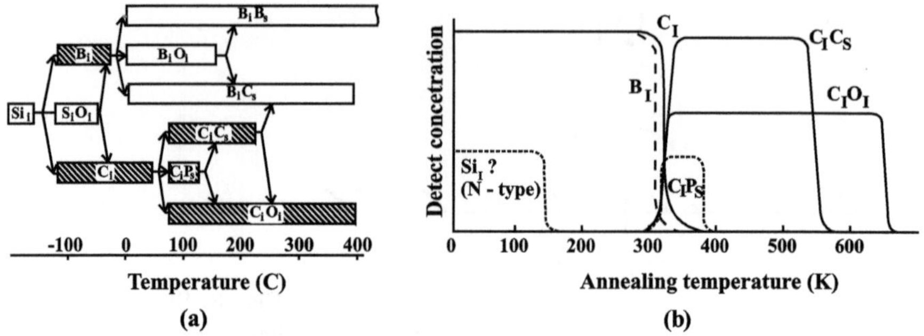

Fig. 9. (**a**) Interstitial-type defect formation and dissociation as a function of the annealing temperature and (**b**) annealing behavior of some of the known I-type defects

become mobile and pair with C, O or P. A further increase in the sample temperature causes the dissociation of defect complexes like the P_sC_i (100°C) or the B_iO_i (150°C). The released B_i or C_i migrate to form extra B- or C-related complexes. When the temperature reaches 250°C the C_iC_s dissociates, while it is necessary to reach 400°C to have the dissociation of the C_iO_i or the B_iC_s complexes. B_iB_s dissociation is achieved over 450°C [77]. In Fig. 9b, the defect concentration, in arbitrary units, is plotted as a function of the annealing temperature for B_i and for all of the C_i-related defects. In the same figure, the supposed annealing curve for the Si-self interstitial in n-type Si is also shown as a dashed line.

2.2 Point-Defect Generation: Electron Irradiation vs. Ion Implantation and Role of Impurities

Defects producted by electron irradiation have been widely investigated, while a comparable detailed investigation is missing for ion-implanted Si. Ion implantation has some inherent differences compared to electron irradiation. Several investigations [6, 84–86] showed that the same electrically active defects are produced by both ions and electrons, in the as-implanted state, while a direct comparison of the damage created in the same Si sample by the two processes [87] has shown interesting results.

It is known that the defect profile for ion-implanted samples is not spatially uniform, in contrast to electron-irradiated samples, but the damage is mainly localized in a well-defined peak at the end of range. The ion-implantation straggling is a function of the implantation energy and, for a light ion, of the defect effective migration length [87–90]. Also, the density of collisional events is different in the two cases, being several orders of magnitude higher for ions than for electrons. As a result, denser collision cascades where RT annihilation and defect clustering can be greatly favored, are obtained in ion-implanted samples. Finally, while electron irradiation produces equal numbers of Vs and Is, ion implantation produces an imbalance between Vs and Is caused by the presence of the incorporated extra ion. This last observation has noticeable consequences and provides the basis for the phenomenological "+1" model [41], being the basis for the description of I clustering and of all the related phenomena occurring for high-dose implantation and at high annealing temperature. For all the reasons listed above a significant difference between the damage evolution in ion-implanted and in electron-irradiated Si is expected.

To apply the knowledge derived from investigations on electron-irradiated samples to ion implantation, it is fundamental to compare directly the damage introduced by the two processes. To this purpose both n-type and p-type Czochralski (CZ) Si samples, P and B doped, respectively, were ion implanted and electron irradiated. In order to have roughly the same total defect concentration in electron-irradiated and ion-implanted Si samples, the relative doses were tuned. In particular, Brotherton et al. [91] determined an introduction rate η of $4.5 \pm 0.4 \times 10^{-2}$ VV/cm for 12-MeV electron irradiation in float zone (FZ) Si to doses in the range 2×10^{13}–5.5×10^{14} e/cm^2, by using DLTS measurements. η is given by the ratio between the trap concentration and the irradiation dose. The volume-averaged defect concentration created by 1.2-MeV Si ion implantation to a dose of 1×10^9 Si/cm^2 in n-type epitaxial Si was measured, and a value of $\sim 2 \times 10^{14}$ VV/cm^3 found by DLTS. To tune the electron irradiation versus Si ion implantation, the following relation was used:

$$\Phi = \frac{N_\mathrm{T}}{\eta}, \tag{2}$$

where Φ is the electron irradiation dose, N_T is the defect concentration value estimated from Si ion implantation and η is the VV introduction rate for electron irradiation. An irradiation dose of $\sim 4.4 \times 10^{15}$ e/cm^2 was estimated. The ion-implantation energy, 1.2 MeV, was chosen to place the defect concentration peak within the DLTS measurement volume. Furthermore, Si was chosen as implanted species since it is a light and chemically inert ion. This approach allows one a direct comparison of the two different damaging procedures on the same substrate, thus fixing the impurities (C and O) and dopant (either P or B) concentrations. Ion implantation and electron irradiation on the same set of samples were performed (see [87] for the experimental details). The samples were measured by DLTS. Reverse biases and filling pulses were chosen in order to include the entire ion-damage distribution within the measurement region.

The comparison between the damage structure of electron irradiation and ion implantation is shown in Figs. 10a and b for n-type and for p-type CZ Si samples [86]. The DLTS spectra of both ion-implanted (dashed line) and electron-irradiated (solid line) samples are shown in the figure, thus providing information on defects having deep levels in both halves of the Si bandgap.

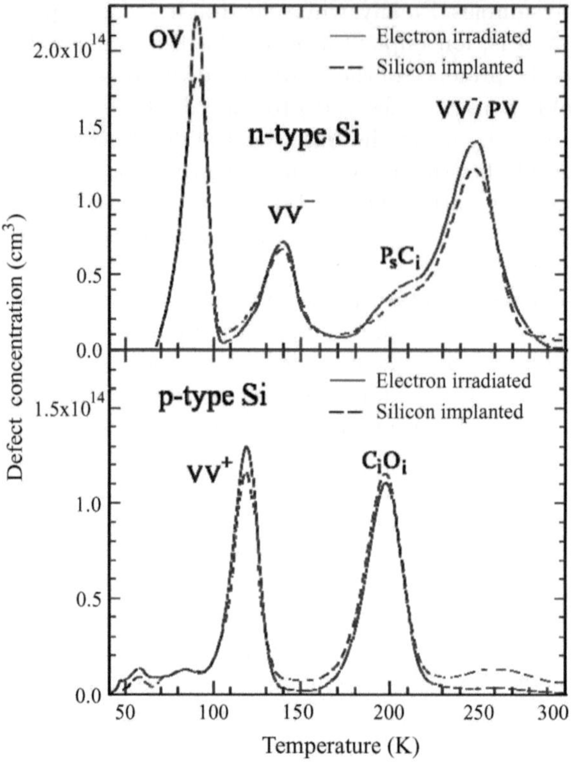

Fig. 10. Comparison of DLTS spectra resulting from 9.2-MeV, $3.5 \times 10^{15}\,\mathrm{cm}^{-2}$ electron irradiation (*solid line*) and 1.2-MeV, $1 \times 10^{9}\,\mathrm{cm}^{-2}$ Si ion implantation of (*dashed line*) in (**a**) n-type and (**b**) p-type CZ Si

In both p-type and n-type Si the spectra are almost identical, regardless of the implanted species, demonstrating that the same defect structures are introduced by ion implantation and electron irradiation. Remarkably, no measurable differences arise as a consequence of the denser ion-collision cascade, even if the Si collision cascade is seven orders of magnitude denser than the electron one. In n-type Si, see Fig. 10a, the signatures of the OV at $E_C - 0.17\,\mathrm{eV}$, the $VV^{=}$ at $E_C - 0.23\,\mathrm{eV}$, the VV^{-} at $E_C - 0.40\,\mathrm{eV}$, and the PV at $E_C - 0.44\,\mathrm{eV}$ deep levels, are observed. In addition, the signature of the P_sC_i at $E_C - 0.29\,\mathrm{eV}$, is also observed as a shoulder between the two VV signatures. This is in agreement with DLTS studies on electron-irradiated Si samples reported in the literature [6, 8, 57, 64].

The DLTS spectrum of p-type Si (Fig. 10b) shows the signatures of the third charge state of the divacancy, VV^{+} at $E_V + 0.23\,\mathrm{eV}$ and the C_iO_i at $E_V + 0.36\,\mathrm{eV}$. Once again the electron-irradiated and ion-implanted samples exhibit the same damage structures, having very similar average concentrations, suggesting that, once the implantation and irradiation doses are tuned, the defects form in the same relative concentration. The situation changes dramatically as the ion-implantation mass increases. In fact, the implanted ion during its motion in the lattice displaces many Si atom from their position, providing them enough energy to create secondary collision cascades. As a result, many point defects are generated very close to each other and this effect is enhanced if the ion mass is increased, due to the higher energy deposited in nuclear collisions. It has been amply studied in the literature by Svensson and coworkers [6, 84, 85]. The relative concentration of the generated defect complexes changes with the ion mass. In particular, the OV concentration strongly decreases by increasing the ion mass (not shown) due to the higher clustering effect. Free Vs find more likely other Vs to pair with and to form VV, or even more complex V-clusters, than O atoms to be trapped by. The trapping of Vs by impurities, like O, plays a big role only if the collision cascade is diluted [92]. Finally, a strong deviation from unity is observed in the ratio between the concentration of the singly ionized and the doubly ionized VV ($[VV^-]/[VV^=]$). When the ion mass is increased, a big difference in the relative concentration of the RT-stable defect complexes generated by the ion beam is observed [93]. In particular, a strong increase in the VV^- concentration compared with the $VV^=$ and the OV concentration in n-type Si is observed. The strong reduction in the $VV^=$ peak intensity is explained by considering that heavier ions cause a higher lattice strain. Since the $VV^=$ characteristic peak occurs at lower temperatures than the VV^- peak, the defect cannot easily rearrange itself in the presence of high lattice strain: it is more sensitive to lattice strain than VV^- [6, 84, 85]. Hence, the increase in the strain produced by heavier ions, causes a progressive reduction of the $VV^=$ signature until, for heavy ions (such as Ge), the level is barely visible in DLTS.

Another interesting effect occurring when the ion mass is increased is shown in Fig. 11, where the number of Vs stored into stable complexes per implanted ion as a function of the energy deposited in nuclear collision for

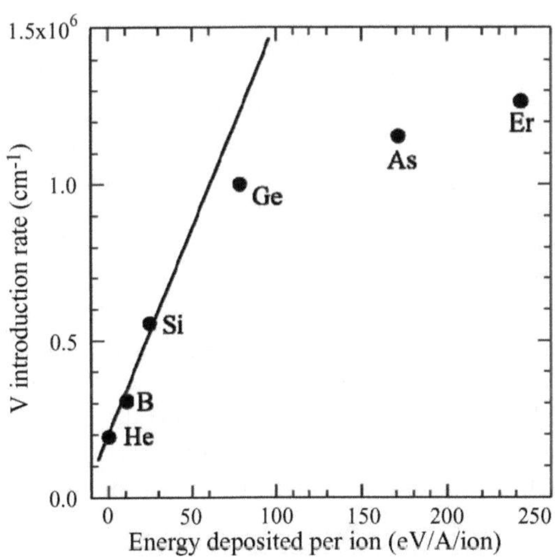

Fig. 11. V-type defect concentration as obtained from DLTS data (●) and TRIM simulation (*solid line*) as a function of the deposit of energy

ions with different masses, is plotted. The introduction rate has been obtained by adding the VV and the OV concentrations of a DLTS spectrum, while the energies deposited per ion are obtained from TRIM (transport of ions in matter) [94], a Monte Carlo program. A linear relation (solid line) exists between the number of Vs introduced per implanted ion and the nuclear-energy deposition for light ions, such as He or Si. A Si ion, implanted with an energy of 3 MeV in Si, deposits 28 times the energy deposited by 1-MeV He according to TRIM, hence, it produces 28 times more V and I per ion. Therefore, it is reasonable to assume that a defect concentration a factor 28 higher than for He ion implantation is obtained in the Si case, as actually occurs. On the other hand, an increase in the ion mass, e.g. by implantation of Ge or Er, produces a decrease in the defect introduction rate and a strong deviation from the simulation values. An increase in the ion mass causes an increase in the density of energy deposited into nuclear collisions but it does not cause a proportional increase in the point-defect concentration. The experimental results, clearly in contrast with simulations, are explained by observing that heavier ions produce denser collision cascades, and the generated point defects are much closer than in a light-ion collision cascade. Under this assumption, the recombination probability is enhanced. Moreover, since the point defects are generated very close to each other, the probability of clustering is greatly enhanced when the implanted ion mass is increased. Many studies [92, 95] confirm that the efficiency in the production of point defects decreases by increasing the ion mass, due to the denser collision cascades and more complex defects form in the as-implanted state. Such defects act as sinks for free Vs and Is, resulting in the strong reduction of the point-defect generation. This effect is not taken into account in simulation programs like TRIM, hence these programs fail in the defect-production simulation due to heavy ions.

Once verified that light-ion implantation and electron irradiation produce the same kind of damage, the impurity effect on the damage structure must be studied, reducing the impurity content of the substrate, as occurs in commercial wafers. It is well known that impurities are effective traps for Frenkel pairs generated by the beam. The introduction rates and annealing kinetics of the known point-defect pairs depend critically on the relative concentrations of dopant atoms, O and C of the starting material [53–55, 96]. It is reasonable to expect that the impurity content reduction might cause a different defect evolution. To this purpose He-ion implantations were performed on Si substrates, both p- and n-type, with different impurity contents [86]. In n-type Si, Fig. 12a, the impurity effect is only slightly visible, resulting in a modification of the relative defect concentrations. In particular, a variation in the sample C content, from $1 \times 10^{16}\,\mathrm{C/cm^3}$ (solid line) to $5 \times 10^{17}\,\mathrm{C/cm^3}$ (dashed line), produces a large enhancement in the P_sC_i concentration. The VV^-, $VV^=$ and OV dominate the DLTS spectra in n-type Si for low C concentrations. It is well known that P_sC_i has a branching ratio much higher, a

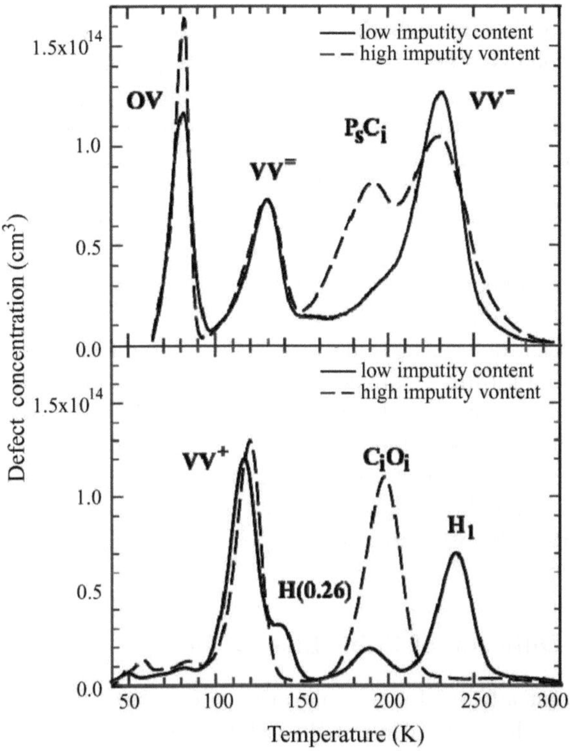

Fig. 12. DLTS spectra measured on n-type (**a**) and (**b**) p-epitaxial (*solid line*) and high-carbon CZ (*dashed line*) Si implanted with $5 \times 10^8\,\mathrm{cm}^{-2}$ 1-MeV He ions

factor 10, than C_iC_s and C_iO_i [57]. As a result, C_i is preferentially trapped by P.

On the contrary, a variation in the impurity concentration produces dramatic effects on the damage structures measured in p-type Si (see Fig. 12b). A large difference in the DLTS spectra of low (solid line) and high (dot-dashed line) impurity Si samples is observed. The results show that there is a very large difference in the defect production as a function of substrate impurity content. The results just shown clearly prove that the impurity content has a much larger effect than the collision cascade in determining the RT-stable defect structures [86]. The wafers used have very low impurity concentrations, causing a lower C_iO_i concentration than in high impurity content (CZ) Si. Furthermore, two additional defect electrical signatures are detected and labeled as H(0.26) (at $E_V + 0.26\,\mathrm{eV}$), and H_1 (at $E_V + 0.44\,\mathrm{eV}$). Such levels are not detected in the as-implanted CZ samples at RT but do appear also in this sample after heat treatment at 150°C 30 min [96]. Detailed studies of H(0.26) annealing behavior allowed us to conclude that it is the electrical signature of the B_iC_s complex.

The difference in the B_iC_s formation temperature for the two sets of samples is a consequence of the difference in the C content. If the C concentration is lower than the B content (epi Si), the B branch of the reaction is favored, the Is generated by irradiation will preferentially form B_i instead of C_i (see Fig. 9a). The mobile B_i will then cluster with either the dopant or the impurities forming B-related defect complexes, such as B_iC_s [53–55]. Thus, while in low-impurity Si B-related complexes directly form in the as-implanted state, in high-impurity Si they form only when Is are released by the dissociation of C_i-related complexes. We cannot be conclusive about the nature of H_1, but many pieces of evidence suggest it is an I-type defect. In fact, it forms when the C_iO_i complex cannot form, confirming that it competes with C in storing Is. Moreover, since both C and O concentrations are quite low in epitaxial Si samples, this complex is likely to embody a B atom [96, 97].

A detailed comprehension of defect–impurity and defect–dopant interactions is crucial for the understanding of phenomena such as TED and for the validation of the simulation tools used for their description. It is known that defect–dopant interactions, responsible of TED, and defect–defect interactions, responsible for extended defect formation, are strongly affected by impurities. In particular, it has been shown that C, interacting strongly with Si self-Is can be successfully used to suppress TED [43] and reduce extended-defect formation [98].

2.3 Room Temperature Diffusion of Point-Like Defects

The substrate impurity content affects also the migration properties of I and V. Ion-induced defects undergo fast migration at RT, which is interrupted by the formation of stable complexes with dopants or impurities [87–90]. Point defects injected at the surface by low-energy ion implantation can migrate to distances as long as 3 μm in highly pure epi Si. Information on defects RT diffusion can be obtained by studying the depth profiles of V- and I-type RT stable complexes. Since ion-implantation damage is well localized at the end of range of the implanted ion it can be used to monitor RT recombination and migration of Vs and Is [99–101]. For such studies, Si samples with various C and O concentrations were He implanted. Helium was used since it has a very low atomic mass (4 a.m.u.) and it creates very diluted collision cascades. Therefore, it is possible to minimize the direct-clustering effect within the single cascade. Furthermore, MeV He-ion implantation produces a very sharp Frenkel pair distribution at the ion end of range, providing an ideal system to study point-defect migration. Finally, very low implantation doses, e.g., $5 \times 10^8\, \mathrm{He/cm^2}$, avoid ion-track overlapping. Si ion implantation was used to test the effect of a denser collision cascade. The implanted samples were characterized by DLTS measuring the depth profiles of impurity–point-defect pairs (OV and P_sC_i) and VV. Depth profiles of the VV (●) and OV (■) complexes for a CZ Si sample (high impurity content) are shown in Fig. 13 [87].

A very high OV concentration was measured in this sample: indeed the OV profile in the figure has been divided by a factor 1.5 to allow the direct comparison with the VV^- profile. This confirms that O is a very efficient trap for migrating Vs. The solid line represent the depth profile of the I–V pairs

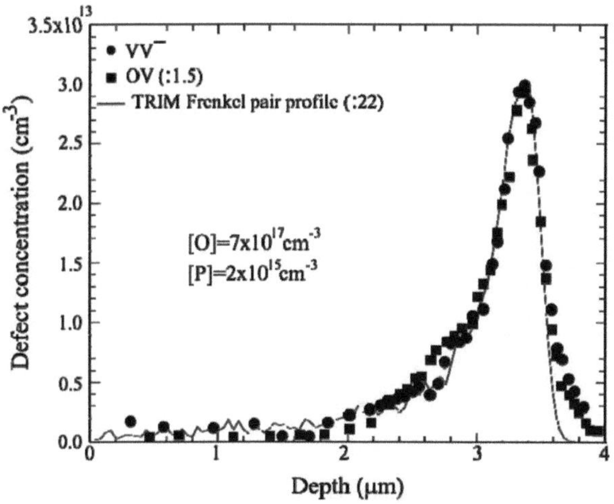

Fig. 13. Depth profiles of VV^- (●), and OV (■) for a CZ sample ($4\,\Omega$ cm, [O] $= 7 \times 10^{17}\,cm^{-3}$ and [C] $= 1 \times 10^{16}\,cm^{-3}$) implanted with 1-MeV He, $5 \times 10^8\,cm^{-2}$. The profiles are compared with a TRIM simulation of the implantation damage (*solid line*)

generated by the beam, as calculated by TRIM. The simulated generation curve has been divided by a factor of 22 in order to allow comparison. A large fraction of the defects generated by the beam anneals during the implant, or immediately after. Such extensive recombination occurs within the single-ion cascade, as the implantation dose is well below the threshold for cascades overlapping. The concentration of residual electrically active defects is only ~16% of the total defect concentration generated by the beam. This result was obtained by adding the OV concentration and the VV concentration derived from the depth profiles, multiplied by a factor 2 since two Vs are stored in each VV. In spite of the extensive recombination, the depth profile precisely mirrors the Frenkel-pair distribution predicted by TRIM. This implies that the point defects escaping recombination cannot migrate to long distances due to the formation of complexes with other Vs or pairing with O or P atoms, confirming the high impurity efficiency in trapping point defects [87, 99]. The relative strength of defect clustering and trapping at the impurities can be changed by varying the impurity content in the material and/or changing the implanted ion mass.

Is and Vs migration properties were explored by comparing the effect of the same implant on samples having different O and P contents. The resulting depth concentration profiles are shown in Figs. 14a and b for OV and VV^-, respectively. The reduction of the O content from 7×10^{17} O/cm^3 (Fig. 13) to $2 \times 10^{16}\,cm^{-3}$ (FZ sample, ■) results in a significant broadening of the depth profile of both defects with respect to the simulated defect profile (solid lines). This effect is even further enhanced in the epi Si (~10^{15} O/cm^3, ●) [87].

The clustering probability is low when the collision cascade is diluted. If the trap concentration is low, the trapping at impurities is also not very efficient. The generated V and I must migrate in order to be stored in RT-stable defects, or to anneal. The resulting depth profile will be broader than

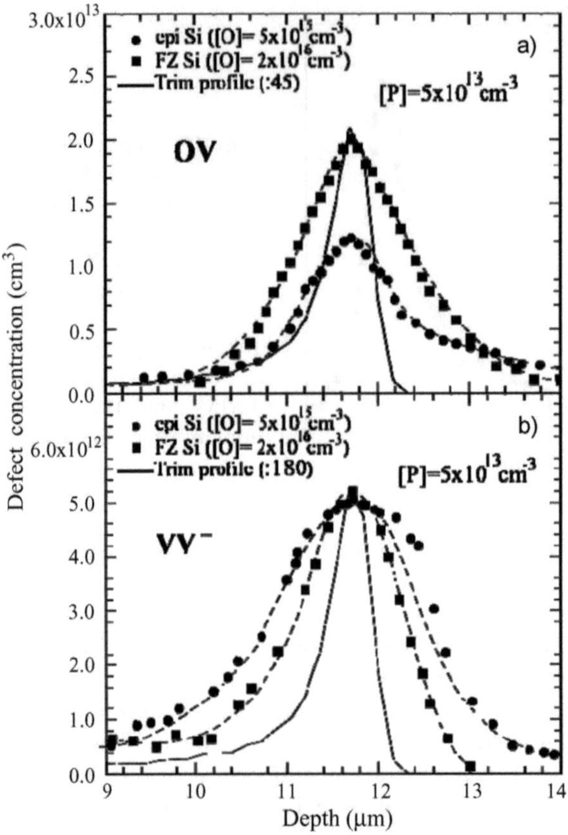

Fig. 14. Depth profiles of OV (**a**), and VV (**b**) in samples implanted with 3-MeV He, $5 \times 10^8\,\mathrm{cm}^{-2}$ on epi Si, $60\,\Omega\,\mathrm{cm}$, $[\mathrm{O}] \leq 5 \times 10^{15}\,\mathrm{cm}^{-3}$ and $[\mathrm{C}] < 1 \times 10^{16}\,\mathrm{cm}^{-3}$, (●); on FZ Si $60\,\Omega\mathrm{cm}$, $[\mathrm{O}] \sim 1 \times 10^{16}\,\mathrm{cm}^{-3}$ and $[\mathrm{C}] \leq 1 \times 10^{16}\,\mathrm{cm}^{-3}$, (■). *Solid lines* are TRIM simulations of the generated Frenkel pairs

the initial profile of ion-beam-generated defects. The strong reduction of the impurity content also causes a strong reduction in the concentration of V-type defects stable at RT. In CZ Si a concentration of ∼16% of the vacancies generated by the beam is stored in RT-stable complexes. This fraction reduces to ∼3% and ∼2.6% in FZ and epi Si, respectively. The reduction occurs since V and I can migrate for longer distances before being trapped and the probability of annihilation increases.

The experimental depth profiles show a symmetric broadening compared to simulations. Since the peak of the defect distribution lies at a large depth, ∼12 μm, any preferential defect recombination at the surface can be ruled out. The observed broadening has therefore to be attributed only to the longer defect migration length in highly pure material [87–90, 99, 101]. The similar depth dependence found for both VV and VO depth profiles suggests that most of the VV are not formed directly in the collision cascade but result from clustering of two migrating V [87]. Finally, a higher concentration of VV is formed in epi than in FZ Si, see Fig. 14b. In fact, the maximum peak concentration is the same in the two profiles, but the half-maximum width is much larger in epi Si, hence the total number of VV is higher in this sample. In epi Si the probability of VV formation is higher than the probability of OV formation, while it is equally probable in the two samples to form PV complexes. If the O concentration in epi Si is enhanced the half-maximum width of the defect depth profiles becomes very sharp, close to the CZ sample. This confirms that the O presence prevents RT defect migration [87]. The big discrepancy between TRIM simulations and experimental data for samples with a low trap concentration arises from the fact that TRIM does not take into account defect migration and trapping at impurities. It is possible to simulate RT diffusion by using programs that take into account defect migration, recombination and interaction with the sample impurities [89]. In such a program, the initial distribution of point defects (V and I) is obtained by a traditional ion-implantation simulation program, MARLOWE [102], a Monte Carlo code. In this way the initial positions of the I and V generated by the ion cascade, are determined. The defect evolution proceeds during the time interval between an ion and the successive one in the same area, at a rate determined by the implantation conditions. The defect diffusion is achieved by allowing I and V to jump randomly, at a rate derived by their RT diffusivity. When a defect, i.e., a V, jumps recombination will take place if there is a free I within a distance of ~5 Å, the second-neighbor distance. If there is a trap (P or O) or another V within such radius, it is stored in a stable defect, according to the following equations:

$$\mathrm{V}+\mathrm{I} \Rightarrow \varnothing, \tag{3}$$

$$\mathrm{V}+\mathrm{V} \Leftrightarrow \mathrm{VV}, \tag{4}$$

$$\mathrm{P}+\mathrm{V} \Leftrightarrow \mathrm{PV}, \tag{5}$$

$$\mathrm{O}+\mathrm{V} \Leftrightarrow \mathrm{OV}. \tag{6}$$

The opposite reactions, detrapping and re-emission from clusters (VV) are also included. Finally, the surface is assumed to act as a perfect sink for migrating defects. The results are in good agreement with the experimental data for both cases of low trap and high trap concentration samples implanted with He [100].

A similar study on the I-type-defect depth profiles can be performed to obtain information on the RT I diffusion process. The same kind of process, a trap-limited diffusion process, is observed to occur for both V and I. In particular, the P_sC_i complex, detectable by DLTS in n-type Si, provides similar information. It experiences a large broadening as the impurity concentration in the sample is reduced.

The profile broadening is observed to increase as a function of the implantation dose [87]. These data can be explained by introducing a critical dose. If the implantation dose is below the critical dose there is no overlap between the defect profiles from different cascades and the measured depth profile mirrors the defect distribution in the single cascade, according to the results just shown. When the dose is above the critical value, due to the point-defect

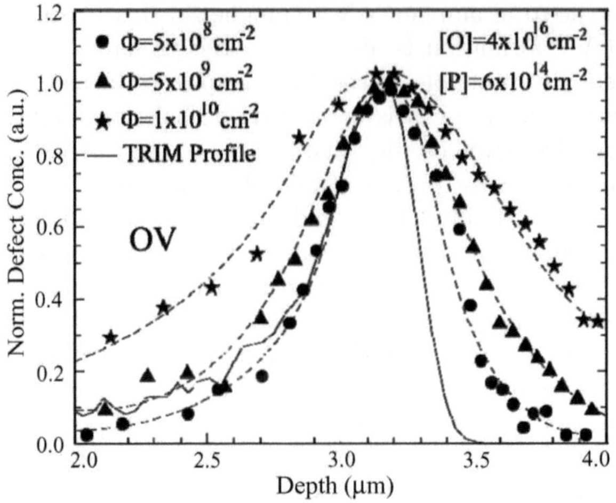

Fig. 15. Normalized concentration depth profile of the OV complexes, as obtained from DLTS, for doses of $5 \times 10^8\,\mathrm{cm}^{-2}$ (●) $5 \times 10^9\,\mathrm{cm}^{-2}$ (▲); $1 \times 10^{10}\,\mathrm{cm}^{-2}$ (⋆). The *dashed lines* are only guides to the eye, while the *solid line* is the TRIM defect depth profile

migration, the traps confined in the space within two ion-collision cascades start to be filled. If another ion reaches the same region, as certainly occurs for higher doses, the defects will be generated in a region with a trap concentration lower than expected. Most of the traps are unable to interrupt the free migration of newly generated defects. Point defects will migrate to longer distances before trapping or recombination occur. As a result, the final depth profile exhibits an increase in the half-maximum width as shown in Fig. 15 [87, 99] where the OV depth profile as a function of the implanted dose is plotted. He was implanted to doses from $5 \times 10^8\,\mathrm{cm}^{-2}$ (●) up to $1 \times 10^{10}\,\mathrm{cm}^{-2}$ (⋆). The dashed lines are only guides to the eye, while the solid line is the TRIM defect depth profile. On increasing the ion dose a broadening of the defect depth profile is observed since the defects need to migrate to longer distances before trapping.

3 Evolution from Point to Secondary Defects

As observed before, our knowledge on defect annealing behavior in c-Si mostly relies on studies performed on electron-irradiated samples [5, 8, 16, 53–55]. Since the Si impurity content plays a big role in defect formation, it might produce a difference also in their annealing behavior. For these reasons, the annealing behavior of electron-irradiated and Si ion-implanted samples was compared by monitoring samples after isochronal anneals, with 50°C steps, from 100–600°C, as a function of the sample impurity content. In particular, the defect evolution upon annealing in ion-implanted and electron-irradiated epitaxial Si samples were monitored. DLTS spectra taken on p-type and n-type epi Si are plotted in Fig. 16. Data for as-implanted samples (solid line) as well as for samples annealed at 150°C (dashed line) or 350°C (dot-dashed

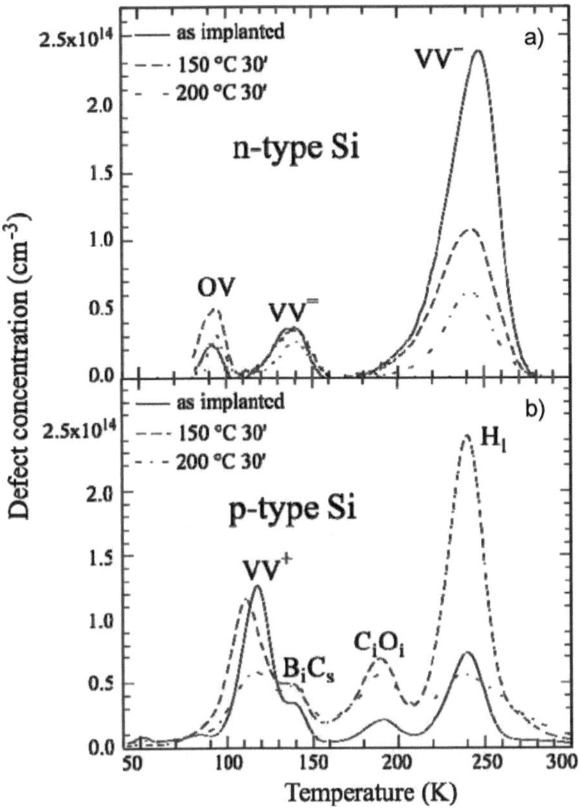

Fig. 16. DLTS spectra measured on n-type (**a**) and p-type (**b**) epitaxial Si samples implanted with 1.2-MeV Si to a dose of $1 \times 10^9\,\mathrm{cm}^{-2}$. Spectra are reported for the as-implanted sample (*solid lines*) and for samples annealed at 150°C (*dashed lines*) and 200°C (*dot-dashed lines*) for 30 min

line) for 30 min are shown. In n-type Si, the as-implanted spectrum comprises the well-known OV at $E_C - 0.16\,\mathrm{eV}$; $VV^{=}$ at $E_C - 0.21\,\mathrm{eV}$ and VV^{-} at $E_C - 0.41\,\mathrm{eV}$. The VV^{-} peak also includes the contribution of the PV at $E_C - 0.46\,\mathrm{eV}$ whose signal occurs at the same spectral position [8]. Annealing at 150°C causes the VV^{-} peak reduction due to the PV dissociation [8, 67], as the invariance in the $VV^{=}$ concentration confirms: part of the released V form additional OV. Annealing above 200°C causes the VV^{-} dissociation [67]. Finally, after annealing at 350°C, the OV also dissociate [64] and only a low residual damage concentration ($\leq 10^{11}\,\mathrm{cm}^{-3}$) is left [96]. Recent studies, performed on silicon on insulator (SOI), allowed characterization of the V-type complexes reducing the I-type defect contribution in the annealing behavior. In particular, it was shown that V-type defects are more stable with temperature (above 400°C for the VV) if Is do not contribute to their dissociation [103].

In p-type Si, the defect structures exhibit a complex annealing kinetics. The as-implanted spectrum includes signatures of both V-type, VV^{+} at $E_V + 0.21\,\mathrm{eV}$, and I-type defects such as the C_iO_i, at $E_V + 0.36\,\mathrm{eV}$, the B_iC_s at $E_V + 0.26\,\mathrm{eV}$, and H_1 at $E_V + 0.45\,\mathrm{eV}$ [54, 55, 96]. After 200°C annealing a strong increase in the H_1 concentration and a much lower increase in the C_iO_i concentration is observed due to the dissociation of the B_iO_i complex [54, 55, 96]. The Is released by the defect dissociation increase the other I-type defect concentrations. As previously mentioned the C_iO_i peak in epi Si is not dominant in the DLTS spectrum, due to the low C concentration of epi Si. Annealing at 350°C results in a reduction of about one order of magnitude of the total defect concentration. An identical annealing behavior for each of the defect signals detected in the DLTS spectra is observed for both electron irradiation and ion implantation between RT and 300°C.

The annealing behavior of RT-stable complexes allows their identification. The defect annealing behavior analysis allowed the identification of the B_iC_s complex. In fact, this defect is known [53, 55] to compete with the B_iO_i complex. The B_iO_i complex introduces a donor level in the upper part of the Si bandgap, not detectable by DLTS measurements on p-type Schottky barriers. It dissociates between 100°C and 150°C [53–55] and the B_i released migrates to form other B-related complexes, such as B_iC_s. The B_iO_i presence inhibits the formation of B_iC_s complex at RT. Indeed, it forms at lower temperatures if the O concentration is low. Annealing of both electron-irradiated and ion-implanted CZ samples (not shown) reveal that B_iC_s is not present as-implanted Si but it forms upon annealing at 150°C, as reported in the literature [54]. A similar behavior has been observed for H_1. The concentrations of B_iC_s and H_1 first increase ($T < 150$–200°C) and then eventually decrease, while the VV^+ and C_iO_i concentrations decrease monotonically upon annealing [86]. The H_1 annealing behavior does not allow one to identify it, but it is obvious that it competes with the C_iO_i. Moreover, its annealing behavior is similar to the B_iC_s complexes. These evidences, and its strong increase at temperatures of ~150°C, when the B_iO_i complex dissociates, strongly suggest that it is a B-related complex that embodies one I [96]. Further measurements indicate that it is an I-type complex involving H and B interstitials [97].

Point-defect annealing behavior does not depend, at least at low temperatures (≤300°C), on the damage-creation method. On the other hand, the reduction in the impurity content not only changes the stable defect structures at RT, but also their annealing rate and, hence, their thermal stability. In particular, defect annealing is the result of either thermal dissociation or annihilation by point defects released by the dissociation of less-stable defect structures. The dependence of the annealing temperature on the impurity content was already known for some defect complexes, such as the VV [103] and B_iC_s [53].

Also, the C_iO_i annealing strongly depends on the sample impurity content. Its isochronal anneal (30 min) was measured by changing the impurity content, and the results are shown in Fig. 17. The epi (●) and CZ (■) Si samples used have a C concentration, $\leq 10^{16}\,\mathrm{cm}^{-3}$, lower than the CZ Si (▲) samples used in a previous study [54], $6\times10^{16}\,\mathrm{cm}^{-3}$. For the last sample, the

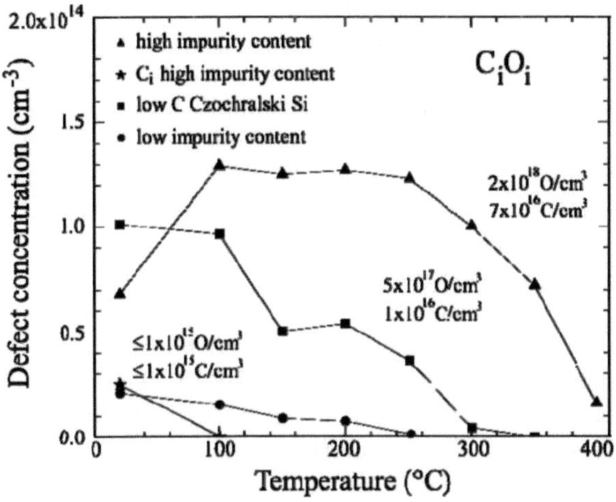

Fig. 17. Carbon–oxygen (C_iO_i) concentration, as obtained from DLTS, as a function of annealing temperature for epitaxial (●) and CZ (■) Si. In addition, the annealing behavior of C_iO_i (▲) and carbon–interstitial C_i (⋆) defects in a CZ Si sample with a higher C content [70] are shown for comparison. All the samples were irradiated with 9.2-MeV electrons to a dose of $3.5 \times 10^{15}\,\mathrm{cm}^{-2}$

C_i annealing behavior (⋆) is also plotted. This defect was not detected in the other samples, due to their low C content.

The annealing curve of the C_iO_i complex in the high-C CZ sample shows the behavior typically observed for this defect. At 100°C, its concentration increases as the C_i, stable at RT, diffuses and is trapped by O, present in high concentration, producing more C_iO_i. The C_iO_i concentration remains stable up to 250°C and decreases slightly in the range 250–350°C. This decrease is probably due to annihilation by the free V released by the VV dissolution that occurs in this temperature range. Finally, complete dissolution of the C_iO_i occurs at 400°C. The C_iO_i defect stability is a consequence of both the high binding energy of the two impurity atoms and of the strong backreaction that competes with dissociation. Indeed, in Si with high O and C concentrations, when a C_iO_i is dissociated the released C_i has a strong chance to pair again with O. In epi Si, the C_iO_i concentration is greatly reduced since most of the Is are expected to be stored into different, B-related complexes. B_i-related complexes dissociate at $T < 400$°C, releasing free B_i [53–55]. They can recombine with V-type defects, as the VV. Their dissolution releases free Vs that migrate and, finally, cause the C_iO_i defects dissolution. In addition, the dissociation of a C_iO_i is not balanced by a backreaction, since the released C_i has low probability of being trapped again by an O atom in the epi material. Complete C_iO_i annealing is achieved at 250°C. The C_iO_i annealing temperature in the low-carbon CZ Si sample is intermediate to that of the high-carbon CZ and of the epi Si samples. Once more, C and B compete to capture Is and then to form stable defects. The presence of B_i stored in RT-stable defects will again result in a different annealing behavior of the C_iO_i complex [105].

Simulations of damage evolution [34] predict that ion implantation produces an imbalance between V and I in the residual damage, due to the extra incorporated ion, as predicted by the "+1" [41] phenomenological model previously discussed. Such an imbalance is not detectable in the as-implanted state since the total concentration of defects generated by the beam is very high compared to the implanted-ion concentration. To monitor the "+1" effect, the I-type and V-type defects annealing in ion-implanted and electron-irradiated epi Si samples was followed. The Vs and Is concentration can be quantitatively measured by DLTS as a function of annealing. Hence, the number of I and V stored in these defects can be counted. The DLTS measurements in Fig. 16 and similar provided the total concentration of V and I. The total V concentration $[N_{\rm V}]$ value was obtained adding the VV and the OV concentration. This last contribution is the only one derived from n-type Si spectra since the OV signature is not detectable by DLTS in p-type Si samples. The total interstitial concentration $[N_{\rm I}]$ is obtained by adding the C_iO_i, B_iC_s and H_1 concentrations [96]. To summarize:

$$[N_{\rm V}] = 2[{\rm VV}] + [{\rm OV}], \tag{7}$$

$$[N_{\rm I}] = [{\rm C_iO_i}] + [{\rm B_iC_s}] + [{\rm H_1}]. \tag{8}$$

The calculation was performed as a function of the annealing temperature and the results are plotted in Fig. 18. The concentration of both I (■, ⋄) and V (▲, ⋆) stored in the defects, is given for ion-implanted (first symbols) and electron-irradiated samples (second symbols) isochronally (30 min) annealed in the range 100–550°C. The lines are guides to the eye. The right-hand scale of the plot is the number of defects per implanted ion, N, calculated as:

$$N = \frac{C_{\rm d} x}{\Phi}, \tag{9}$$

where $C_{\rm d}$ is the average defect concentration in the region probed during the DLTS measurements, x is the thickness of this region (~1 μm) and Φ is the implanted dose. Although each ion produces ~2500 Frenkel pairs (as

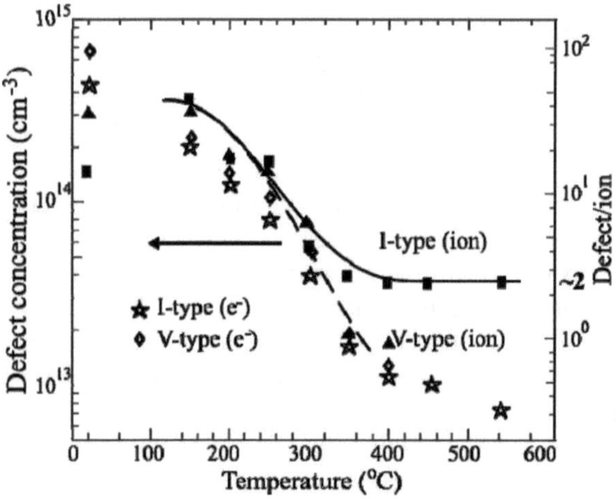

Fig. 18. Total concentration of I and V stored in defects, as a function of annealing temperature for epi Si implanted with 1.2-MeV Si, $1\times10^9\,{\rm cm^{-2}}$, I (■) V (▲), and electron irradiated with $3.5 \times 10^{15}\,{\rm cm^{-2}}$, 9.2 MeV, I (⋆) and V (⋄). The *lines* are guides to the eye for ion-implantation-generated defects

calculated by TRIM), only ~60 per ion escape recombination and form RT stable defects. This result is in perfect agreement with the results described previously. Annealing at temperatures up to 300°C produces a concomitant reduction of I- and V-type defects. This behavior is explained by assuming that the V pairs dissociation produces free Vs that migrate to I-type complexes causing their dissociation, and vice versa. This process results in defect recombination in the bulk, thus maintaining the balance between I and V. Indeed, an equal number of V-type and I-type defects remain although their absolute concentrations have been reduced by approximately one order of magnitude [96]. This confirms that, as expected from Monte Carlo simulations [106], most of the defect recombination occurs in the bulk. If I or V surface annihilation had been dominant, the balance between the two defects would not have been maintained.

At temperatures above 300°C all of the V-type clusters annealed out (the V data points at $T \geq 300°C$ are upper limits given by measurement sensitivity), but a measurable number of interstitials (~2–3 per implanted ion) is still present in the samples. This imbalance between I- and V-type defects is due to the extra incorporated ion that only becomes experimentally detectable when most of the defects have recombined [105, 107].

The same counting experiment was performed on identical n- and p-type electron irradiated samples. The results are shown in the same figure. The as-implanted DLTS spectra are identical, as previously observed. The imbalance in the two concentrations observed for ion implantation at temperatures ≥350°C, is not detected in electron-irradiated samples. I- ($\star$) and V-type ($\diamond$) defects continue to anneal concomitantly. Despite the fact that the two samples have the same as-implanted spectrum, after annealing at 350°C the difference is experimentally detectable [96, 108].

The results clearly demonstrate that the imbalance between Vs and Is in ion-implanted samples has to be attributed to the extra ion introduced in the implantation process. This is an important experimental validation of the idea already predicted by the phenomenological model for the description of TED and by Monte Carlo simulations [106].

It should be recalled that such studies were performed by low-dose implants. DLTS measurements cannot be carried out in the as-implanted state for samples implanted to high doses since deterioration of the junction characteristics occurs. This is why, traditionally, DLTS was not used to detect the electrically active damage in the "defect in cluster" region of Fig. 2. However, after annealing at 400°C, most of the damage recovers and DLTS measurements can be performed again. At annealing temperatures of 350–500°C, after all the V-type defects anneal, *interstitial point-like defects* form [108, 109]. These point defects store the I "excess" caused by the extra ion presence. Point defects form complexes with impurities (see the "*defects in traps*" region of Fig. 2). In fact, the residual damage concentration increases when the sample impurity concentration increases. In particular, when the annealing proceeds (e.g., 400°C, 30 min), the residual damage concentration, is higher in high impurity content Si (not shown). More defects are trapped by impurities and cannot recombine with other defects or at the surface.

To follow the defect evolution in ion-implanted Si, Si ions were implanted in a wide dose range, 1×10^{9}–$5 \times 10^{13}\,\mathrm{cm}^{-2}$, and annealed in the range 400–750°C. The residual damage left after annealing was monitored using both electrical (DLTS) and optical (PL) measurements. The comparison of the two techniques allowed us to demonstrate that the damage left after low-dose implantation and annealing at temperatures below 600°C is formed mainly by secondary defects. These defects have the typical characteristics of point-like defects but do not form in the as-implanted state. Implants to doses $\leq 10^{11}\,\mathrm{cm}^{-2}$ cause the formation of roughly the same residual damage. The same DLTS signatures, within the experimental error (below ± 0.01 eV), are observed, regardless of the implantation dose. In this regime, on increasing the dose a linear increase in the defect concentration is achieved, suggesting that the collision cascades generated by different ions do not markedly overlap. The final damage configuration depends only on the defect density and the defect interactions within the single-ion track. In fact, the DLTS signatures at these doses show large variations only as a function of the annealing temperature. Totally different is the residual damage spectrum after implantation at intermediate dose (1×10^{12}–$1 \times 10^{13}\,\mathrm{cm}^{-2}$). In this case, interactions between defects belonging to different collision cascades occur, dramatically changing the damage.

The residual damage features were also monitored by PL measurements. PL is sensitive to optically active centers, on which the electron–hole recombination occurs, producing photons. To reduce as much as possible the non-radiative recombination paths, all the measurements were carried out at 17 K. The PL spectra of Si samples implanted with Si and annealed at 400°C for 30 min are compared in Fig. 19.

After $1 \times 10^{9}\,\mathrm{cm}^{-2}$ Si implantation (dashed line) the optically active damage is barely visible and the band to band carrier recombination is clearly visible at 1121 nm (Si band edge). On increasing the implantation dose the Si band edge is no longer visible, confirming the results already obtained by DLTS. The damage is present in higher concentration and most of the radiative recombinations occur at the damage. We identified the PL lines and found they belong to I-type defects. The W line [16, 17] at 1218 nm (1.018 eV) is the dominant center for doses $\geq 1 \times 10^{11}\,\mathrm{cm}^{-2}$. This line has been recently associated with small I-rich clusters [16, 17, 110]. Moreover, the shoulder observed in the spectra at 1279 nm (0.969 eV) is probably the G-line, due to the C_iC_s [16, 17] complex. The spectra also exhibit the characteristic peak at 1570 nm (0.789 eV) associated with the C_iO_i complex, the well-known C line [16, 17]. Since PL is not quantitative, it is not possible to extract information on the defect concentration. An increase in the PL signal of these lines is observed, increasing the implantation dose. The comparison

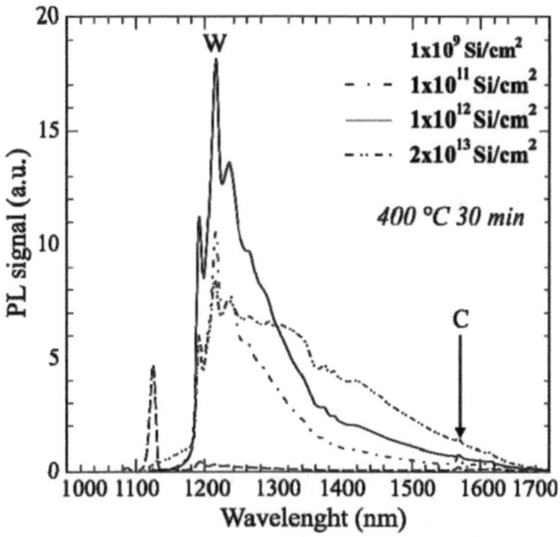

Fig. 19. PL spectra of CZ Si samples implanted with 1.2-MeV Si to doses of $1 \times 10^9\,\mathrm{cm}^{-2}$ (*dashed line*), $1 \times 10^{11}\,\mathrm{cm}^{-2}$ (*dot-dashed line*), $1 \times 10^{12}\,\mathrm{cm}^{-2}$ (*solid line*), $2 \times 10^{13}\,\mathrm{cm}^{-2}$ (*double dot-dashed line*). All samples were annealed at 400°C 30 min

with the DLTS data suggests that the implantation damage increases. PL can be really useful to monitor samples having a high damage concentration, where the DLTS fails. In fact, it is possible to monitor the residual damage after $2 \times 10^{13}\,\mathrm{Si/cm^2}$ implantation (double-dot-dashed line). In this case the W line is the dominant feature of the spectrum, but a pronounced shoulder between 1300 nm and 1400 nm appears, probably due to the high degree of lattice disorder.

The damage evolution was followed as a function of both the implantation dose and the annealing temperature using both DLTS and PL. Between the two techniques, only the DLTS can provide quantitative information on the defect concentration. In fact, assuming that every complex involves a single point defect, we added the defect concentration obtained from the DLTS spectra, using the counting procedure previously described ((7) and (8)) [96, 109]. The results obtained for 1.2-MeV Si implantation on epi Si samples are summarized in Fig. 20. In this figure, the total defect concentration (Fig. 20a) and the number of defects per ion (Fig. 20b, obtained by dividing the defect concentration by the ion dose) are plotted as a function of the temperature for isochronal (30 min) annealing in samples implanted to doses from $1 \times 10^9\,\mathrm{cm}^{-2}$ (■), to $1 \times 10^{12}\,\mathrm{cm}^{-2}$ ($\star$). The solid and dashed lines are used as a guide to the eye. An increase in the ion dose up to $1 \times 10^{11}\,\mathrm{Si/cm^2}$ ($\diamond$) determines a roughly linear increase in the total defect concentration. However, for 600°C annealing, a strong reduction in the defect concentration is produced due to the damage recovery. The lattice recovery for annealing at 600°C in samples implanted to doses up to $10^{11}\,\mathrm{cm}^{-2}$ is confirmed by the PL measurements that exhibit only the band-edge recombination line (not shown).

Ion implantation to larger doses ($\geq 1 \times 10^{12}\,\mathrm{cm}^{-2}$) reveals a quite different behavior. First, the DLTS spectra of samples annealed at temperatures

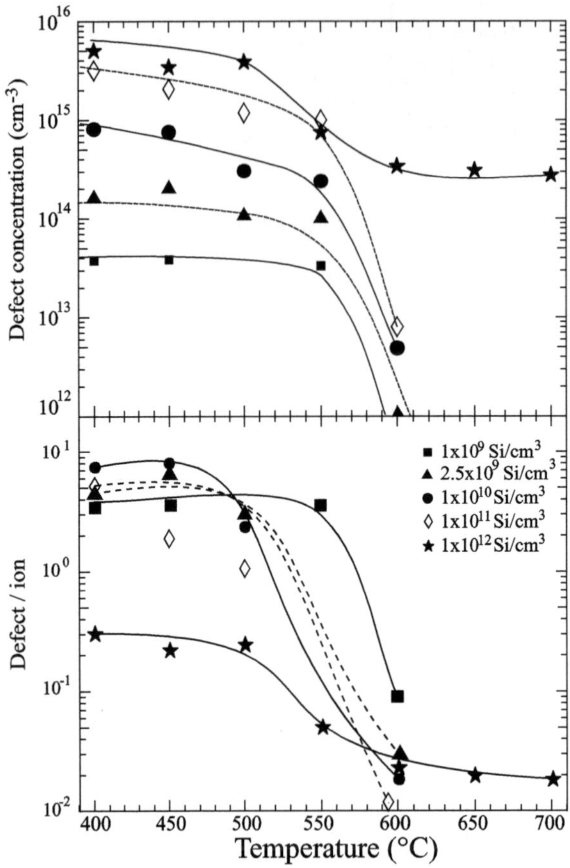

Fig. 20. (**a**) Total defect concentration and (**b**) defect per ion, obtained from DLTS measurements, as a function of isochronal (30 min) annealing for p-type epi Si samples implanted with 1.2-MeV Si to doses of $1 \times 10^9\,\mathrm{cm}^{-2}$ (■), $2.5 \times 10^9\,\mathrm{cm}^{-2}$ (▲), $1 \times 10^{10}\,\mathrm{cm}^{-2}$ (●), $1 \times 10^{11}\,\mathrm{cm}^{-2}$ (⋄) and $1 \times 10^{12}\,\mathrm{cm}^{-2}$ (⋆). The *lines*, solid or dashed, are only guides to the eye

≤500°C exhibit a residual damage dramatically different from that observed in samples implanted to lower doses. Furthermore, it is still detectable in DLTS after annealing at temperatures ≥600°C, as shown in Fig. 20. These results strongly suggest that a more stable class of defect is formed. We believe that this difference arises from the higher I supersaturation obtained for higher-dose implants that causes the formation of more complex defects already in the early stages of annealing.

The modification in the defect structure is visible in Fig. 20b, where the number of defects per ion is plotted. For doses up to $10^{11}\,\mathrm{cm}^{-2}$ the number of defects per ion is roughly 3 regardless of the dose. For implants at $10^{12}\,\mathrm{cm}^{-2}$ the number of defects per ion is much lower, suggesting that the simple counting procedure used for lower doses cannot be applied.

4 Formation and Annihilation of I Clusters and Extended Defects

4.1 Evolution from Secondary Defects to Interstitial Clusters

The simple point-defect pairs formed after low-dose implants anneal out already at temperature below 500°C. On the other hand, extended defects formed at high implantation doses are particularly stable, up to 800°C. It is therefore interesting to characterize the residual damage obtained for Si implants $\geq 1 \times 10^{12}\,\mathrm{cm}^{-2}$. To this purpose, once again, both PL and DLTS were used. DLTS analyses show that the same class of defects is formed regardless of the annealing conditions after high-dose implants. In particular, the DLTS spectra of a sample implanted with 145-keV Si to a dose of $2 \times 10^{13}\,\mathrm{cm}^{-2}$ on p-type CZ Si and annealed at 680°C 1 h (solid line) and of a sample implanted with 1.2-MeV Si to a dose of $1 \times 10^{12}\,\mathrm{cm}^{-2}$ on p-type epi Si and annealed at 600°C 30 min (dashed line, multiplied by 10) are compared in Fig. 21.

The residual damage for the two samples is mainly given by two signatures, with activation energies of $E_\mathrm{V} + 0.33\,\mathrm{eV}$ and $E_\mathrm{V} + 0.52\,\mathrm{eV}$ and labeled in the figure as B_1 and B_2 (B-lines), respectively [107–109]. The same kind of defect was detected regardless of the implantation energy, dose and impurity content of the sample. O and C content are, respectively, 3 and 1 orders of magnitude higher in CZ than in epi Si. Also, the dopant concentration does not affect the damage, since the CZ Si wafers used have about one order of magnitude more B than the epi Si ones. Moreover, differently from low-dose residual damage, the annealing is not the dominant factor in determining the final defect characteristics. The two samples underwent thermal treatments different in temperature and time. It should be mentioned that DLTS measurements exhibit small differences, within $\pm 0.05\,\mathrm{eV}$ ($\sim$10%), in

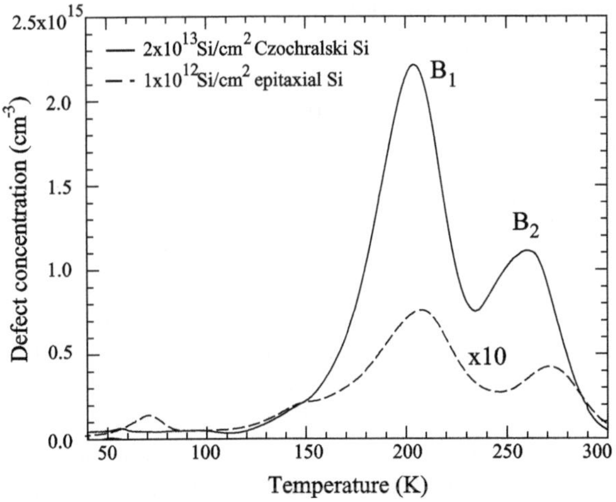

Fig. 21. DLTS spectra of p-type CZ Si implanted with 145-keV Si to a dose of $2 \times 10^{13}\,\mathrm{cm}^{-2}$ and annealed at 680°C 1 h (*solid line*); epi Si implanted with 1.2-MeV Si to a dose of $1 \times 10^{12}\,\mathrm{cm}^{-2}$ and annealed at 600°C 30 min (*dashed line*, multiplied by a factor of 10)

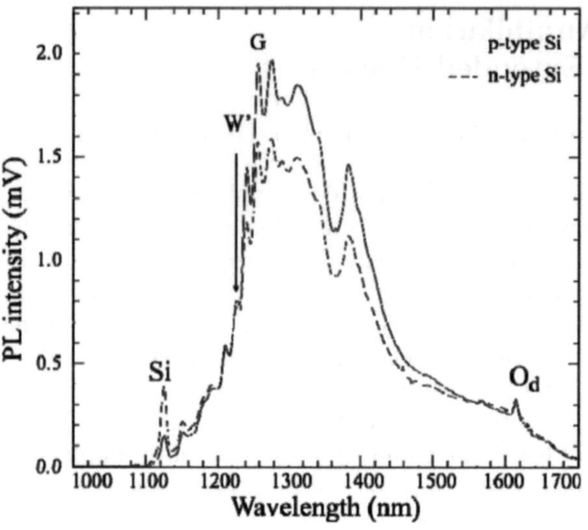

Fig. 22. PL signal for p-type (*solid line*) and n-type (*dashed line*) Si implanted with 1.2-MeV Si, $1 \times 10^{13}\,\mathrm{cm}^{-2}$ annealed at 600°C for 30 min

the activation-energy values for the two lines (B_1 and B_2), but they have not been successfully correlated to any difference in the sample characteristics, implantation conditions or thermal treatment. These results and more extensive studies [109] show that neither impurities nor the dopant are the main constituents of these defects. The evidence that the defect structures after high-dose implantation do not depend on the impurity concentration for wide dose, temperature and time ranges confirms that they are particularly stable, probably clusters of intrinsic defects.

The PL spectra of p-type (solid line) and n-type (dashed line) Si samples implanted to a dose of $1 \times 10^{13}\,\mathrm{Si/cm^2}$ and annealed at 600°C for 30 min are compared in Fig. 22. The same optically active centers are observed in both samples under the same measurement conditions. The optically active residual damage is the same regardless of the dopant (P and B, respectively). This result shows that the dopant does not influence the residual damage features, as previously observed for the electrically active residual damage. The PL spectra present two main features in the 1100–1400 nm range consisting of two broad peaks centered at 1320 nm (0.94 eV) and 1390 nm (0.89 eV). These two features are broad luminescence bands and do not arise from the convolution of narrower peaks [111]. The broadening in the PL peak has been associated [112] to the quantum confinement of carriers in regions with a high-strain region surrounding the defects (see below). Several sharp lines in the range 1200–1280 nm are superimposed to these broad peaks. All of them have been associated to point-like defect and defect–impurity complexes formed as a consequence of ion-beam irradiation. In particular, a line W′ at 1233 nm (1.0048 eV) which is a perturbed form of the W line (1218 nm, 1.018 eV) [16, 17] has been identified. As previously mentioned, this line is associated with small I-rich clusters [17, 110]. The line at 1279 nm (0.969 eV) is the

G-line, associated with the CiCs complex. The spectra also exhibit the well-known C line at 1570 nm (0.789 eV) associated to the C_iO_i complex [16, 17] and two peaks at 1620 nm (0.765 eV) and 1660 nm (0.7466 eV) associated to oxygen thermal donors [113].

Although the comparison of DLTS and PL measurements cannot allow us to conclude that the optically active and the electrically active centers belong to the same defect clusters both datasets can be easily associated to the residual damage and in both cases the dopant is not responsible for the final defect characteristics. Both measurements show the presence of a stable class of damage not depending on the impurity or dopant concentration or type.

Measurements were carried out in wide dose (1×10^{12}–$1 \times 10^{14}\,cm^{-2}$), temperature (550–750°C) and time (10 min–15 h) ranges and the same signatures are always detected. The two features at 1320 nm (0.94 eV) and 1390 nm (0.89 eV) are unchanged as a function of the ion-implantation dose. Finally, TEM analyses revealed that no extended defects are formed in these samples even after at $2 \times 10^{13}\,cm^{-2}$ Si ion implantation [111]. Broad features in PL spectra have been usually observed [112] in samples having extended defects, such as in O-precipitated, Sb-precipitated and hydrogen-plasma-treated Si. Broadening is associated with the quantum confinement of carriers in regions with the high-strain region surrounding the defects. Since no extended defects are detected in our samples, we believe that these signatures are associated with the carrier recombination in the strained region surrounding the small I clusters embedded in the Si matrix [111, 114]. Also, the cluster electrical signatures (B-lines) exhibit a broadening that could be associated with the defects strain (see later).

The results so far shown suggest that I-rich regions, with I-type point defects (such as C–O and C_iC_s), small I complexes (the PL W line) and I clusters, are present. Since we did not observe any V-type defect signature we believe that the I excess observed is the direct consequence of the extra implanted ion introduced during implants.

The I-cluster defect-formation rate is strongly nonlinear as the implantation dose is increased and a strong reduction in the defect concentration at $5 \times 10^{13}\,Si/cm^2$ implantation is observed. It can be attributed to the formation of extended {113} defects that compete with the B-lines in storing the I excess [109]. These results confirm the I nature of the clusters, since their concentration reduce when I-type extended defects form.

Another evidence of the complex nature of these defects is provided by the B-lines depth concentration profile, as determined by DLTS. The depth-profile distribution of a point-like interstitial defect, the C_iO_i (●), and of B_1 (■) are shown in Fig. 23a and b, respectively, and are compared with the Frenkel-pair (dashed line) and the extra-ion (solid line) distributions simulated by MARLOWE. The C_iO_i experimental profile, obtained after a low-dose implant, precisely mirrors the simulated Frenkel-pair profile. The

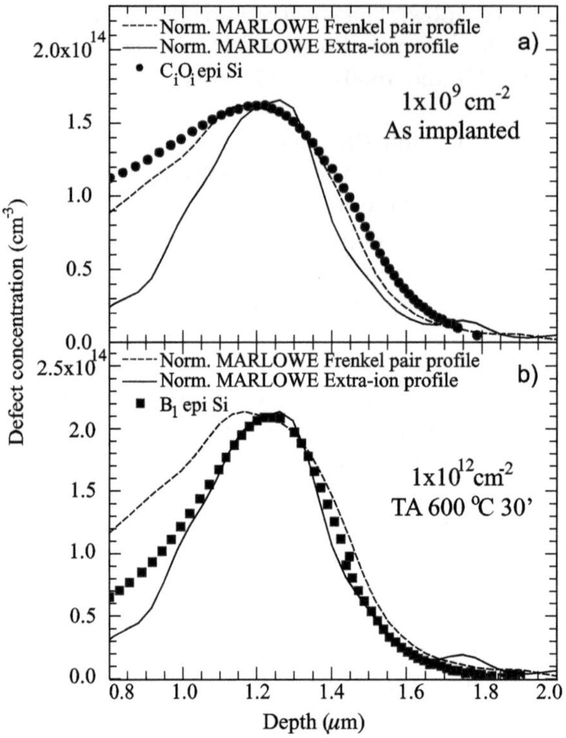

Fig. 23. Depth concentration profile of: (**a**) C_iO_i defect pairs in p-type epi Si as implanted with 1.2-MeV Si, $1 \times 10^9\,cm^{-2}$; (**b**) B_1 line in p-type epi Si implanted with 1.2-MeV Si, $1 \times 10^{12}\,cm^{-2}$, and annealed at 600°C, 30 min. The lines are the MARLOWE depth profile simulation of the implanted ion (*solid line*) and of the Frenkel pairs generated by the beam (*dashed line*)

extra-ion profile simulation shows a narrower and slightly deeper peak. On the other hand, the B_1 depth profile, obtained after high-dose implants and annealing at 600°C for 30 min, having the maximum at ~1.3 μm, precisely mirrors the extra implanted-ion simulation profile at this dose (divided by a factor 100 to allow comparison). Similar depth profiles were measured for B_2 (not shown). These considerations and the result that the B-lines defect structures form in the end-of-range region, the region that experiences the maximum I supersaturation, support the conclusion that these defects are Si I clusters.

Finally, the B-line signatures depth profiles confirm that B is not a main constituent of the defect clusters. We cannot exclude that a single B atom is involved in the cluster structure but it is unlikely that a big number of atoms participate to the cluster formation. In fact, the maximum peak concentration in the B_1 depth profile is $\sim 2 \times 10^{14}\,cm^{-3}$. A peak concentration value of $\sim 1.5 \times 10^{14}\,cm^{-3}$ was measured for B_2. Hence, the total B-lines concentration is 3.5×10^{14} clusters/cm^3. If more than two B atoms were stored in such defects, becoming inactive, a detectable reduction of the shallow doping concentration, in the implanted region, would have been observed by capacitance–voltage (CV) measurements. However, CV measurements show a constant shallow doping concentration throughout the implanted region.

Moreover, B clustering is also unlikely for simple probability considerations. The monitored samples have a B concentration of $\sim 7 \times 10^{15}\,\mathrm{B/cm^3}$, and the probability that more than two B_i could migrate and find each other during the annealing process is very low. In fact, even the B_iB_s complex formation is not observed for concentrations below $5 \times 10^{16}\,\mathrm{B/cm^3}$.

4.2 Interstitial Cluster Formation and Dissociation

The next issue to be addressed is the I-cluster formation kinetics. The existence of a threshold dose for their formation and its dependence on the impurity content is of great interest. We monitored the residual damage as a function of the implantation dose and of the material impurity concentration. The DLTS spectra on epi Si show that the B-line forms only after Si implantation at doses $\geq 1 \times 10^{12}\,\mathrm{cm^{-2}}$, while in CZ Si samples B_1 is present already after $1 \times 10^{10}\,\mathrm{Si/cm^2}$ implants and both signatures are already visible [109] at $10^{11}\,\mathrm{cm^{-2}}$, even if their concentration is quite low ($<1 \times 10^{13}\,\mathrm{cm^{-3}}$). Hence, the impurity presence plays a role in the I-clusters formation only for implantation doses below $10^{11}\,\mathrm{cm^{-2}}$. The experimental data are explained assuming that only clusters above certain dimension can survive annealing. As a result, very small clusters tend to dissociate for times shorter than 30 min, e.g., in epi Si, where the I can be stored only in small clusters. A different scenario is proposed for CZ Si, where the impurity presence favors the I trapping. In fact, impurities, in particular C, are efficient traps for I [97, 115] storing them, thus preventing the formation of bigger clusters [116] or extended defects. C and I might form mobile CI complexes [117] followed by the nucleation and growth of CI agglomerates. The impurities reduce the I diffusion length by trapping and preventing their annealing to the surface. As a result, a local high I supersaturation in the end-of-range region is experienced and the cluster structures rapidly grow above the critical size to be stable. This effect is visible only at low doses. As the implantation dose exceeds a critical value the I supersaturation allows the growth of I-clusters. Our interpretation of the data is confirmed by the evidence that extended defects are known [98, 115] to form at a threshold dose increasing with the sample impurity content. Moreover, I-cluster thermal stability strongly depends on the implantation dose [118–122].

In order to verify if the I stored in small I-clusters might be responsible for the TED phenomena at low implantation dose, we studied their annealing behavior, by calculating the dissociation energy as a function of the annealing time and temperature. The data analysis at different annealing times for a given temperature allows one to obtain the characteristic annealing time τ_0 at that temperature. The values were determined by fitting the data obtained for annealing at temperatures ranging from 550°C (30 min–15 h) to 700°C (10–40 min). The characteristic time (τ_0) is given by:

$$C(t) = C_0 \exp\left(-\frac{t}{\tau_0}\right), \tag{10}$$

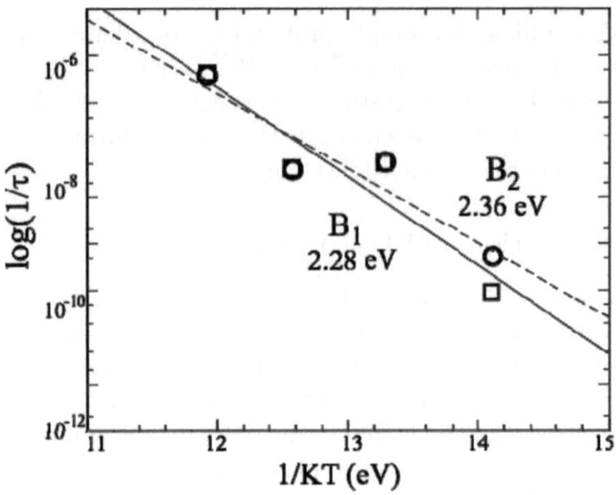

Fig. 24. Arrhenius plot of the B_1 (□) and B_2 (○) characteristic times. The *solid lines* are linear fits of the data

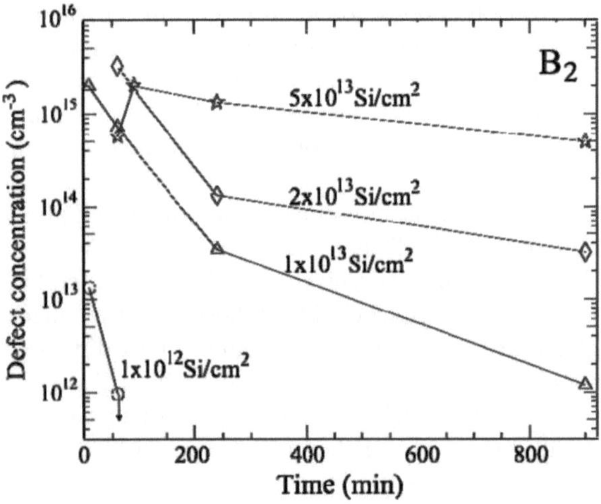

Fig. 25. Defect concentration values, as obtained from DLTS measurements, for p-type epi Si samples implanted with 145-keV Si to doses of $1 \times 10^{12}\,\mathrm{cm}^{-2}$ (○), $1 \times 10^{13}\,\mathrm{cm}^{-2}$ (△), $2 \times 10^{13}\,\mathrm{cm}^{-2}$ (◇) and $5 \times 10^{13}\,\mathrm{cm}^{-2}$ (⋆). All samples were annealed at 680°C

where C_0 is the initial cluster concentration.

The τ_0 values in the temperature range explored are summarized in the Arrhenius plot of Fig. 24 for both B_1 and B_2. The data best fits are plotted as a solid and dashed line for B_1 and B_2, respectively. The fit slopes provide an estimation of the activation energy for dissociation. Dissociation energy values of 2.28 eV, and 2.36 eV were obtained for B_1 and B_2, respectively. These values, equal within the experimental error (∼15%), are consistent with the TED characteristic energy value in the absence of extended defects [28, 123].

The B-line dissociation was monitored as a function of the implantation dose. The results for B_2 are plotted in Fig. 25, being similar those for B_1. The samples were implanted with Si to doses from $10^{12}\,\mathrm{cm}^{-2}$ (○) up to $5 \times 10^{13}\,\mathrm{cm}^{-2}$ and annealed at 680°C for times ranging from 10 min to 15 h. At the lowest implantation dose the defect concentration is below the sensitivity limit of the DLTS system after 1 h annealing, as indicated by the arrow in the figure. On increasing the dose the time needed for dissociation increases and a high defect concentration is present even after 15 h annealing. I-clusters are more stable with temperature for higher implantation doses. Since a higher dose implies a higher I supersaturation, the results suggest that bigger clusters are formed in this case.

4.3 Interstitial Cluster Characterization

As already pointed out, the I-clusters PL spectra present broad features peculiarly different from the extremely sharp lines observed for point-like defects. Also, the DLTS spectra of I-clusters exhibit much broader features than those expected for point-like defects. As an example, the DLTS spectrum of a sample implanted with Si to a dose of $1 \times 10^{12}\,\mathrm{cm}^{-2}$ and annealed at 700°C for 20 min is compared, in Fig. 26, with the simulation of a DLTS spectrum (dashed line). The simulation was performed assuming two point-like defects having the same activation energy and capture cross section experimentally

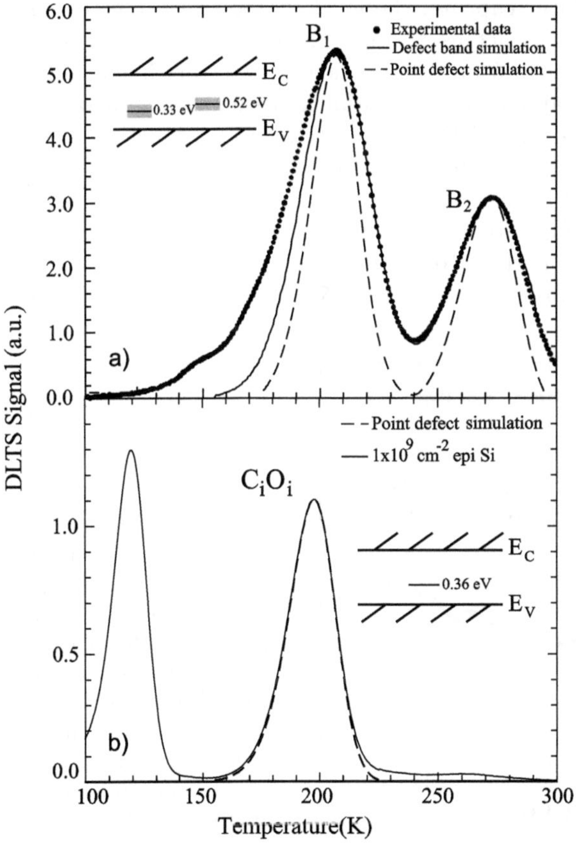

Fig. 26. DLTS spectra of p-type epi Si samples (**a**) implanted with 1.2-MeV Si, $1 \times 10^{12}\,\mathrm{cm}^{-2}$, and annealed at 700°C for 20 min (•), (**b**) as-implanted with 1.2-MeV Si, $1 \times 10^{9}\,\mathrm{cm}^{-2}$ (*solid line*). The experimental data were compared with the simulations obtained assuming point defects (*dashed lines*) and defect bands (*solid line* in **a**) signatures. The *inserts* show the deep levels schematic, levels in (**a**) are reproduced by small bands

determined for B-line I-clusters. The point-defect simulation is narrower than the experimental data, that are in good agreement with the simulation curve (solid line) obtained with the procedure described later in the text. The same procedure was used to reproduce the DLTS spectrum of the C_iO_i (solid line) plotted in Fig. 26b. The simulation (dashed line) was performed by assuming a single deep level in the gap having the same activation energy and capture cross section as the C_iO_i complex. A bandgap schematic with the point-like defect level is shown in the insert. One of the causes of the observed broadening could be the convolution of signatures quite close in the DLTS spectrum. To minimize the contribution arising from defects less stable with temperature, samples annealed at 700°C were monitored and the broadening measured.

Broadening of DLTS peaks may have different explanations. In order to rule out most of them, and to get more insight into the I-cluster nature, different measurements were performed. A peak broadening could result from a temperature dependence of the defect carrier capture kinetic, the ability of a defect to trap a carrier as a function of time. A simple point-like defect, such as C_iO_i, exhibits exponential capture kinetics since only one carrier can be trapped at each level. After capture, the level charge state changes and the center is no longer able to act as a trap. Therefore, the DLTS signal exponentially approaches a saturation value (when all traps are filled) as the pulse width increases. An example of exponential capture kinetic is shown as a solid line in Fig. 27, where the DLTS signal intensity as a function of the logarithm of the filling pulse duration is plotted.

The carrier capture kinetics related to an extended defect, e.g., the {113} electrical marker, exhibits fully logarithmic behavior, as shown in Fig. 27. This last measurement was performed by monitoring the {113} electrical signature at $E_V + 0.50\,\mathrm{eV}$ observed in DLTS. The electrical and structural

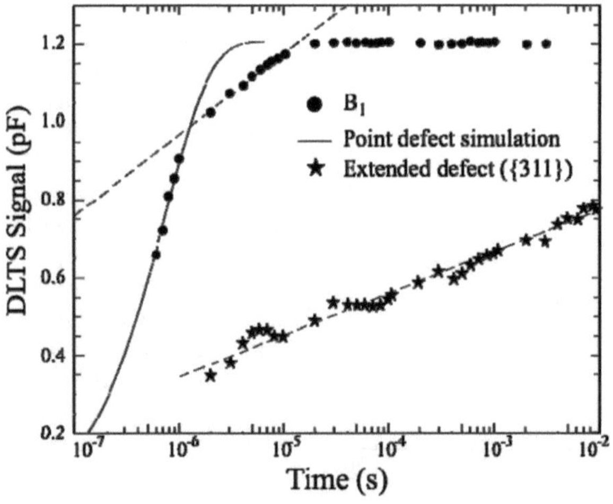

Fig. 27. Capture cross section measurements of B_1 (●) and {113} (⋆) electrical markers. The *solid line* is the expected capture cross section for a point-like defect. The two *dashed lines* are linear fits of the data. For B_1, only some of the points were fitted

characterization are discussed in Sect. 4. The straight line in the figure is a guide to the eye and shows that the capture kinetics is fully logarithmic in this case. This behavior is a direct consequence of the extended nature of the defect, which is able to accommodate more than one carrier. The capture of a single carrier by the defect does not produce a net change in its charge state, since the carrier is "delocalized" either in defects "surrounding" the extended defect surface, or in the extended defect core [114, 123]. As a consequence of the charge storage at the defect, a Coulombic repulsion significantly reduces the rate at which other carriers are trapped on the same defect. The DLTS signal increases logarithmically with the pulse width and does not reach saturation. Comparison of the capture cross section of the I-cluster signature, of point defects and of extended defects yields some hints about defect structure. The data refers to B_1 and are measured from an epi Si sample implanted with 1.2 MeV Si to a dose of $1 \times 10^{12}\,\mathrm{cm}^{-2}$ and annealed at 700°C for 20 min. Similar behaviors have been detected for all the samples implanted with 1×10^{12} Si/cm^2 annealed in the range 600–750°C. The I-cluster capture cross section has a behavior that is a combination of exponential and logarithmic trends. For short filling pulses, until 1×10^{-7} s, the measured trend can be well fitted by an exponential. An increase in the filling pulse duration results in a linear region well fitted by a straight line (dashed line in the figure), and corresponding to a logarithmic capture kinetics. When the pulse duration is above $\geq 10^{-5}$ s, a saturation in the DLTS signal is observed. This behavior is not due to a temperature dependence of the capture cross section, as demonstrated by capture kinetics measurements in the temperature range 190–230 K. The data are explained by assuming that, since the cluster is bigger than a point defect, it can accommodate more than one carrier during the trapping process, as an extended defect does. However, since they are smaller than an extended defect, there is an upper limit to the number of carriers that can be trapped by a single level, and the signal, eventually, saturates.

Another possible cause of peak broadening could be a field dependence of the trap emissivity [124]. However, the measurements show that the emissivities do not depend on the electric field, since the Poole–Frenkel effect was not detected. We believe that the broad DLTS signatures are related to a distribution of defect energy states in the Si bandgap rather than to field- or capture-kinetic-related effects on a single level. The broadening could be associated with a spread in the emission energies from the deep levels. Examples reported in the literature are highly dislocated [125, 126] Si and the EL2 level [127] in $GaAs_{1-x}P_x$. Omling et al. [125] analyzed broad DLTS peaks and extracted the deep-level real concentration and energy spread. The analysis is based on the assumption that the deep levels in the gap that produce the broad peaks are not associated with a single level with activation energy E_0 but to a narrow band of levels that form a Gaussian distribution having E_0 as the mean value and with a broadening described by a factor S. The data are fitted by varying the broadening factor S. Using this procedure, Ayres et al. [128] obtained an energy spread of ~35 meV for the DLTS signature of end-of-range extended defects in preamorphized crystalline Si.

An increase in the S factor causes the broadening and the intensity reduction of the measured DLTS peaks. For this reason an equivalent defect concentration, N_S, is defined as the concentration that would be obtained from the DLTS measurements if the broadening factor S were zero. Moreover, the incorporation of a defect bigger than a point-like one, causes a large reconstruction of the lattice around it. As a result, the number of levels introduced in the gap by each defect can be much lower than the number of defects (i.e., I) stored in it. This condition has been verified for extended defects [120, 125, 128] and it is reasonable to assume that it holds also for defect clusters. Therefore, the measured DLTS concentrations are related to the number of clusters rather than to the Is stored in them.

Finally, the data can be reproduced assuming a symmetric (Gaussian) broadening of the two energy levels. The results are plotted as a solid line in Fig. 26a. There is a good agreement between the simulation and the experimental data except at low temperatures, the difference arising from a contribution by an additional peak at ~150 K and/or to a nonperfectly Gaussian distribution of the defect states [128]. An S value of 19.5 meV and 17 meV was obtained for B_1 and B_2, respectively, to fit the data of Fig. 26a. Since the measured peak heights are lowered as a consequence of the broadening, the fitting procedure allowed us to also extract the correct defect concentration N_S.

It is interesting to note that the measured spectra are very similar, both in energy and temperature location to those determined by Ayres et al. [129] for the DLTS signature of end-of-range extended defects in preamorphized crystalline Si, although the energy spread is much smaller, ~18 eV, than that, ~35 meV. This difference, observing that the B-lines are signatures of small I-clusters rather than of extended defects, would suggest that the S parameter is sensitive either to the defect dimension, increasing as the size increases, or to the stress around it. This hypothesis is in perfect agreement with the PL data shown in the previous section. In fact, a broadening in the PL spectra and the shift of the lines related to point-like defects (e.g., the W line) are associated to a lattice strain due to the defect presence.

The data so far reported clearly indicate that similar structures are found in regions having a variable local environment and/or a larger stress. The same structure can exhibit slightly different carrier emission energies either due to the lattice stress surrounding the defect, or as a consequence of the interaction between carriers trapped on clusters close to each other. Two possible explanations fit our data. Upon annealing, clusters less stable with temperature dissociate, releasing the Is stored. An average decrease of the number of clusters (N_S) is expected. At the same time, the released Is will join clusters more stable with temperature, increasing the stress around them and the associated S value. The other possible scenario is that, as the annealing proceeds, small clusters that survive dissociation can coalesce and come closer to each other, as suggested by Monte Carlo simulations carried out by La Magna et al. [129, 130]. The presence of a larger number of clusters in a smaller area produces a higher local stress. The S variation could be due either to a variation in the local stress around the defect or to the interaction of carriers trapped at different sites very close to each other. This S dependence from the stress and/or trapped carrier interactions would be responsible for the high values reported for extended defects [123].

Measurements on epi Si samples implanted at 145 keV Si to a fluence of $2 \times 10^{13}\,\mathrm{cm}^{-2}$ and annealed at 680°C for 4 h, offer further insights into the physical meaning of the S values. The reduced range of 145-keV Si ions lie within the zero-bias depletion region, and only the tails of the damage distribution can be monitored by DLTS. In the insert of Fig. 28 the depth concentration profiles of B_1 and B_2 are compared to the extra implanted-ion depth profile (solid line, divided by a factor 100) simulated with MARLOWE.

The comparison clearly shows that no long-range Is migration is involved in the defect-cluster formation and regions of the sample at different depths experience different Is supersaturation. At the same time, a larger stress and a closer spatial defect distribution are expected in the higher supersaturation region. If S is related to one of the two effects a different peak broadening in different regions is expected. DLTS measurements were performed in two

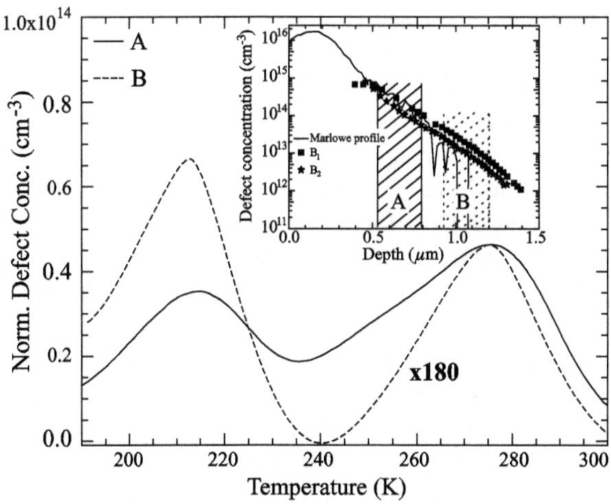

Fig. 28. DLTS spectra of p-type epi Si implanted with 145-keV Si, $2 \times 10^{13}\,\mathrm{cm}^{-2}$, annealed at 680°C for 4 h. Spectra were measured in region A (*solid line*, $V_{\mathrm{Rev}} = 3\,\mathrm{V}$, $V_{\mathrm{Fill}} = 0\,\mathrm{V}$) and region B (*dashed line*, $V_{\mathrm{Rev}} = 7\,\mathrm{V}$, $V_{\mathrm{Fill}} = 4\,\mathrm{V}$). This last spectrum was multiplied by a factor of 900. In the insert the depth concentration profiles for B_1 (■) and B_2 (⋆) are plotted. The *solid line* is the simulation of the implanted ion depth distribution according to MARLOWE

different regions labeled as A (solid lines area in the insert, measured using $V_{\mathrm{Rev}} = +3\,\mathrm{V}$ and $V_{\mathrm{Fill}} = 0\,\mathrm{V}$) and B (dashed lines area in the insert, measured using $V_{\mathrm{Rev}} = +7\,\mathrm{V}$ and $V_{\mathrm{Fill}} = +4\,\mathrm{V}$). The resulting DLTS spectra are plotted in the figure as a solid and a dashed line for regions A and B, respectively. Once again, the same signatures B_1 and B_2 are present in both cases, but the DLTS spectrum of region B has been smoothed and multiplied by a factor of 900 to allow comparison. The calculated values for S are $\sim$27 meV in region A and $\sim$5 meV in region B. Since region A experienced a higher I supersaturation during implantation than B, both a larger stress and a denser clusters distribution are expected in this region. Thus, the S dependence may result from both effects. The cluster density varies with depth due to the I supersaturation distribution during implant, hence both the stress of the lattice around the defects and the defect dimensions are larger in the region that experienced the highest I supersaturation.

On the basis of these results, the S data should be regarded as an average of the cluster distribution, or of the stress distribution. However, the contribution of the clusters in the tail of the ion distribution is expected to be very small due to their low concentration. In addition, their contribution will decrease at longer annealing times.

4.4 Extended Defect Characterization

At high implantation doses, $\geq 5 \times 10^{13}\,\mathrm{Si/cm^2}$, and annealing temperatures $\geq 680°\mathrm{C}$, a new regime is entered, characterized by the formation of extended defects, and {113} rod-like defects [26] (*extended defect region*, Fig. 2), which have been widely characterized by TEM analysis. In order to compare the DLTS and PL results with the data known from the literature, the analyses were performed in this dose–temperature range.

A new electrical signature is found in the residual damage spectrum measured by DLTS when {113} extended defects form. Si was implanted to doses in the range $1\text{–}5 \times 10^{13}\,\mathrm{cm^{-2}}$, on p-type Czochralski Si substrates, heavily doped ($\sim 5 \times 10^{16}\,\mathrm{B/cm^3}$). The doping level was chosen in order to have a doping concentration higher than the defect concentration at each implantation dose. All the samples were annealed at 680°C for times ranging from 1 h to 15 h. The DLTS residual damage spectra after 1 h annealing are plotted in Fig. 29. The I-cluster concentration increases more than linearly with increasing ion-implantation dose from $1 \times 10^{13}\,\mathrm{Si/cm^2}$ (solid line) to $2 \times 10^{13}\,\mathrm{Si/cm^2}$ (dashed line). The first spectrum has been multiplied by a factor 6 to allow comparison. A further increase in the ion-implantation dose to $5 \times 10^{13}\,\mathrm{cm^{-2}}$ (dot-dashed line), does not cause a corresponding increase in the I-clusters concentration as observed for lower doses, but their signals abruptly decrease. Beside the I-clusters signatures, still present at high doses, a shoulder, marked with an arrow in the figure, is visible in the high-temperature part of the spectrum [108, 109]. Deconvolution of the spectral peaks yields a tentative energy

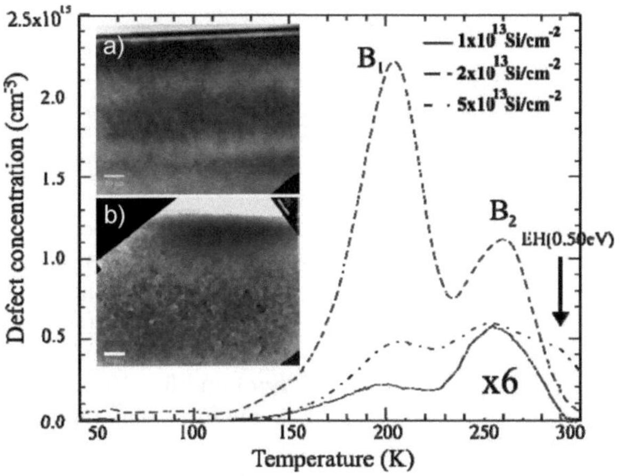

Fig. 29. DLTS spectra of residual damage after 145-keV Si ion implantation at $1 \times 10^{13}\,\mathrm{cm}^{-2}$ (*solid line*), $2 \times 10^{13}\,\mathrm{cm}^{-2}$ (*dashed line*) and $5 \times 10^{13}\,\mathrm{cm}^{-2}$ (*dot-dashed line*) on p-type Si annealed at 680°C 1 h. In the *insert* TEM, dark-field, plan-view of the samples implanted at $2 \times 10^{13}\,\mathrm{Si/cm}^2$ (*upper image*) and $5 \times 10^{13}\,\mathrm{Si/cm}^2$

for thermally stimulated carrier emission of $E_\mathrm{V} + 0.50\,\mathrm{eV}$ (EH). Its logarithmic capture kinetic (Fig. 27) suggests that this DLTS peak is the electrical marker of an extended defect. In order to confirm the relation between the DLTS marker and extended-defect, plan-view, dark-field, TEM micrographs were taken on this sample, and one of them is shown in the inset of Fig. 29 (lower image). It clearly shows that {113} defects are present in the samples where the EH (0.50 eV) signature is observed by DLTS. Moreover, the {113} defects are not visible in the lower-dose samples, which display only the signals related to I-clusters, as shown in the upper image of Fig. 29. This suggests that the shoulder in the DLTS spectra is a {113} defect signature. Support to this idea is achieved by comparing the formation and dissociation kinetics of the DLTS signature with TEM observations.

Detailed DLTS studies of EH evolution as a function of the annealing time compared to TEM analyses on the same samples show that the DLTS signal is a {113} defects electrical signature [106]. The Si ion-implantation regime for doses ranging from $5 \times 10^{12}\,\mathrm{Si/cm}^2$ to $1 \times 10^{14}\,\mathrm{Si/cm}^2$ has been extensively studied by TEM. The damage evolves into {113} defects after heat treatments of 670–815°C [19]. The {113} concentration decreases while their size increases in an Ostwald ripening process, accompanied by the injection of Is. The {113} density seen in plan-view TEM data was compared with the DLTS measurement of the EH (0.50 eV) concentration for the same samples. The data are consistent with a linkage between the extended defect signals in DLTS and the {113} defects measured by TEM and provide the first observations of an electrical signature related to the {113} presence. Detailed TEM analysis of {113} defects confirmed that, as previously reported [115], extended defects are formed at lower doses when the impurity content is low. I-cluster formation is inhibited by the presence of extended defects. This is a

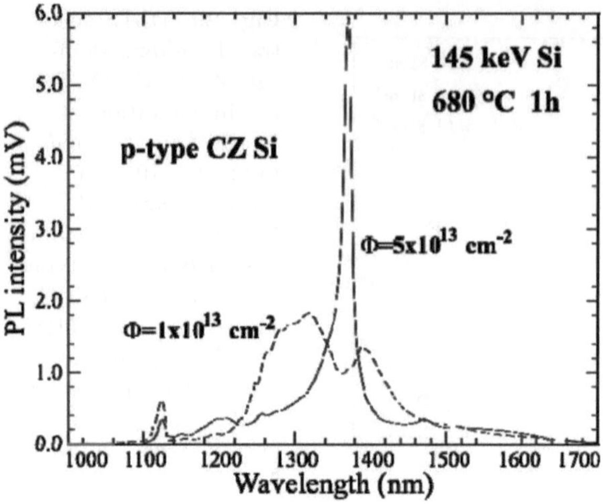

Fig. 30. PL spectra taken at 17 K on n-type Si samples implanted with 145-keV Si to doses of $1 \times 10^{13}\,\mathrm{cm}^{-2}$ (*dashed line*) and $5 \times 10^{13}\,\mathrm{cm}^{-2}$ (*solid line*). The samples were annealed at 680°C for 1 h

further confirmation of the B lines I-cluster nature, since they compete with {113} in storing the I excess.

When the annealing temperature is increased to 680°C major modifications also occur in the residual damage optical properties. The PL spectra of p-type CZ Si samples implanted with Si to a dose of $1 \times 10^{13}\,\mathrm{cm}^{-2}$ (dashed line) and $5 \times 10^{13}\,\mathrm{cm}^{-2}$ (solid line) and annealed at 680°C 1 h are plotted in Fig. 30. These samples are the same as these used to record the DLTS spectra shown in Fig. 29. The PL spectrum of the lower-dose sample exhibits the typical I-clusters spectrum, already observed, while a totally different scenario emerges from the PL spectrum of the sample containing {113}: a sharp peak at 1376 nm (0.9007 eV) dominates the spectra. TEM analysis on all of the samples implanted at doses $\geq 5 \times 10^{13}\,\mathrm{cm}^{-2}$ revealed the presence of {113} planar defects [111, 114]. A one-to-one correlation exists between the observation of the PL line at 1376 and the presence of {113}, allowing us to associate this line to an optical transition occurring at or close to {113} defects.

4.5 Transition from Defect Clusters to Extended Defects

The data reported in the previous section, as well as many literature data, show that ion implantation to doses, $\geq 5 \times 10^{13}\,\mathrm{Si/cm}^2$, and annealing temperatures $\geq$680°C causes extended-defect formation, in particular {113} rod-like defects, widely characterized by TEM analysis. We have so far shown that DLTS and PL measurements are really powerful in exploring the extended-defect region.

In this section, we will show that both DLTS and PL measurements can be used to monitor the transition from I-cluster to extended defects. When the annealing temperature is above 680°C a new regime is entered, characterized by major modifications in the optical and structural properties of I defects.

Following the same arguments used in the previous sections, it is reasonable to assume that the sample impurity content plays a relevant role in the extended-defect formation kinetics. To this purpose we used CZ Si samples with a low impurity concentration, [C] $\sim 1 \times 10^{16}\,\mathrm{cm}^{-3}$ and $[\mathrm{O}] = 7 \times 10^{17}\,\mathrm{cm}^{-3}$. At the lowest dose ($1 \times 10^{12}\,\mathrm{Si/cm}^2$, dotted line) the damage is fully recovered and the spectrum only reveals the 1121 nm signature due to Si band-edge phonon-assisted recombination [16]. However, at doses $\geq 10^{13}\mathrm{cm}^{-2}$ in samples with a low C concentration the sharp peak at 1376 nm (0.9007 eV) typical of {113} defects dominates the spectra. The width of this peak slightly increases with dose and a broad band centered at 1550 nm is developed at the highest dose ($5 \times 10^{13}\,\mathrm{cm}^{-2}$). The TEM analysis on all of these samples revealed the {113} defects presence [111]. A one-to-one correlation exists between the observation of the PL line at 1376 and the presence of {113} defects in the sample, and this line has been associated with optical transition occurring at or close to {113} defects. This result suggests PL as a promising technique to fully characterize the {113} defect formation. In fact, the {113} presence in DLTS is determined by a level at $E_V + 0.50\,\mathrm{eV}$ in a spectrum in which the I-cluster electrical markers are still present, that lie very close to the {113} peak. When {113} defects form the optical properties of the sample are totally modified and their presence is clearly visible by the presence of the 1376 nm line.

The results indicate the presence of a threshold dose for the formation of {113} defects. Finally, measurements we performed on samples with a carbon content $\geq 5 \times 10^{13}\,\mathrm{cm}^{-3}$ showed a threshold dose for extended defect formation $> 2 \times 10^{13}\,\mathrm{Si/cm}^2$ [131]. Si with low C ($\sim 10^{16}\,\mathrm{cm}^{-3}$) exhibits a much lower threshold, certainly lower than $1 \times 10^{13}\,\mathrm{cm}^{-2}$. The formation of {113} defects at relatively low dose, $1 \times 10^{13}\,\mathrm{Si/cm}^2$, in these samples is in agreement with the literature data, confirming the presence of a lower threshold dose in highly pure materials. In fact, it is known [26] that the threshold dose for {113} formation reduces with the impurity concentration. As previously mentioned, this has been attributed to the fact that C stores Is, thus preventing the formation of big self-I clusters and, eventually, extended defects. If the C concentration in the sample is strongly reduced the threshold dose for {113} formation is expected to change.

We characterized the extended-defect formation as a function of the annealing temperature and time. To this purpose samples implanted to doses of $1 \times 10^{12}\,\mathrm{cm}^{-2}$ (dotted line), $1 \times 10^{13}\,\mathrm{cm}^{-2}$ (solid line), $2 \times 10^{13}\,\mathrm{cm}^{-2}$ (dashed line), $5 \times 10^{13}\,\mathrm{cm}^{-2}$ (dot-dashed line) were annealed at 600°C for times up to 15 h. The PL spectra measured at 17 K on samples annealed at 600°C for 4 h are compared in Fig. 31. It is noteworthy to observe that although the spectra intensity increases with ion dose, the major features remain unchanged. The intensity of these lines is reduced when the dose is increased: this suggests

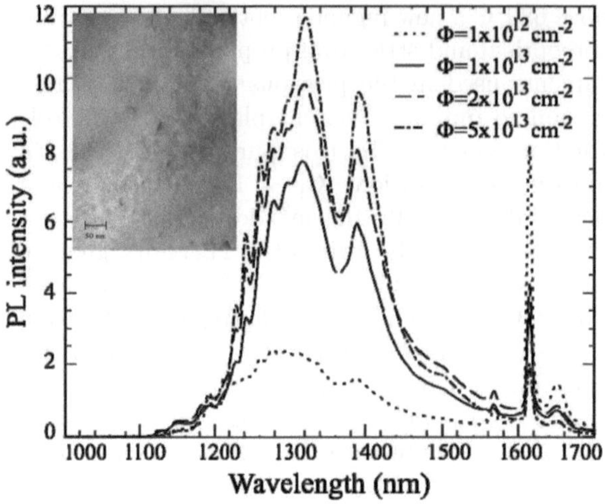

Fig. 31. PL spectra at 17 K of n-type Si samples implanted with 1.2-MeV Si to doses of $1 \times 10^{12}\,\mathrm{cm}^{-2}$ (*dotted line*), $1 \times 10^{13}\,\mathrm{cm}^{-2}$ (*solid line*), $2 \times 10^{13}\,\mathrm{cm}^{-2}$ (*dashed line*) and $5 \times 10^{13}\,\mathrm{cm}^{-2}$ (*dash-dot-dashed line*). All samples were annealed at 680°C for 1 h

that the centers responsible for these features compete with those responsible of the 1200–1400 nm features in trapping and recombine excess carriers.

TEM analyses performed on these samples revealed that no extended defects are formed in these samples even at the highest dose ($5 \times 10^{13}\,\mathrm{cm}^{-2}$). The cross section analysis on this sample is shown in Fig. 32: only a weak contrast at the projected range (at ~1.35 μm from the surface, at the end-of-range of 1.2-MeV Si) is observed, demonstrating that a heavily damaged region, probably consisting of small defects with a large strain, is present in the sample. DLTS measurements (not shown) only reveal the two well-defined signatures characteristic of the Si I-clusters. They increase in concentration when the ion dose is increased.

PL measurements performed on samples annealed at 600°C for times of 30 h do not show {113} defects, thus confirming the presence of a threshold temperature. For $T \leq 600$°C and independently of ion dose, {113} defects are not formed at all: the I stored in the clusters are eventually released and anneal out, probably at the surface.

Finally, we analyzed in more detail the PL signatures in the region 1200–1450 nm, performing high-resolution (0.5 nm) PL analysis on a sample implanted to a dose of $5 \times 10^{13}\,\mathrm{cm}^{-2}$ and annealed at 600°C for 4 h. In Fig. 33 the high-resolution spectrum is plotted as a solid line, while the dashed line spectrum is that taken at low resolution (3.2 nm). The set of sharp peaks in the range 1340–1410 nm are due to the stretching modes of the OH molecules [124, 132] in the air and hence do not reflect any feature of the sample. On the other hand, the high-resolution spectrum reveals that the sharp lines in the 1200–1280 nm presents well-defined structures. All of them are associated with point-like defect-defect and defect–impurity complexes formed as a consequence of ion-beam irradiation. In particular, we can unambiguously identify: (a) a line W′ at 1233 nm (1.0048 eV) that is a perturbed form of the

Fig. 32. TEM cross section of n-type CZ Si sample implanted with 1.2-MeV Si to a dose of 5×10^{13} Si/cm^2 and annealed at 600°C for 4 h

W line (1218 nm, 1.018 eV) observed in regions with high strain [16, 17]. The two features at 1320 nm (0.94 eV) and 1390 nm (0.89 eV) are unchanged in the high-resolution spectra; they are indeed broad luminescence bands and do not arise from the convolution of narrower peaks. Broad features in PL spectra have been usually observed [112] in samples where extended defects are observed and were associated with the quantum confinement of carriers in regions with the high-strain region surrounding the defects. Since no extended defects are detected in our sample, it is tempting to associate these signatures with the carriers recombination in the strained region surrounding the small I-clusters embedded in the Si matrix. The presence of this strain is also evident in the TEM picture of Fig. 32. In summary, the results suggest that the material contains I-rich regions where I-type point-defect complexes (such as C–O and C_iC_s), small I-complexes and I-clusters are present. This I excess is the direct consequence of the extra implanted ion introduced during implants.

The results reported indicate that major transformations in the optical, electrical and structural properties of Si occur when {113} are formed. It is

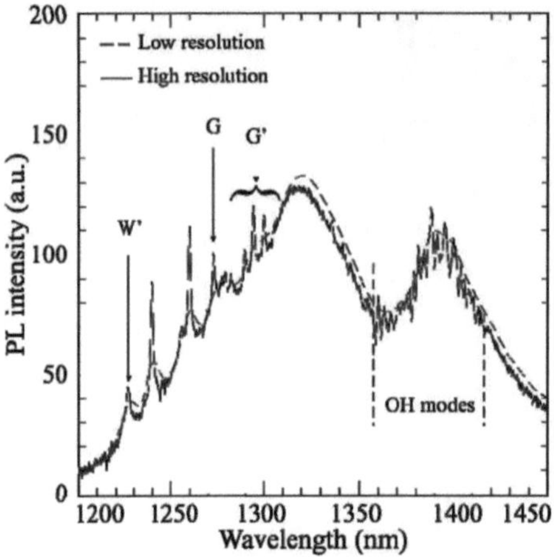

Fig. 33. High-resolution PL spectrum of a n-type CZ Si sample implanted with 1.2-MeV Si to a dose of 5×10^{13} Si/cm^2 and annealed at 600°C 4 h

therefore tempting to speculate that the early stage of I nucleation produces small I-clusters whose structure differs significantly from that of the planar {113} defects. A transition between the two structures requires some morphological transformation. At low annealing temperatures, the probability to overcome this nucleation barrier is small and most of the I will remain stored in the I clusters until they are annealed out: these clusters dominate the optical, electrical and structural properties. On the other hand, when temperature is increased and the dose is high enough to provide large I-clusters, the probability of transition is very large and {113} defects are soon formed. These results show that the I-clusters evolution into extended defects cannot occur through a simple Ostwald ripening mechanism, in perfect agreement with the Monte Carlo simulations on a lattice that will be discussed in the next section. A structural transformation of the I-clusters has to occur before they can grow into extended defects. Only after this transformation occurs does the "traditional" OR take place and the extended-defect regime is entered. Our results also fit with the explanation proposed by Cowern [28]. In fact, they assume a very stable I-clusters configuration (with about 8 I) and a large potential barrier before the OR can take place. We believe that the "magic number" they find arises from the fact that at a certain size the clusters must undergo a structural transformation in order to grow bigger.

4.6 Simulation of Defect Evolution

The Technological Roadmap of the Semiconductor Industry indicates that technology computed-aided design (TCAD) tools will allow, in the next decade, a 30% saving of the overall costs dedicated to the research and development of future devices [133]. However, the process simulators actually used are based on continuum models, that will make them soon obsolete to the process development of the metal-oxide-semiconductor (MOS) leading-edge technology. Indeed, continuum models fail at feature sizes below 100 nm (typical of the next-generation MOS devices). Their manufacturing requires a material modification control pushed to the atomic level, whilst the nanostructural evolution occurs in a far from equilibrium regimes. Therefore, the future process simulators in TCAD tools should be based on atomistic simulators. To this aim many attempts have been devoted in the last years in projecting simulation tools able to be used in the submicrometer regime [130, 134].

Moreover, a definitive understanding of the experimental findings in terms of micro-structural evolution, as observed before, is still lacking. Indeed, the studies on the stability of I-type complexes performed using quantum-mechanical calculation (QMC) [135, 136] are not able per se to predict the behavior of any evolving system. Statistical methods could satisfy the demand for an atomic-level investigation of system evolution, if the relation between structures and stability, derived by QMC, were correctly recovered in such codes. Lattice kinetic Monte Carlo (LKMC) constitutes a suitable framework to recover the statics derived by QMC in studying defect evolution. In fact, this innovative simulation approach has been successfully used to follow V [137, 138] and I [129] systems in c-Si. In this code defect–defect effective interaction models are used instead of a fixed form for the aggregate binding energy. The mechanism driving the ripening of V-clusters in Si was studied by means of lattice kinetic Monte Carlo simulations using different binding models. A modified Ising model, also taking into account second-neighbor interaction, the data of V-cluster energetics, results in quantitative agreement with tight-binding molecular-dynamics calculations [135]. They show that, when this model is used, the ripening process is also driven by the migration of small V-clusters, and not solely by free V. This produces a faster vacancy agglomeration and a strong modification of the cluster-size distribution.

LKMC was also applied to investigate the evolution of an I supersaturation in c-Si [129]. The model allowed one to map the results of QMC using local interactions between defects that evolve over a superlattice matrix including both regular and tetrahedral sites. Using the knowledge of I-type defects structure and stability acquired using tight-binding molecular-dynamics (TBMD) [135] and first-principles calculations in the local density approximation (LDA) [136]. In stable small I-type defects ($n = 1, 2, 3$) a lattice atom is strongly displaced from its reference site so that it cannot be distinguished from the added atoms in the defect structure, e.g., the dumbbell ($n = 1$) configuration. The defective atoms sit in four quasi-equivalent positions at small distances from the four tetrahedral sites nearest neighbor to the reference lattice site. The atom coordination z is $z = 4, 5, 6$ when $n = 1, 2, 3$, respectively, with $z = 6$ an upper limit for ions involved in the I-type defects in Si. A comparative analysis of available QMC on I-type defects reveals substantial structural differences between small stable aggregates and extended defects (e.g., {113} rod-like defects). The simplified assumption that precursors have the same morphology as extended defects seems to be invalidated by the large positive difference between the formation energy of small-size I-chain in {113} like configuration and I-clusters. Indeed, according to, e.g., TBMD calculations, a two-member chain has $E_\mathrm{f} = 4.8\,\mathrm{eV}$ formation energy per I, whilst the same quantity is $E_\mathrm{f} = 2.4\,\mathrm{eV}$ for a two-member compact cluster [136, 139]. However, nominally infinite I-chains are more stable than I-cluster since its estimated formation energy per I is $E_\mathrm{f} = 1.7\,\mathrm{eV}$ [139], while in the case of chain-like structures formed by I-clusters an average value of $E_\mathrm{f} \sim 2.2\,\mathrm{eV}$ has been calculated [136].

Compact cluster energetics is modeled by the following two-body local defective atoms interaction: a bond forms, with binding energy E_b, between two ions sitting in tetrahedral sites next neighbor to the same empty regular site. A defective atom can also bind with more defective atoms, however, the overcoordination energetic cost has to be considered. Then, the two parameters E_5 and E_6 must satisfy the constraint $E_6 > E_5 > 0$. Finally, a single defect cannot belong to a chain-like defect when it is in an overcoordinated state (i.e., when it forms a E_b bond). The system evolves according to two elementary displacements of Si atoms: from a regular site to an empty tetrahedral site (or vice versa); from an occupied tetrahedral site to an empty one. If no more defective atoms interact with the moving one, the former movement leads to the formation of an unstable $\mathrm{I_T}$–V complex with an activation energy E_IV, whilst the latter takes place between two degenerate configurations. There is a one-to-one correspondence between the six model parameters $(E_\mathrm{IV}, E_\mathrm{b}, E_5, E_6, E_\mathrm{ex}, E_\mathrm{in})$ and six key parameters, extracted by QMC, since they can be related through the following close relations [129]:

$$
\begin{aligned}
&E_\mathrm{IV} = E_\mathrm{IV}^\mathrm{MD}; \qquad E_\mathrm{b} = E_\mathrm{f}(\mathrm{I_T}) - E_\mathrm{f}(\mathrm{I_D}) + E_\mathrm{IV};\\
&E_5 = (1/3)\left[E_\mathrm{f}(\mathrm{I_2}) - 2E_\mathrm{f}(\mathrm{I_T}) - E_\mathrm{IV} + 3E_\mathrm{b}\right];\\
&E_6 = (1/4)\left[E_\mathrm{f}(I_3) - 3E_\mathrm{f}(\mathrm{I_T}) - E_\mathrm{IV} + 6E_\mathrm{b}\right];\\
&E_\mathrm{ex} = E_\mathrm{f}^\mathrm{ch}(n=2)/2 - E_\mathrm{f}(\mathrm{I_T});\\
&E_\mathrm{in} = E_\mathrm{f}^\mathrm{ch}(n\ \mathrm{large})/n - E_\mathrm{f}(\mathrm{I_T}),
\end{aligned}
$$

where key parameters in the second members are the formation energies of the following complexes: $\mathrm{I_T}$–V complex ($E_\mathrm{IV}^\mathrm{MD}$), the compact I defects with $n = 1, 2, 3$ ($E_\mathrm{f}(\mathrm{I_D}), E_\mathrm{f}(\mathrm{I_2}), E_\mathrm{f}(\mathrm{I_3})$, respectively), the I chain with $n = 2$ $E_\mathrm{f}^\mathrm{ch}(n = 2)$, the long I chain $E_\mathrm{f}^\mathrm{ch}(n$ large). Using the key parameters calculated by TBMD or LDA [123–126] two set of LKMC parameters are derived:

$$
\begin{aligned}
&E_\mathrm{IV} = 4.0; \quad E_\mathrm{b} = 4.6; \quad E_5 = 2.0; \quad E_6 = 4.3;\\
&E_\mathrm{ex} = 0.4; \quad E_\mathrm{in} = -2.7 \quad (\mathrm{TBMD})\\
&E_\mathrm{IV} = 4.0; \quad E_\mathrm{b} = 4.5; \quad E_5 = 2.2; \quad E_6 = 4.2;\\
&E_\mathrm{ex} = -0.9; \quad E_\mathrm{in} = -2.8 \quad (\mathrm{LDA}).
\end{aligned}
$$

All values are in eV units. Note that the two QMC schemes give similar estimates of LKMC parameters apart from the E_{ex} value.

The simulations show [129] that small-size compact clusters form below the microsecond timescale. During this stage the spatial distribution of compact clusters may change, if it is allowed by the thermal budget, whilst the aggregates maintain the same structural identity. In order to understand this behavior it must be assumed that an energy gain follows the inclusion of an approaching free I in a structure containing one or more adjacent compact clusters. The system needs extra energy, during the capture process, to achieve a germinal chain-like structure. As a consequence, the system remains frozen in a metastable disordered phase (t_{clust}). This phase is characterized by a real evolution (occurrence of dissolution/capture processes) but its effective consequence is the formation of a droplet consisting of closer compact clusters, which can be considered as a local high density of added atoms stored in these substructures. Droplet formation promotes the transition to chain-like defects. Indeed, unstable chain-like nuclei, which form inside a droplet due to thermal fluctuations, can soon stabilize capturing ions in the neighborhood, abruptly reaching a configuration energetically favored with respect to an equal number of I stored in compact clusters. At the microstructural level, this transition is driven by the recovery of four-fold coordination for I in the chain, and by back displacements to regular sites for the lattice ions sharing, with added ones, the compact cluster defect structure. These transitions lead to an energy gain E_{IV}.

A useful parameter to monitor this transition is the rate $s = t_{free}/t_{sample}$ between the time, t_{free}, the system spends to perform free I diffusive transitions, over a suitable sampling time t_{sample}. s is proportional to the supersaturation level maintained by the aggregates. In Fig. 34 $s(t)$ is shown as a function of time (full circles and line). Two abrupt changes occur, related to the compact cluster formation and to the disordered–ordered transition, respectively. At this temperature, compact clusters maintain a supersaturation level more than three orders magnitude larger than chain-like defects. No midlevel plateau of s is observed when compact cluster energetics is not considered (open circles and dots), while this plateau extends indefinitely if chain-like defects energetics is not considered (open squares and large dashes).

These results suggest an appealing qualitative explanation of available experimental findings on the evolution of I-type defects in c-Si. The PL experimental results obtained after Si ion implantation ($2 \times 10^{13}\,\mathrm{cm}^{-2}$) and annealing at 400°C (dashed line), 600°C (dot-dashed line) and 680°C (solid line) (a), are compared with the simulations obtained after ~0.1 s of simulated evolution at the same temperature: 400°C (b), 600°C (c), and 680°C (d) in Fig. 35.

After annealing at low temperature (400°C) the PL spectra show typical features of small I-type defects (see above). Consistently, simulations show a state blocked after the compact clusters nucleation. Dissolution events occur

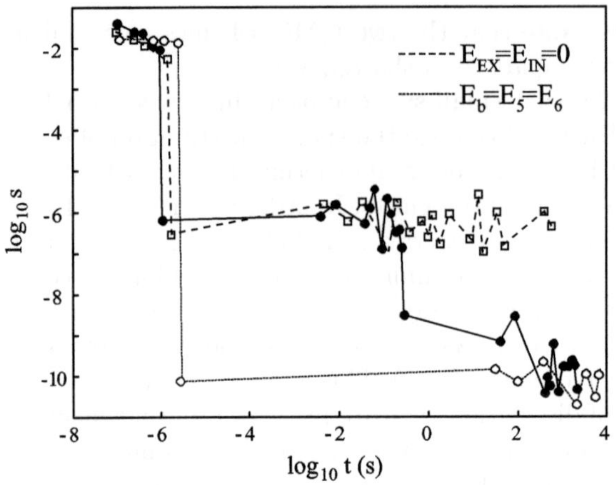

Fig. 34. s parameter plotted as function of time for a system with $N_{\mathrm{sites}} = 32{,}768$, $N_{\mathrm{added}} = 16$ and $T = 650°\mathrm{C}$ using LDA parameters with $E_{\mathrm{ex}} = -0.1\,\mathrm{eV}$ (*solid line* and *circles*), using LDA parameter with $E_{\mathrm{ex}} = E_{\mathrm{in}} = 0\,\mathrm{eV}$ (*open circles* and *dashes*), using $E_{\mathrm{ex}} = -0.1\,\mathrm{eV}$, $E_{\mathrm{in}} = -2.8\mathrm{eV}$, $E_{\mathrm{b}} = E_6 = E_5 = 0$ (*open squares* and *points*). In this last case diffusivity is set $D = 0.003\exp(0.5/kT)\,\mathrm{cm}^2/\mathrm{s}$

over a large timescale. The optical signatures change when the annealing temperature is increased (600°C). This change has been related to the formation of larger I aggregates that induce a large stress field. The simulation shows droplet formation Finally, the PL spectra at 680°C arises from {113} defects emission. In this case, chain-like defects dominate the simulated evolution of this ideal I system.

5 Conclusion

In this chapter we reviewed the fundamental properties of point defects (I and V), as well as their interactions with impurities and dopant atoms in c-Si. The structure of RT-stable defect complexes and their annealing kinetics was found to depend on the sample impurity content. RT-stable defect structures produced by electron irradiation and ion implantation in Si and their low-temperature annealing kinetics (≤300°C) are the same. Moreover, impurities play a major role in determining the final depth distribution of V-type and I-type defects. These results open a new area of study and strongly suggest that caution is imperative in the comparison of new experiments with previously published data whenever the impurity concentration in the sample is not known or strongly differs from the value reported for the substrate used in the literature data.

The study of defect evolution upon annealing provided several insights into the mechanisms of damage accumulation the evolution of I-type defects from simple pairs to extended {113} defects in c-Si. Defect recombination

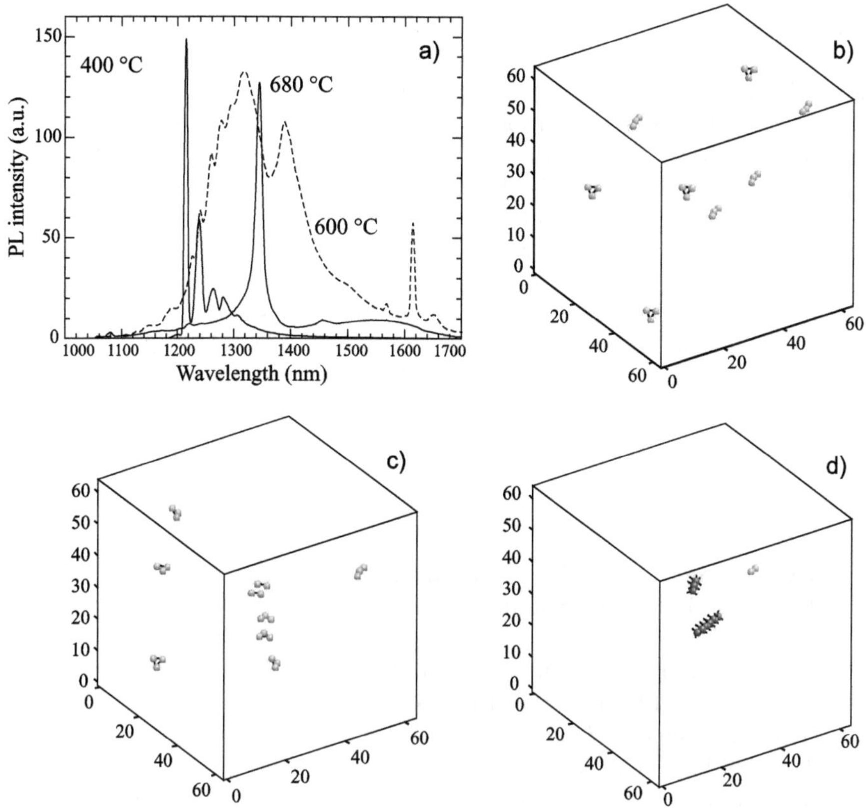

Fig. 35. (**a**) PL spectra of 1.2-MeV Si implanted samples to a dose of $2 \times 10^{13}\,\mathrm{cm}^{-2}$, and annealed at 400°C (*dashed line*), 600°C (*dot-dashed line*), 680°C (*solid line*). Snapshots of I-aggregates obtained by LKMC simulations ($t \sim 0.1\,\mathrm{s}$) for a closed system with $N_{\mathrm{sites}} = 32{,}768$, $N_{\mathrm{added}} = 16$ at: (**b**) 400°C, (**c**) 600°C, (**d**) 680°C

occurs preferentially in the bulk. Evidences of I-cluster formation, evolution and annealing have been provided. Implantation to doses $\leq 10^{11}\,\mathrm{Si/cm^2}$ results in the formation of secondary-order point-defect dissociating at temperatures ≤550°C. Increasing the implantation dose, 10^{12}–$10^{13}\,\mathrm{Si/cm^2}$ and annealing temperature, 550–700°C, I-clusters are formed. Their signatures are detectable in DLTS and PL. I-clusters are present in a dose and temperature regime at which TED of B occurs without the formation of extended defects. Their dissociation energy varies as a function of the implantation dose, but the measured dissociation energy value for I-clusters at $1 \times 10^{12}\,\mathrm{Si/cm^2}$, ∼2.3 eV, is in good agreement with those observed for low-dose TED.

High implantation doses, $\geq 5 \times 10^{13}\,\mathrm{Si/cm^2}$, and annealing temperatures, ≥680°C 1 h, cause the formation of extended defects, such as {113}, observed by TEM and correlated to an electrical signature, at $E_{\mathrm{V}} + 0.50\,\mathrm{eV}$, and an optical signature at 1376 nm. Extended defects compete with small clusters in storing the I-excess. The transition from I-clusters to {113} defects was explored determining a threshold dose (1×10^{13} cm^{-2}) and a threshold temperature (600°C) for {113} formation. A structural transformation accompanies the transition from I cluster to {113} defects.

Finally, new insights into the dynamics of the I agglomeration process were found using LKMC simulations. I agglomeration can follow two pathways having a completely different structural identity. The competition of these two classes of structures leads to a peculiar kinetic behavior, since a metastable disordered phase occurs after the early annealing times and characterizes the status of the system over a period It critically depends on the annealing temperature and the I supersaturation. After this time, the transition to the ordered phase dominated by the structures rod-like defect takes place.

References

1. J.W. Corbett, J.C. Bourgoin, in *Point Defects in Solids*, ed. by J.H. Crawford, L.M. Slifkin (Plenum, New York, 1975), p. 1 147, 149
2. P.M. Fahey, P.B. Griffin, J.D. Plummer, Rev. Mod. Phys. **61**(2), 289 (1989) 147, 148, 149, 153, 156, 160
3. H.G. Van Bueren, *Imperfections in Crystals* (North-Holland, Amsterdam, 1961) 147, 149
4. R. Car, P.J. Kelly, A. Oshiyama, S.T. Pantelides, in *Thirteenth International Conference on Defects in Semiconductors*, ed. by L.C. Kimerling, J.M. Parsey Jr. (AIME, Warrendale, 1985), p. 269 148, 149, 160
5. G.D. Watkins, in *Radiation Damage in Semiconductors* (Dunod, Paris, 1964), p. 97 148, 149, 150, 154, 156, 160, 172
6. B.G. Svensson, C. Jagadish, A. Hallen, J. Lalita, Nucl. Instrum. Methods B **106**, 183 (1995) 148, 149, 154, 162, 164, 165
7. L. Rubin, Semicond. Int., April 1997 148, 149
8. L.C. Kimerling, in *Radiation Effects in Semiconductors*, ed. by N.B. Urli and J.M. Corbett, Inst. of Phys. Conf. Ser., vol. 31 (London, 1977), p. 221. And references therein 149, 150, 151, 154, 156, 157, 158, 164, 172, 173
9. G.D. Watkins, in *Electronic Structure and Properties of Semiconductors*, ed. by W. Schröter, R.W. Cahn, P. Haasen, E.J. Kramer. Material Science and Technology, vol. 407 (VHC, Weinheim, 1991) 149, 150, 154
10. M. Servidori, R. Angelucci, F. Cembali, P. Negrini, S. Solmi, P. Zaumseil, J. Appl. Phys. **61**, 1834 (1987) 149
11. K.S. Jones, J. Liu, L. Zhang, V. Krishnamoorthy, R.T. DeHoff, Nucl. Instrum. Methods B **106**, 227 (1995) 149

12. D.J. Eaglesham, P.A. Stolk, H.-J. Gossmann, J.M. Poate, Appl. Phys. Lett. **65**, 2305 (1994) 149
13. D.V. Lang, J. Appl. Phys. **45**, 3014 (1974) 150
14. D.V. Lang, J. Appl. Phys. **45**, 3023 (1974) 150
15. G.L. Miller, D.V. Lang, L.C. Kimerling, Rev. Annu. Mater. Sci. **377** (1977) 150
16. G. Davies, Phys. Rep. (Rev. Sector Phys. Lett.) **176**, 83 (1989) 150, 154, 160, 161, 172, 178, 182, 183, 195, 197
17. G. Davies, S. Hayama, L. Murin, R. Krause-Rehberg, V. Bondarenko, A. Sengupta, C. Davia, A. Karpenko, Phys. Rev. B **73**, 165202 (2006) 150, 154, 160, 161, 178, 182, 183, 197
18. E.V. Monakhov, J. Wong-Leung, A.Yu. Kuznetsov, C. Jagadish, B.G. Svensson, Phys. Rev. B **65**, 245201 (2002) 151, 158
19. S. Takaeda, Jpn. J. Appl. Phys. **30**, L639 (1991) 151, 193
20. M.-J. Caturla, T. Diaz de la Rubia, G.H. Gilmer, J. Appl. Phys. **77**, 3121 (1995) 151
21. P.W. Voorhees, M.E. Glicksman, Acta Metall. **32**, 2001 (1984) 151
22. H.P. Strunk, in *Analysis of Microelectronic Materials and Devices*, ed. by M. Grasserbauer, H.W. Werner (Wiley, New York), p. 327 151
23. P. Børgesen, in *Analysis of Microelectronic Materials and Devices*, ed. by M. Grasserbauer, H.W. Werner (Wiley, New York), p. 513 151
24. A. Bourret, Inst. Phys. Conf. Ser. **87**, 39 (1987) 151
25. I.G. Salisbury, M.H. Loretto, Philos. Mag. A **39**, 317 (1979) 151
26. P.A. Stolk, H.-J. Gossmann, D.J. Eaglesham, D.C. Jacobson, C.S. Rafferty, G.H. Gilmer, M. Jaraiz, J.M. Poate, H.S. Luftman, T.E. Haynes, J. Appl. Phys. **81**, 6031 (1997) 151, 152, 153, 192, 195
27. T.Y. Tan, Philos. Mag. **44**, 101 (1981) 151, 152
28. N.E.B. Cowern, G. Mannino, P.A. Stolk, F. Roozeboom, H.G.A. Huizing, J.G.M. van Berkum, F. Cristiano, A. Claverie, M. Jaraíz, Phys. Rev. Lett. **82**, 4460 (1999) 152, 153, 186, 198
29. P. Alippi, S. Coffa, L. Colombo, A. La Magna, Solid State Phenom. **85–86**, 177 (2002) 152, 157, 158
30. J. Kim, F. Kirchhoff, J.W. Wilkins, F.S. Khan, Phys. Rev. Lett. **84**, 503 (2000) 152
31. J.P. Goss, P.R. Briddon, T.A.G. Eberlein, R. Jones, N. Pinho, A.T. Blumenau, S. Öberg, Appl. Phys. Lett. **85**, 4633 (2004) 152
32. G.M. Lopez, V. Fiorentini, Phys. Rev. B **69**, 155206 (2004) 152
33. R. Jones, T.A.G. Eberlein, N. Pinho, B.J. Coomer, J.P. Goss, P.R. Briddon, S. Oberg, Nucl. Instrum. Methods Phys. Res. Sect. B, Beam Interact. Mater. At. **186**, 10 (2002) 152
34. P.B. Griffin, S.T. Ahn, W.A. Tiller, J.D. Plummer, Appl. Phys. Lett. **51**, 115 (1987) 153, 154, 175
35. N.E.B. Cowern, Appl. Phys. Lett. **64**, 2646 (1994) 153
36. J. Zhu, T. Diaz de la Rubia, L.H. Yang, C. Mailhiot, G.H. Gilmer, Phys. Rev. B **54**, 4741 (1996) 153

37. A.E. Michel, W. Rausch, P.A. Ronsheim, R.H. Kastl, Appl. Phys. Lett. **51**, 487 (1987) 153
38. A.E. Michel, W. Rausch, P.A. Ronsheim, R.H. Kastl, Appl. Phys. Lett. **50**, 416 (1987) 153
39. S. Solmi, F. Baruffaldi, R. Cantier, J. Appl. Phys. **69**, 2135 (1991) 153
40. P.A. Packan, Ph.D. Dissertation, Dept. Elect. Eng., Stanford University, 1991 153
41. M. Giles, J. Electr. Soc. **138**, 1160 (1991) 153, 163, 176
42. H.-J. Gossmann, P. Asoka-Krumer, T.C. Leng, B. Nielsen, K.G. Lynn, F.C. Unterwals, L.C. Feldman, Appl. Phys. Lett. **61**, 540 (1992) 154
43. H.G. Huizing, Ph.D. Dissertation, University of Delft, The Netherlands, 1996, and references therein 154, 168
44. A.J. Smith, N.E.B. Cowern, R. Gwilliam, B.J. Sealy, B. Colombeau, E.J.H. Collart, S. Gennaro, D. Giubertoni, M. Bersani, M. Barozzi, Appl. Phys. Lett. **88**, 082112 (2006) 154
45. M.T. Asom, J.L. Benton, R. Sauer, L.C. Kimerling, Appl. Phys. Lett. **51**, 256 (1987) 154, 155
46. J.L. Benton, M.T. Asom, R. Sauer, L.C. Kimerling, Mater. Res. Soc. Symp. Proc. **104**, 85 (1988) 154
47. G.D. Watkins, J.W. Corbett, Phys. Rev. **121**, 1001 (1961) 154, 157
48. P.M. Mooney, L.J. Cheng, M. Sull, J.D. Gerson, J.W. Corbett, Phys. Rev. B **15**, 3836 (1977) 154
49. J.W. Corbett, G.D. Watkins, Phys. Rev. A **138**, 555 (1965) 154, 156
50. J. Adey, R. Jones, D.W. Palmer, P.R. Briddon, S. Öberg, Phys. Rev. B **71**, 165211 (2005) 154
51. G.D. Watkins, Phys. Rev. B **12**, 5824 (1975) 155, 159
52. C.A. Londos, J. Grammatikakis, Phys. Stat. Sol. (a) **109**, 421 (1988) 155
53. P.J. Drevinsky, C.E. Caefer, L.C. Kimerling, J.L. Benton, in *Proceedings of the International Conference on the Science and Technology of Defect Control in Semiconductors*, ed. by K. Sumino (Elsevier, North-Holland, Amsterdam, 1990), p. 341 155, 161, 166, 168, 172, 174, 175
54. P.J. Drevinski, C.E. Caefer, S.P. Tobin, J.C. Mikkelsen Jr., L.C. Kimerling, in *Mater. Res. Soc. Symp. Proc.*, vol. 104, ed. by M. Stavola, S.J. Pearton, G. Davies (Materials Research Society, Pittsburgh, 1988), p. 167 155, 161, 166, 168, 172, 173, 174, 175
55. P.J. Drevinsky, H.M. DeAngelis, in *Thirteenth International Conference on Defects in Semiconductors*, ed. by L.C. Kimerling, J.M. Parsey Jr. (The Metallurgical Society of AIME, Warrendale, 1985), p. 807 155, 161, 166, 168, 172, 173, 174, 175
56. J. Hermansson, L.I. Murin, T. Hallberg, V.P. Markevich, J.L. Lindström, M. Kleverman, B.G. Svensson, Physica B **302–303**, 188 (2001) 155
57. J.W. Corbett, *Electron Radiation Damage in Semiconductors and Metals* (Academic Press, New York, 1966) 156, 161, 164, 167

58. G.D. Watkins, Mater. Res. Soc. Symp. Proc. **469**, 139 (1997). And references therein 156, 157
59. G.D. Watkins, Phys. Rev. Lett. **33**, 223 (1964) 156
60. J.C. Brabant, M. Pugnet, J. Barbolla, M. Brousseau, J. Appl. Phys. **47**, 4809 (1976) 156
61. G.D. Watkins, J.R. Troxell, A.P. Chatterjee, in *Defect and Radiation Effects in Semiconductors*, Inst. of Phys. Conf. Ser., vol. 46 (1979), p. 16 156
62. G.D. Watkins, Physica B **117–118**, 9 (1983) 156
63. J.R. Troxell, A.P. Chatterjee, G.D. Watkins, L.C. Kimerling, Phys. Rev. B **19**, 5336 (1979) 156
64. G.D. Watkins, in *Lattice Defects in Semiconductors*, ed. by F.A. Huntley, London, Inst. Phys. Conf. Ser., vol. 23 (1975), p. 1 156, 161, 164, 173
65. J.A. Van Vechten, Phys. Rev. B **10**, 1482 (1974) 156
66. G.D. Watkins, Phys. Rev. Lett. **33**, 223 (1964) 157
67. G.D. Watkins, J.W. Corbett, Phys. Rev. A **138**, 543 (1965) 157, 158, 159, 160, 173
68. K.L. Brower, in *Radiation Effects in Semiconductors*, ed. by J.W. Corbett, G.D. Watkins (Gordon and Breach, New York, 1971), p. 189 157
69. O.O. Awadelkarim, B. Monemar, J. Appl. Phys. **65**, 4779 (1989) 157, 159
70. L.C. Kimerling, Inst. Phys. Conf. Ser. **43**, 113 (1979) 175
71. Y.-H. Lee, J.W. Corbett, Phys. Rev. B **8**, 2810 (1973) 157
72. Y.-H. Lee, J.W. Corbett, Phys. Rev. B **9**, 4351 (1974) 157
73. L. Colombo, A. Buongiorno, T. Diaz de La Rubia, Mater. Res. Soc. Symp. Proc. **469**, 205 (1997) 157
74. L.C. Kimerling, P. Blood, W.M. Gibson, Inst. Phys. Conf. Ser. **46**, 273 (1979) 158
75. G.D. Watkins, in *Deep Centers in Semiconductors*, ed. by S. Pantelides (Gordon and Breach, New York, 1986), p. 147 159
76. R.D. Harris, J.L. Newton, G.D. Watkins, J. Electron. Mater. **14**, 799 (1987) 159
77. L.C. Kimerling, M.T. Asom, J.L. Benton, P.J. Drevinsky, C.E. Caefer, Mater. Sci. Forum **38–41**, 141 (1989) 159, 161, 162
78. R.D. Harris, G.D. Watkins, J. Electron. Mater. **14**, 799 (1984) 159
79. C.Z. Wang, K.M. Ho, Compos. Mater. Sci. **2**, 93 (1994) 159
80. L. Colombo, Annu. Rev. Comput. Phys. **IV** (1996) 159, 160
81. M. Tang, L. Colombo, J. Zhu, T. Diaz de la Rubia, Phys. Rev. B **V**, 5717 (1997)
82. J. Zhu, L. Yang, C. Mailhiot, T.D. de la Rubia, G.H. Gilmer, Nucl. Instrum. Methods B **102**, 29 (1995) 160
83. J.C. Bourgoin, J.W. Corbett, Phys. Lett. A **38**, 135 (1972) 160
84. J. Lalita, N. Keskitalo, A. Hallen, C. Jagadish, B.G. Svensson, Nucl. Instrum. Methods B **120**, 27 (1996) 162, 165

85. A. Hallen, B.U.R. Sundqvist, Z. Paska, B.G. Svensson, M. Rosling, J. Tiren, J. Appl. Phys. **64**, 1266 (1990) 162, 165
86. S. Libertino, J.L. Benton, D.C. Jacobson, D.J. Eaglesham, J.M. Poate, S. Coffa, P.G. Fuochi, M. Lavalle, Appl. Phys. Lett. **70**, 3002 (1997) 162, 163, 166, 167, 174
87. S. Libertino, S. Coffa, V. Privitera, F. Priolo, Mater. Res. Soc. Symp. Proc. **438**, 65 (1997) 162, 163, 168, 169, 170, 171, 172
88. V. Privitera, S. Coffa, F. Priolo, K. Kyllesbech Larsen, S. Libertino, A. Carnera, Nucl. Instrum. Methods Phys. Res. B **120**, 9 (1996) 162, 168, 170
89. S. Coffa, S. Libertino, Appl. Phys. Lett. **73**, 3369 (1998) 162, 168, 170, 171
90. S. Coffa, S. Libertino, A. La Magna, V. Privitera, G. Mannino, F. Priolo, Mater. Res. Soc. Symp. Proc. **532**, 93 (1998) 162, 168, 170
91. S.D. Brotherton, P. Bradley, J. Appl. Phys. **53**, 5720 (1982) 163
92. A. Hallén, B.G. Svensson, Rad. Eff. Defects Solids **128**, 179 (1994) 165, 166
93. B.G. Svensson, B. Mohadjeri, A. Hallén, J.H. Svensson, J.W. Corbett, Phys. Rev. B **43**, 2292 (1991) 165
94. J. Ziegler, http://www.srim.org 166
95. B.G. Svensson, C. Jagadish, J.S. Williams, Phys. Rev. Lett. **71**, 1860 (1993) 166
96. S. Libertino, J.L. Benton, D.C. Jacobson, D.J. Eaglesham, J.M. Poate, S. Coffa, P.G. Fuochi, M. Lavalle, Appl. Phys. Lett. **71**, 389 (1997) 166, 167, 168, 173, 174, 176, 177, 179
97. S. Fatima, C. Jagadish, J. Lalita, B.G. Svensson, A. Hállen, J. Appl. Phys. **85**, 2562 (1999) 168, 174, 185
98. S. Nishikawa, A. Tanaka, J. Yamall, Appl. Phys. Lett. **60**, 2270 (1992) 168, 185
99. S. Coffa, V. Privitera, F. Priolo, S. Libertino, G. Mannino, J. Appl. Phys. **81**, 1639 (1997) 168, 169, 170, 172
100. L. Enriquez, M. Jaraiz, J. Hernandez, J. Barbolla, S. Libertino, S. Coffa, in *Actas de la 1ª Conferencia de Dispositivos Electronicos*, ed. by R. Alcubilla, J. Pons, Barcelona, Spain, 20–21 February, 1997, pp. 545–550 168, 171
101. E. Rimini, S. Coffa, S. Libertino, G. Mannino, F. Priolo, V. Privitera, Defect Diff. Forum **153–155**, 137 (1998) 168, 170
102. M.T. Robinson, J.M. Torrens, Phys. Rev. B **9**, 5008 (1974) 171
103. J.L. Benton, V. Venezia, R.D. Swain, L. Pelaz, Oral communication at the Mater. Res. Symp. Spring Meeting, San Francisco, CA, April 5–9, 1999 173, 174
104. L.J. Cheng, J.C. Corelli, J.W. Corbett, G.D. Watkins, Phys. Rev. **152**, 761 (1966)

105. S. Libertino, J.L. Benton, S. Coffa, D.C. Jacobson, D.J. Eaglesham, J.M. Poate, M. Lavalle, P.G. Fuochi, Mater. Res. Soc. Symp. Proc. **469**, 187 (1997) 175, 177
106. M. Jaraiz, G.H. Gilmer, J.M. Poate, T. Diaz de la Rubia, Appl. Phys. Lett. **68**, 409 (1996) 177, 193
107. J.L. Benton, S. Libertino, P. Kringhøi, D.J. Eaglesham, J.M. Poate, S. Coffa, J. Appl. Phys. **82**, 120 (1997) 177, 181
108. S. Libertino, J.L. Benton, S. Coffa, D.J. Eaglesham, in *Mater. Res. Soc. Symp. Proc.*, vol. 504, ed. by J.C. Barbour, S. Roorda, D. Ila (Materials Research Society, Warrendale, 1998), p. 3 177, 181, 192
109. J.L. Benton, S. Libertino, S. Coffa, D.J. Eaglesham, in *Mater. Res. Soc. Proc.*, vol. 469, ed. by S. Coffa, T. Diaz de la Rubia, P.A. Stolle, C.S. Rafferty (Materials Research Society, Warrendale, 1997), p. 193 177, 179, 181, 182, 183, 185, 192
110. M. Nakamura, Appl. Phys. Lett. **72**, 1347 (1998) 178, 182
111. S. Coffa, S. Libertino, C. Spinella, Appl. Phys. Lett. **76**, 321 (2000) 182, 183, 194, 195
112. H. Weman, B. Monemar, G.S. Oehrlein, S.J. Jeng, Phys. Rev. B **42**, 3109 (1990) 182, 183, 197
113. N.S. Minaev, A.V. Mudryi, Phys. Status Solidi **68**, 561 (1981) 183
114. D.C. Schmidt, B.G. Svensson, M. Seibt, C. Jagadish, G. Davies, J. Appl. Phys. **88**, 2309 (2000) 183, 189, 194
115. P.A. Stolk, H.-J. Gossmann, D.J. Eaglesham, D.C. Jacobson, H.S. Luftman, Mater. Res. Soc. Proc. **354** (1995) 185, 193
116. J.L. Ngau, P.B. Griffin, J.D. Plummer, J. Appl. Phys. **90**, 1768 (2001) 185
117. L.W. Song, G.D. Watkins, Phys. Rev. B **42**, 5759 (1990) 185
118. J.L. Benton, K. Halliburton, S. Libertino, D.J. Eaglesham, S. Coffa, J. Appl. Phys. **84**, 4749 (1998) 185
119. S. Libertino, S. Coffa, C. Spinella, J.L. Benton, D. Arcifa, Mater. Sci. Eng. B **71**, 137 (2000) 185
120. S. Libertino, J.L. Benton, S. Coffa, D.J. Eaglesham, Nuovo Cimento D **20**(10), 1529 (1998) 185, 190
121. S. Libertino, S. Coffa, J.L. Benton, K. Halliburton, D.J. Eaglesham, Nucl. Instrum. Methods Phys. Res. B **148**, 247 (1999) 185
122. S. Libertino, S. Coffa, J.L. Benton, Phys. Rev. B **63**, 195206 (2001) 185
123. V.V. Kveder, Yu.A. Osipyan, W. Schroeter, G. Zoth, Phys. Status Solidi (a) **72**, 701 (1982) 186, 189, 191, 200
124. P. Blood, J.W. Orton, *The Electrical Characterization of Semiconductors: Majority Carriers and Electron States* (Academic Press, London, 1992), ed. by N.H. March 189, 196, 200
125. P. Omling, L. Samuelson, H.G. Grimmeiss, J. Appl. Phys. **54**, 5117 (1983) 189, 190, 200
126. L. Samuelson, P. Omling, Phys. Rev. B **34**, 5603 (1986) 189, 200

127. P. Omling, E.R. Weber, L. Montelius, H. Alexander, J. Michel, Phys. Rev. B **32**, 6571 (1985) 189
128. J.R. Ayres, S.D. Brotherton, J. Appl. Phys. **71**, 2702 (1992) 190
129. A. La Magna, S. Coffa, S. Libertino, in *Mater. Res. Soc. Symp. Proc.*, vol. 210, ed. by A. Agarwal, L. Pelaz, H. Vuong, P. Packan, M. Kase (Materials Research Society, Warrendale, 2000), p. B11.5.1 190, 191, 199, 200, 201
130. A. La Magna, S. Coffa, Compos. Mater. Sci. **17**, 21 (2000) 191, 199
131. P.A. Stolk, H.-J. Gossmann, D.J. Eaglesham, D.C. Jacobson, J.M. Poate, H.S. Luftman, Appl. Phys. Lett. **66**, 568 (1995) 195
132. G. Davies, K.T. Kun, T. Reade, Phys. Rev. B **44**, 12146 (1991) 196
133. International Technology Roadmap for Semiconductors, 1999 edition, http://public.itrs.net 198
134. A. La Magna, P. Alippi, L. Colombo, M. Strobel, Comput. Mater. Sci. **27**, 10 (2003) 199
135. A. Bongiorno, L. Colombo, F. Cargnoni, C. Gatti, M. Rosati, Europhys. Lett. **50**, 608 (2000) 199
136. J. Kim, F. Kirchhoff, J.W. Wilkins, F.S. Khan, Phys. Rev. Lett. **84**, 503 (2000) 199, 200
137. A. La Magna, S. Coffa, L. Colombo, Phys. Rev. Lett. **82**, 1720 (1999) 199
138. S. Libertino, S. Coffa, A. La Magna, Nucl. Instrum. Methods Phys. Res. B **186**, 265 (2002) 199
139. J. Kim, F. Kirchhoff, J.W. Wilkins, F.S. Khan, Phys. Rev. B **55**, 16186 (1997) 200

Index

Point Defect Kinetics and Extended-Defect Formation during Millisecond Processing of Ion-Implanted Silicon

K. Gable and K.S. Jones

Department of Materials Science and Engineering, University of Florida,
Gainesville, FL 32611, USA, e-mail: kjones@eng.ufl.edu

Abstract. One challenge in successfully scaling the dimensions of the MOSFET transistor is in maintaining a highly activated ultrashallow p-type source/drain extension (p-SDE) region under the gate. Ion implantation introduces significant levels of damage and dopant that is not electrically active into the lattice. The thermal processing necessary to activate this dopant, while also limiting the dopant motion, is evolving. A high-power arc-lamp design has enabled millisecond annealing as an alternative to conventional rapid thermal processing, which operates on the timescale of seconds for ultrashallow junction formation. This chapter summarizes some investigations into the use of millisecond annealing to form a highly activated ultrashallow junction, while simultaneously minimizing diffusion. In order to minimize dopant motion the various mechanisms that lead to diffusion during millisecond annealing are discussed. Examples of how changing the proximity of the damage to the dopant, allow one to understand these mechanisms and the role of annealing temperature are presented. Finally, the mechanisms of postimplantation dopant activation are discussed. It is shown that it is critical to understand the interplay between dopants, point defects, extended defects and processing if one is to understand the evolution of ultrashallow p-type junctions.

Silicon technology development is supported by the significant advantages obtained by following the trend known as Moore's law, which suggests that the average geometrical dimensions and fabrication cost of a transistor will decrease by a factor of two every 18 to 24 months [1]. Figure 1 shows a cross section of a single planar p-type enhancement mode metal-oxide-semiconductor field effect transistor (p-MOSFET), which is the most common device used in current electronics manufacturing [2]. This device has been modified through the use of SiGe source strain regions to lower contact resistance and increase hole mobility by stressing the channel. In addition, the use of high-k dielectrics and metal gates is gaining widescale adoption for some applications. The continued scaling of this transistor offers the ability to produce higher-speed/lower-power devices capable of increasing the functionality and applicability of the resulting product.

One challenge in successfully scaling the dimensions of the MOSFET transistor is in maintaining a highly activated ultrashallow p-type source/drain extension (p-SDE) region under the gate. The SDE of a MOSFET transistor

H. Bernas (Ed.): Materials Science with Ion Beams,
Topics Appl. Physics **116**, 213–226 (2010)
DOI: 10.1007/978-3-540-88789-8_7,

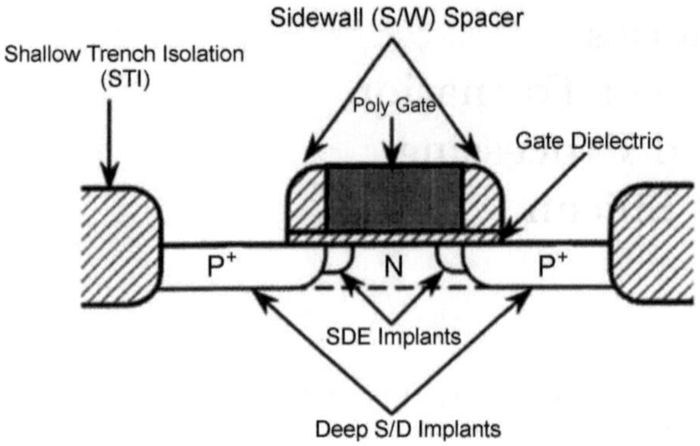

Fig. 1. Cross section of a single planar p-type enhancement mode metal-oxide-semiconductor field effect transistor (p-MOSFET). Reprinted with permission from [2]

is responsible for reducing short-channel effects, which occur when the electric field of the drain begins to enter the channel region and starts affecting the potential barrier between the source and channel areas of the device. The result is a device with a drain current that cannot be effectively controlled by the gate. Ion implantation is commonly used in complementary metal-oxide-semiconductor (CMOS) technology to dope the source/drain extension (SDE) region of the device [3].

The ion-implantation process is known to create a large amount of interstitial–vacancy (I–V) (i.e., Frenkel) pairs caused by the nuclear collisions among the primary ions and recoiled atoms with the host atoms of the substrate. Many of these Frenkel pairs recombine during relaxation of the collision cascade produced by the implanted ion, after which the primary damage generated by the incident ion can be considered stable [2, 4]. The probability of the recombination of a Frenkel pair is dependent upon the separation distance of the interstitial and vacancy, temperature, and the concentration of point-defect traps. For nonamorphizing implants, the stable damage is primarily small defect clusters, dopant–defect complexes, and some isolated Frenkel pairs. It is well known that continuous amorphous layers can be formed by ion implantation, and that these layers are capable of preventing ion channeling associated with the implantation of low-mass species (e.g., B). These layers extend from the substrate surface down to a depth dependent on the implant conditions. When considering amorphous-Si (α-Si) it can be said that the lattice maintains some short-range order, although it is significantly disordered and consists of atoms with unsatisfied bonds that exhibit large tetrahedral bond-angle distortions [5]. The threshold damage density for the formation of an amorphous layer is often taken to be 10% of the Si lattice density [6]. It was shown that damage accumulation saturates after an amorphous state is reached [2]. It is well known that α-Si has a melting temperature and atomic density of approximately 225 ± 50°C and 1.8 ± 0.1% below that of crystalline Si (c-Si), respectively [7–9]. In addition, it was shown that α-Si consists of a covalently bonded continuous random network (CRN) that can exist as either an as-implanted or structurally relaxed state [10–15]. The structurally relaxed α-Si differs from the as-implanted case in that the number of large-angle bond distortions and defect complexes produced during the preamorphization implant are reduced, typically by a low-temperature relaxation anneal (e.g., 500°C for 60 min) [16]. Regardless of the structural state of the α-Si, a large population of excess interstitials are transmitted through the amorphous layer and remain below the original amorphous/crystalline (α/c) interface until postimplant thermal processing.

Postimplant thermal processing is required to induce solid-phase epitaxial growth (SPEG) of the implantation-induced amorphous layer, which regrows the amorphous layer and activates a relatively high concentration of the implanted dopants by establishing them on substitutional sites where they are able to contribute their holes (electrons) to the valence (conduction) band. During SPEG of an amorphous layer, the excess interstitials beyond the amorphous/crystalline interface (termed the end-of-range (EOR) region) begin to coalesce into metastable crystallographic defects that have been shown to enhance dopant diffusion [17], assist in incomplete dopant activation [18], and contribute to junction leakage [19, 20]. During subsequent thermal processing these EOR crystallographic defects, evolve into either {311} defects or dislocation loops. The {311} defect is an extrinsic row of interstitials lying on the {311} habit plane, elongated in the $\langle 110 \rangle$ direction. Two different types of dislocation loops have been observed to also form: so-called perfect prismatic loops with a Burgers vector $b = a/2\langle 110 \rangle$ and faulted Frank loops with a Burgers vector $b = a/3\langle 111 \rangle$ [21]. Dislocation loops have been shown to evolve from unfaulting of {311} defects [22, 23] and are more stable than {311} defects [24]. During postimplant thermal annealing, these defects release interstitials and these interstitials give rise to transient enhanced diffusion (TED), which significantly increases the diffusion behavior of dopants such as B and P, which diffuse primarily or in part by an interstitial(cy) mechanism [25]. It was shown that the amount of TED observed during an anneal decreases when the implant damage is annealed out at a higher temperature [3]. This arises from the fact that the interstitial supersaturation due to the presence of extended defects is higher at a lower temperature [26]. This observation influenced the development of single-wafer thermal processes capable of producing a high-temperature ambient with ramp rates on the order of 50–200°C/s, and fast switching times in order to insulate the dopant from a high degree of TED [27].

Rapid thermal processing (RTP) alone was successful in producing junctions with the performance characteristics necessary for the continued scaling of CMOS technology for several generations. Its ability to satisfy these requirements is associated with improved equipment capability in the form of spike annealing, which decreases the effective thermal budget, allowing for higher annealing temperatures to improve activation and reduce the amount of diffusion of the dopants during the thermal process. A spike anneal is characterized as a short thermal-anneal cycle that can be achieved by increasing the ramp-up and ramp-down rates and by minimizing the dwell time at the temperature of interest. The inability of this technique to produce junctions with the performance characteristics required by current and future technology nodes is in the cycle time of the thermal process, which results in an unacceptable amount of dopant diffusion. The minimum cycle times in conventional RTP techniques are limited by the maximum power delivered to the wafer (which determines the ramp-up rate) and the minimum response time of the relatively large thermal mass incandescent tungsten lamps (which determines both the soak time and the ramp-down rate). Without being able to minimize the soak time and the ramp-down rate, increasing the ramp-up rate above 100°C/s results in negligible improvement in terms of forming a highly activated ultrashallow junction [27]. This illustrates the need to investigate advanced annealing technologies that may be able to produce highly activated junctions without being subject to TED.

A high-power arc-lamp design or nonmelt laser annealing has enabled ultrahigh temperature (UHT) annealing as an alternative to conventional RTP for ultrashallow junction formation [28]. The high-powered arc technique uses continuous-wave mode arc-lamp radiation to heat the wafer to an intermediate temperature (e.g., 700°C) before discharging a capacitor bank into flash lamps, which heats the device side of the wafer to a relatively high temperature (e.g., 1350°C) for a few milliseconds [29–31]. This time duration is significantly reduced from those obtained with conventional RTP, which are of the order of 1–2 s within 50°C of the peak temperature. The arc lamp responds more rapidly than tungsten filament lamps due to the reduced thermal mass of the argon gas used in the arc-lamp system. An approximate value for the response time of the arc-lamp system is 50 ms, whereas the switching time constant for tungsten incandescent lamps is of the order of 0.5 s [32]. A second advantage of arc-lamp radiation is that over 95% of the arc radiation is below the 1.2-μm bandgap absorption of Si at room temperature (compared to 40% for tungsten) [28]. Arc-lamp radiation is strongly absorbed in Si due to band-to-band transitions with very low transmission through the wafer [28, 33]. Even though this annealing technique may be a likely candidate to replace conventional RTP to activate the SDE for future technology nodes, the activation and diffusion mechanisms that take place on these timesscales are not well understood. To better understand the physics involved when using a technique such as UHT annealing, a recent experiment will be presented and the most significant differences between conventional RTP annealing and millisecond annealing will be discussed.

Typically, a preamorphization implant is done prior to the SDE implant so as to reduce channeling associated with low-mass ions (e.g., B) and maintain a Gaussian profile for any subsequently implanted species. Most PMOS implant strategies consist of a Ge^+ preamorphization implant followed by a low-energy (i.e., $\leq$5 keV) B^+ implant. Recently reported results show that the preamorphization implant energy can have a profound effect on interstitial release from the EOR damage region, presumably because of the close proximity of the excess interstitials to the substrate surface [34]. As a comparison study, two 200-mm 3–5 Ω cm (100) n-type Czochralski (CZ) grown Si wafers were preamorphized with either 5- or 48-keV Ge^+ implantation to $5 \times 10^{14}/\mathrm{cm}^{-2}$, and subsequently implanted with 3 keV BF_2^+ molecular ions to $6 \times 10^{14}\,\mathrm{cm}^{-2}$. The wafers were then sectioned and annealed at Vortek Industries to investigate the effects of the UHT annealing technique on the resulting junction characteristics. In this experiment, impulse anneals (i.e., iRTP) were performed over the range of 760 to 1100°C and flash annealing (i.e., fRTP) was performed at 1350°C.

Figures 2a and b show the SIMS results for each of the iRTP anneals used in this study for the 48- and 5-keV preamorphization implants, respectively. Each profile shows an increase in junction depth (x_j) when compared to the as-implanted profile. Junction depth is defined as the depth of the profile at a dopant concentration of $1 \times 10^{18}\,\mathrm{cm}^{-3}$. Figure 2a shows that the 760 and 800°C iRTP anneals display similar profiles with a 19.3-nm x_j for the 48 keV preamorphization implant. A SIMS profile for a 585°C furnace anneal for 45 min is included with the data in Fig. 2a, and shows very similar diffusion behavior when compared to the 760 and 800°C iRTP anneals. Figure 2b shows that the 760 and 800°C iRTP anneals produce profiles with x_j of 17.9 and 18.4 nm, respectively, for the 5-keV preamorphization implant. The profiles are comparable to the as-implanted profile above a concentration of $1 \times 10^{19}\,\mathrm{cm}^{-3}$. It should be noted that the diffusion observed for the 760 and 800°C iRTP anneals in Fig. 2b occurs in c-Si, since the motion occurs below

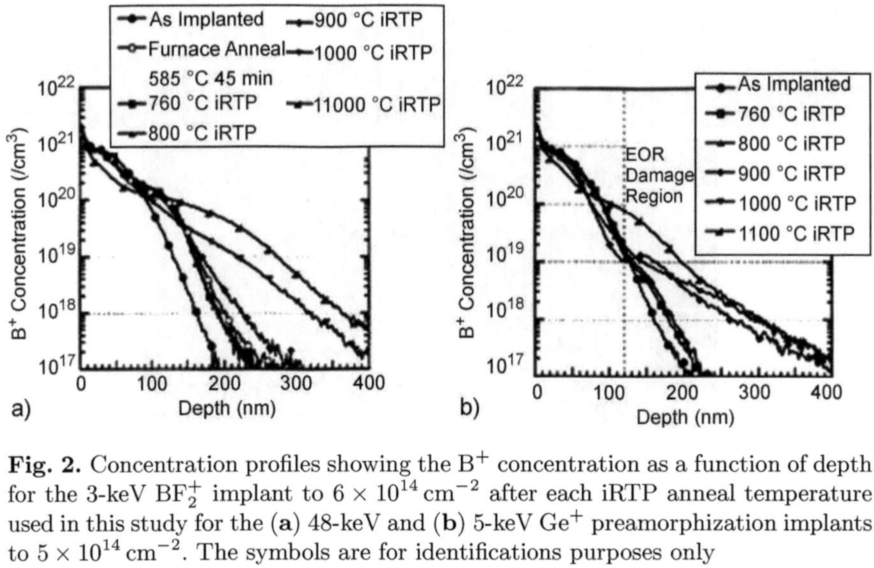

Fig. 2. Concentration profiles showing the B^+ concentration as a function of depth for the 3-keV BF_2^+ implant to $6 \times 10^{14}\,\mathrm{cm}^{-2}$ after each iRTP anneal temperature used in this study for the (**a**) 48-keV and (**b**) 5-keV Ge^+ preamorphization implants to $5 \times 10^{14}\,\mathrm{cm}^{-2}$. The symbols are for identifications purposes only

the original α/c interface produced by the 5-keV preamorphization implant. The SIMS profiles for the 900°C iRTP anneal show that the B diffusion behavior is much larger for the wafer that received the 5-keV preamorphization implant.

Figure 3 shows the corresponding plan-view transmission electron microscopy (PTEM) images of the EOR damage produced by the 48-keV preamorphization implant after each of the iRTP anneals used in this study. It is important to note that similar PTEM investigation showed that no observable defects formed for the 5-keV preamorphization implant. As can be seen by the images, the 760, 800, and 900°C iRTP anneals produce a high density defect structure consisting of defect clusters, which are typically observed after a short anneal at a relatively low temperature [17, 35]. These defect clusters are approximately 4 to 12 nm and 6 to 18 nm in diameter for the 760 and 900°C iRTP anneals, respectively. The PTEM image for the 1000°C iRTP anneal shows that it is sufficient to produce a defect structure mainly consisting of {311} defects and dislocation loops [17]. The {311} defects range from 29 to 88 and average 60 nm in length and the dislocation loops range from 21 to 29 and average 26 nm in diameter. Increasing the iRTP anneal temperature to 1100°C results in a defect structure consisting only of dislocation loops, which shows that {311} defect dissolution is complete

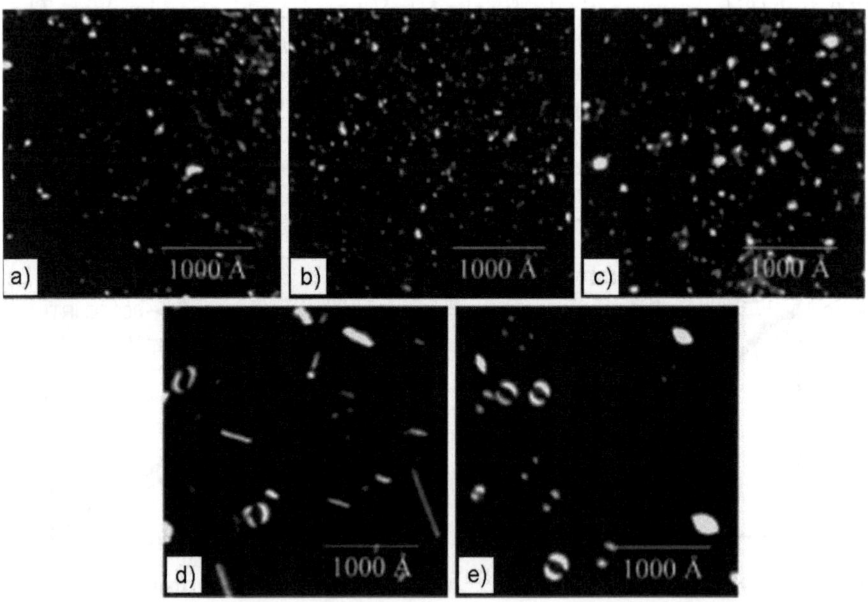

Fig. 3. Plan-view TEM images of the damage produced by the 48-keV Ge^+ preamorphization implant to 5×10^{14} cm^{-2} under a WBDF g_{220} two-beam imaging condition after a (**a**) 760, (**b**) 800, (**c**) 900, (**d**) 1000 and (**e**) 1100°C iRTP anneal

between 1000 and 1100°C. The dislocation loops range from 24 to 32 and average 29 nm in diameter.

The SIMS results for the 760 and 800°C iRTP anneals and the 585°C furnace anneal for 45 min in Fig. 2a showed similar diffusion behavior for the 48-keV preamorphization implant. It was shown that B from both B^+ and BF_2^+ implants into preamorphized Si displayed a similar diffusion enhancement during SPER of an implantation-induced amorphous layer at 550°C [36]. Since B thermal diffusion in c-Si is negligible at 550°C, this diffusion enhancement was attributed to TED (possibly due to the large amount of damage produced by the 20-keV Si^+ preamorphization implant to $5 \times 10^{14}\,\mathrm{cm}^{-2}$). However, the fact that the same diffusion profile was obtained over the temperature range of 585 to 800°C in Fig. 2a suggests that the diffusion enhancement is not TED. The diffusion enhancement observed in Fig. 2a is caused by B diffusion in α-Si before complete recrystallization of the implantation-induced amorphous layer (i.e., not TED).

It was shown that B diffusivity in α-Si at 600°C is more than five orders of magnitude greater than that in c-Si [37]. This was done by growing three narrow B profiles with a peak concentration of $1.3 \times 10^{20}\,\mathrm{cm}^{-3}$ at depths of 170, 338, and 508 nm with respect to the substrate surface. These three B profiles were then implanted at −196°C with 600-keV Si^+ to $5 \times 10^{15}\,\mathrm{cm}^{-2}$, and subsequently implanted with 70-keV Si^+ to $5 \times 10^{14}\,\mathrm{cm}^{-2}$ to produce a continuous amorphous layer extending 900 nm below the substrate surface. The amorphous layer was then recrystallized at 600°C and continuously monitored by time-resolved reflectivity (TRR). The data show that the three B profiles are slightly broadened by the two Si^+ preamorphization implants, and that the three B profiles are further broadened during the SPER of the implantation-induced amorphous layer. The broadening of the three profiles increases with decreasing depth from the substrate surface (i.e., the profiles that spend the most time within the α-Si show the most broadening during SPER of the amorphous layer). This result is inconsistent with TED, which would cause the deepest B profile to broaden the most. This result is also inconsistent with the suggestion that dopant segregation across the advancing α/c interface caused the increase in diffusion behavior [38]. If the diffusion occurred because of mass transfer across the α/c interface, then it would be independent of the amount of time the B spends in α-Si (which was not the case in [37]) [38]. Their results estimate the B diffusivity in α-Si to be approximately $(2.6 \pm 0.5) \times 10^{-16}\,\mathrm{cm}^2/\mathrm{s}$ at 600°C, which matches very well with the calculated (i.e., $2\sqrt{Dt}$) amount of diffusion expected from the 585°C furnace anneal in Fig. 2a (assuming a regrowth velocity of approximately 30 nm/min) [39]. Figure 2b showed no B diffusion during SPER of the implantation-induced amorphous layer, presumably because of the high local concentration of excess interstitials produced by the 5-keV preamorphization implant (which couple with the B atoms and form immobile clusters, preventing diffusion in the amorphous phase). It should be noted that the diffusion coefficient of Ge in α-Si was reported to be very low and is not expected to have a significant impact on the results in Fig. 2 [40].

The profiles after the 760 and 800°C iRTP anneals in Fig. 2a showed approximately 3 nm of diffusion up to a concentration of $1.8 \times 10^{20}\,cm^{-3}$, above which inactive B-cluster formation or precipitation occurred and the B remained immobile during SPER of the amorphous layer produced by the 48-keV preamorphization implant. Similar profiles were observed by Jin et al., who showed that this characteristic is independent of B^+ or BF_2^+ implantation after a 550°C furnace anneal for 40 min [36]. Since additional XTEM results (not shown) revealed that the 76-nm continuous amorphous layer produced by the 48-keV preamorphization implant completely recrystallized during the 760°C iRTP anneal, it can be said that the 760 and 800°C iRTP anneals result in similar dopant profiles due to the fact that the B remains in α-Si the same amount of time before recrystallization of the amorphous layer is complete. This shows that the 760 and 800°C iRTP anneals are insufficient to evolve the excess interstitials to the point where TED begins to affect the overall diffusion profile. The observation of similar dopant profiles for the 760 and 800°C iRTP anneals suggests that there is a temperature range in which the iRTP anneal will result in equivalent dopant profiles without being subject to TED.

Figure 2a showed that the 900°C iRTP anneal increased the x_{j} from 19.3 to 22.5 nm when compared to the 800°C iRTP anneal, for the 48-keV preamorphization implant. FLorida Object Oriented Process Simulator (FLOOPS) simulations estimate that approximately 3 min at 900°C are required to produce the 3.2-nm increase in x_{j} for the 900°C iRTP anneal. Since the entire anneal cycle (i.e., ramping up to 900°C and cooling down to room temperature) was complete on the order of 8–10 s, the increase in diffusion behavior is attributed to TED. The corresponding PTEM results (not shown) revealed that the 760, 800, and 900°C iRTP anneals produce defect structures consisting of a high density of defect clusters. This suggest that either submicroscopic interstitial cluster dissolution and evolution or a nonconservative defect-coarsening process of the EOR damage is responsible for the diffusion enhancement observed for the 900°C iRTP anneal in Fig. 2a.

It can be seen that the 900°C iRTP anneal in Fig. 2a showed less of a diffusion enhancement than in Fig. 2b. This difference is presumably caused by the effect of interstitial release from the EOR damage region. It was shown that the interstitial flux from the EOR damage is approximately an order of magnitude greater into the substrate than toward the surface for overlapping 112-keV and 30-keV Si^+ implants to $1 \times 10^{15}\,cm^{-2}$ performed at (20 ± 1)°C [41]. The decrease in the interstitial flux toward the surface was attributed to the EOR damage acting as interstitial traps, which prevent a significant fraction of the interstitials from diffusing toward the substrate surface. Jones et al. correlated the EOR dislocation loop density with the amount of interstitial backflow toward the surface, which increased with decreasing implant temperature (presumably due to the fact that less EOR damage is available to prevent the interstitials from diffusing toward the substrate surface) [42]. The difference in the interstitial flux for near-room-temperature preamorphization implants offers an explanation for observing increased diffusion enhancement below and the lack of diffusion behavior above the original α/c interface produced by the 5-keV preamorphization implant. It is presumed that the boron–interstitial clusters (BICs) that form for the 5-keV preamorphization implant behave in a similar way to EOR damage in that they are capable of obstructing interstitial backflow toward the surface.

Figure 2a showed that the 900, 1000, and 1100°C iRTP anneals increased the x_j by 3.2, 12.4, and 16.8 nm, respectively, compared to the 800°C iRTP anneal for the 48-keV preamorphization implant. These results show that the largest difference in B diffusion behavior is observed for the 1000°C iRTP anneal. The increase in diffusion behavior for the 1000°C iRTP anneal is most likely due to a significant fraction of the interstitial flux toward the surface, which is capable of reaching the B profile during the 1000°C iRTP anneal but is less pronounced for the 900°C iRTP anneal. Such a significant pulse of TED was shown to occur for 40-keV Si^+ implants to both 2×10^{13} and $5 \times 10^{13}/\,cm^{-2}$ during the first 15 s of annealing at 700°C [36]. This pulse of TED was shown to be in excess of the enhancement caused by {311} defect dissolution, suggesting a different source of interstitials [22, 23]. It is presumed that a similar mechanism is causing the diffusion enhancement for the 1000°C iRTP anneal in Fig. 2a for the 48-keV preamorphization implant, because the corresponding PTEM results in Fig. 3d revealed that {311} dissolution is incomplete following the 1000°C iRTP anneal. Figure 2 showed that the 1000 and 1100°C iRTP anneals increased the x_j much less for the 5-keV preamorphization implant when compared to the 48-keV preamorphization implant, presumably because once the interstitials pass the B profile they no longer affect its diffusion behavior.

Figures 4a and b show the SIMS results for the 1350°C fRTP anneal for the 48- and 5-keV preamorphization implants, respectively. As can be seen in Fig. 4a, the 760 and 800°C intermediate temperatures produce profiles with x_j of 21.3 and 22.4 nm following the 1350°C fRTP anneal for the 48-keV preamorphization implant, respectively. The 900°C intermediate temperature produces a profile with a x_j of 25.0 nm. As can be seen in Fig. 4b, the 1350°C fRTP anneal results in similar profiles independent of the intermediate anneal temperature for the 5-keV preamorphization implant, all of which display an average x_j of 23.3 nm.

Figure 5 shows the corresponding PTEM images of the EOR damage produced by the 48-keV preamorphization implant after the 1350°C fRTP anneal. No observable defects formed for the 5-keV preamorphization implant. The 760°C intermediate temperature produced a defect structure consisting of defect clusters, {311} defects, and dislocation loops. The {311} defects range from 19 to 29 and average 25 nm in length and the dislocation loops

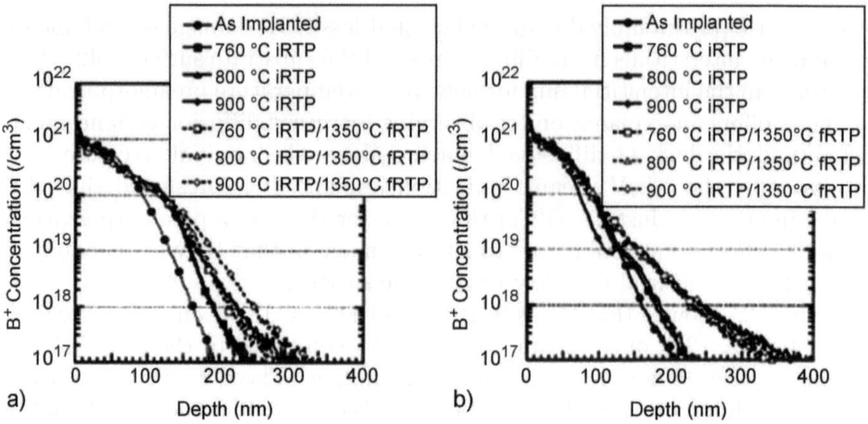

Fig. 4. Concentration profiles showing the B^+ concentration as a function of depth for the 3-keV BF_2^+ implant to $6 \times 10^{14}\,cm^{-2}$ before and after the 1350°C fRTP for the (**a**) 48-keV and (**b**) 5-keV Ge^+ preamorphization implants to $5 \times 10^{14}\,cm^{-2}$. The symbols are for identifications purposes only

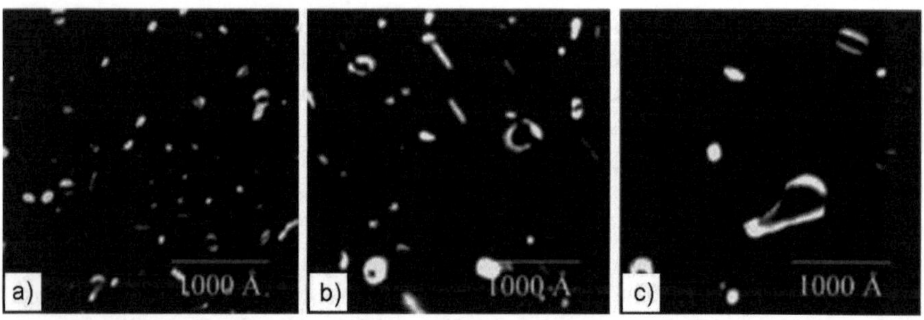

Fig. 5. Plan-view TEM images of the damage produced by the 48-keV Ge^+ preamorphization implant to $5 \times 10^{14}\,cm^{-2}$ under a WBDF g_{220} two-beam imaging condition for the 1350°C fRTP using a (**a**) 760, (**b**) 800 and (**c**) 900°C intermediate temperature

range from 18 to 24 and average 19 nm in diameter. The 800°C intermediate temperature produced a defect structure mainly consisting of {311} defects and dislocation loops. The {311} defects range from 19 to 43 and average 32 nm in length and the dislocation loops range from 19 to 59 and average 32 nm in diameter. The 900°C intermediate temperature produced a defect structure consisting of dislocation loops. The dislocation loops range from 24 to 115 and averaged 62 nm in diameter. The most marked difference between these images is the size and overall evolution of the dislocation loops, which increases with the intermediate annealing temperature. The largest dislocation loops in each of the images are approximately 24, 59, and 115 nm in diameter for the 760, 800, and 900°C intermediate temperatures, respectively. When comparing these images it can be seen that the resulting defect structure after a fRTP anneal is significantly dependent on the intermediate anneal temperature, suggesting that both the intermediate and fRTP anneal temperatures need to be considered when using this UHT annealing technique.

Conclusions

With the significant advantages obtained by following Moore's Law, the semiconductor industry is constantly evaluating novel processing techniques and/or integration schemes to further scale the MOSFET transistor. One challenge in successfully scaling the dimensions of the MOSFET transistor is in maintaining a highly activated ultrashallow p-type source/drain extension (p-SDE) region under the gate. A high-power arc-lamp design has enabled UHT annealing as an alternative to conventional RTP for ultrashallow junction formation. This technique heats the wafer to an intermediate temperature (e.g., 800°C) before discharging a capacitor bank into flash lamps, which heats the device side of the wafer to a relatively high temperature (e.g., 1350°C) for a few milliseconds. This time duration is significantly reduced from those obtained with conventional RTP, which are of the order of 1–2 s within 50°C of the peak temperature. The purpose of this chapter was to investigate the possibility of the UHT annealing technique to form a highly activated ultrashallow junction without being subject to TED. It is shown that, of the 3.6 nm of diffusion that occurs during a 1200°C UHT anneal for a 48-keV preamorphization implant, 3.0 nm is caused by B diffusion in α-Si before complete recrystallization of the implantation-induced amorphous layer. The additional 0.6 nm of diffusion that occurred during the 1200°C fRTP anneal is very close to what would be expected under intrinsic conditions (i.e., 1200°C anneal for 3 ms). Although 3.8 nm of diffusion occurred for the 5-keV preamorphization implant during the same anneal, the diffusion occurs in c-Si and is presumably caused by TED due to the close proximity between the excess interstitials and B atoms. Both preamorphization conditions show approximately 200 Ohm/sq improvement in activation during a 1200°C fRTP anneal (compared to the sample that was only annealed to the intermediate temperature); this improvement in R_s may be caused by the breaking up of initially clustered B atoms, but needs to be better understood. Dopant activation is degraded for the 5-keV preamorphization implant, presumably because the high local concentration of excess interstitials and B atoms results in inactive B–interstitial clusters (BICs) formation during postimplant UHT annealing. This shows that a highly activated ultrashallow p-type junction can be formed without being subject to TED only when the excess interstitials are sufficiently separated from the B atoms implanted near the substrate surface.

References

1. G.E. Moore, Electronics **38**, 114 (1965) 213
2. J.D. Plummer, M.D. Deal, P. Griffin, *Silicon VLSI Technology – Fundamentals, Practice and Modeling* (Prentice-Hall, Upper Saddle River, 2000) 213, 214
3. J.F. Ziegler (ed.), *Ion Implantation Science and Technology* (Ion Implantation Technology Co., Edgewater, 2000) 214, 215
4. F.F. Morehead, B.L. Crowder, Radiat. Eff. **6**, 27 (1970) 214
5. S. Roorda, W.C. Sinke, J.M. Poate, D.C. Jacobson, S. Dierker, B.S. Dennis, D.J. Eaglesham, F. Spaepen, P. Fuoss, Phys. Rev. B **44**, 3702 (1991) 214
6. L.A. Christel, J.F. Gibbons, T.W. Sigmon, J. Appl. Phys. **52**, 7143 (1981) 214
7. M. Thompson, G. Galvin, J. Mayer, P. Peercy, J. Poate, D. Jacobson, A. Cullis, N. Chew, Phys. Rev. Lett. **52**, 2360 (1984) 214
8. J. Poate, J. Cryst. Growth **79**, 549 (1986) 214
9. J.S. Custer, M.O. Thompson, D.C. Jacobson, J.M. Poate, S. Roorda, W.C. Sinke, F. Spaepen, Appl. Phys. Lett. **64**, 437 (1994) 214
10. D.E. Polk, D.S. Boudreaux, Phys. Rev. Lett. **31**, 92 (1973) 215
11. F. Wooten, K. Winer, D. Weaire, Phys. Rev. Lett. **54**, 1392 (1985) 215
12. F. Wooten, D. Weaire, in *Solid State Physics: Advances in Research Applications*, vol. 40, ed. by D. Turnbull, H. Ehrenreich (Academic Press, New York, 1987), p. 2 215
13. R. Car, M. Parrinello, Phys. Rev. Lett. **60**, 204 (1988) 215
14. T. Uda, Solid State Commun. **64**, 837 (1987) 215
15. R. Biswas, G.S. Crest, C.M. Soukoulis, Phys. Rev. B **36**, 7473 (1987) 215
16. A. Polman, D.C. Jacobson, S. Coffa, J.M. Poate, S. Roorda, W.C. Sinke, Appl. Phys. Lett. **57**, 1230 (1990) 215
17. S.C. Jain, W. Schoenmaker, R. Lindsay, P.A. Stolk, S. Decoutere, M. Willander, H.E. Maes, J. Appl. Phys. **91**, 8919 (2002) 215, 218
18. L.S. Robertson, P.N. Warnes, K.S. Jones, S.K. Earles, M.E. Law, D.F. Downey, S. Falk, J. Liu, Mater. Res. Soc. Symp. Proc. **610**, B4.2 (2001) 215
19. P. Ashburn, C. Bull, K.H. Nicholas, G.R. Booker, Solid State Electron. **20**, 731 (1977) 215
20. C. Bull, P. Ashburn, G.R. Booker, K.H. Nicholas, Solid State Electron. **22**, 95 (1979) 215
21. A. Claverie, B. Colombeau, G.B. Assayag, C. Bonafos, F. Cristiano, M. Omri, B. de Mauduit, Mater. Sci. Semicond. Proc. **3**, 269 (2000) 215
22. H.G.A. Huizing, C.C.G. Visser, N.E.B. Cowern, P.A. Stolk, R.C.M. de Kruif, Appl. Phys. Lett. **69**, 1211 (1996) 215, 221
23. J.-H. Li, K.S. Jones, Appl. Phys. Lett. **73**, 374 (1998) 215, 221

24. P.A. Stolk, H.-J. Gossmann, D.J. Eaglesham, D.C. Jacobson, C.S. Rafferty, G.H. Gilmer, M. Jaraíz, J.M. Poate, H.S. Luftman, T.E. Haynes, J. Appl. Phys. **81**, 6031 (1997) 215
25. S.M. Hu, Mater. Sci. Eng. R **13**, 105 (1997) 215
26. H.-J. Gossmann, in *Semiconductor Silicon*, ed. by H.R. Huff, U. Goselle, H. Tsuya, Elect. Chem. Soc. Proc. **98**, 884 (1998) 215
27. A. Agarwal, H.-J. Gossmann, A.T. Fiory, J. Electron. Mater. **28**, 1333 (1999) 215, 216
28. D.M. Camm, B. Lojek, in *Proc. 2nd Int. RTP Conf.* (1994), p. 259 216
29. R.T. Hodgson, V.R. Deline, S. Mader, J.C. Gelpey, Appl. Phys. Lett. **44**, 589 (1984) 216
30. A.T. Fiory, D.M. Camm, M.E. Lefrancois, S. McCoy, A. Agarwal, in *Proc. 7th Int. Conf. Advanced Thermal Processing of Semiconductors* (1999), p. 273 216
31. R.S. Tichy, K. Elliott, S. McCoy, D.C. Sing, in *Proc. 9th Int. Conf. Advanced Thermal Processing of Semiconductors* (2001), p. 87 216
32. A.T. Fiory, D.M. Camm, M.E. Lefrancois, S.P. McCoy, A. Agarwal, in *195th Electrochem. Soc. Symp. Proc.* (1999), p. 133 216
33. T. Sato, Jpn. J. Appl. Phys. **6**, 339 (1967) 216
34. A.C. King, A.F. Gutierrez, A.F. Saavedra, K.S. Jones, D.F. Downey, J. Appl. Phys. **93**, 2449 (2003) 217
35. F. Cristiano, B. Colombeau, B. de Mauduit, C. Bonafos, G. Benassayag, A. Claverie, Mater. Res. Soc. Symp. Proc. **717**, C5.7.1 (2002) 218
36. J.-Y. Jin, J. Liu, U. Jeong, S. Metha, K.S. Jones, J. Vac. Sci. Technol. B **20**, 422 (2002) 219, 220, 221
37. R.G. Elliman, S.M. Hogg, P. Kringhøj, in *1998 Intl. Conf. on Ion Imp. Tech. Proc.* (1998), p. 1055 219
38. W.F.J. Slijkerman, P.M. Zagwijn, J.F. van der Veen, G.F.A. van de Walle, D.J. Gravesteijn, J. Appl. Phys. **70**, 2111 (1991) 219
39. J. Borland, Mater. Res. Soc. Symp. Proc. **717**, C1.1.1 (2002) 219
40. B. Park, F. Spaepen, J.M. Poate, F. Priolo, D.C. Jacobson, Mater. Res. Soc. Symp. Proc. **128**, 243 (1989) 220
41. L.S. Robertson, A. Lilak, M.E. Law, K.S. Jones, P.S. Kringhøj, L.M. Rubin, J. Jackson, D.S. Simons, P. Chi, Appl. Phys. Lett. **71**, 3105 (1997) 220
42. K.S. Jones, K. Moller, J. Chen, M. Puga-Lambers, B. Freer, J. Berstein, L. Rubin, J. Appl. Phys. **81**, 6051 (1997) 221

Index

Magnetic Properties and Ion Beams: Why and How

T. Devolder[1] and H. Bernas[2]

[1] Institut d'Electronique Fondamentale, Université Paris-Sud 11, 91405 Orsay, France
[2] CSNSM, CNRS and Université Paris-Sud 11, 91405 Orsay, France, e-mail: bernas@csnsm.in2p3.fr

Abstract. We show how structural modifications due to ion irradiation affect the magnetic anisotropy, and how this may provide information on, and ultimate control of, basic interactions in magnetic nanostructures. Novel applications are demonstrated.

1 Introduction

This chapter is not a general review of ion-beam effects on magnetism. We deliberately focus on a single property, the magnetic anisotropy, and show in some detail how the combined knowledge of magnetic properties and ion-irradiation physics can provide information on basic interactions, and lead to novel applications. Our example involves magnetic nanostructures in ultra-thin films, not just because they are fashionable and interesting, but because they constitute a prime example of the specific potential advantages of ion-beam treatments in magnetism. The basic electronic structure properties of materials are all related to some characteristic lengths, whose scales depend on which property is considered: e.g., the electron mean-free path or the Fermi wavelength for conductivity, the Debye wavelength for phonons, the domain-wall width or the exchange length for magnetic interactions, the pair-correlation length for superconductivity.... All these are in the range ~0.1 to a few tens of nanometers. Many of the typical lengths involved in thin-film magnetism (e.g., the exchange length, the domain-wall width or Barkhausen jump length) are nanometric, so that reducing the lateral sample size to such scales leads to new insights into nanomagnetism and to potential applications such as ultrahigh density recording via nanopatterned magnetic media.

When the physical dimensions of the sample shrink down to the characteristic length scale of a basic physical property, the latter is radically affected by its size and shape, by the symmetry of its environment and by its coupling (chemical bonds, radiation...) to the latter. This is an example of the "Third Way" [1] (Fig. 1) differing from the familiar "top-down" (involving lithography of some kind) and "bottom-up" (involving chemical, e.g., colloidal) techniques.

How to design the irradiation conditions so that adequate control is obtained over the nature and properties of the nanosized metastable alloy?

H. Bernas (Ed.): Materials Science with Ion Beams,
Topics Appl. Physics **116**, 227–254 (2010)
DOI: 10.1007/978-3-540-88789-8_8, © Springer-Verlag Berlin Heidelberg 2010

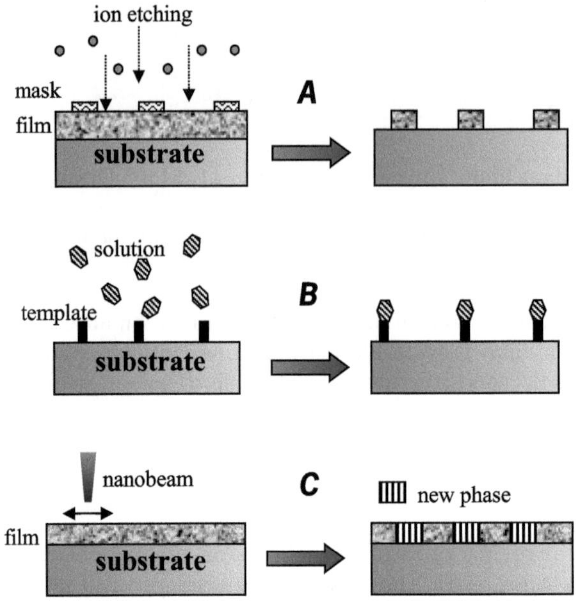

Fig. 1. Nanostructuring techniques. "*Top-down*" (*A*) involves ion or photon treatment through a mask (lithography). "*Bottom-up*" (*B*) involves self-organized deposition of a solution component on a template. The "Third Way" (*C*) involves treating a predeposited film or cluster array by an ion (or electron, or photon) nanobeam in order to induce localized phase changes without requiring further manipulation. The nanobeam may be produced by focusing or via a stencil mask. From [1]

A specific feature of ion beams is that they can *nanopattern the physical properties* as well as the sample geometry. Because of the size effects mentioned above, a detailed understanding of the consequences of irradiation (implantation) on *both* the nanostructure *and* the physical properties to be tailored is required. Typical ion impacts and displacement sequences are on the nanometer scale, and their size and density may be controlled by varying beam conditions. On the other hand, it is also important to stress that ion-beam treatments in nanostructure fabrication generally make sense within a "building strategy" in which a finely tuned feature of ion-beam interactions is one step among others leading to the final nano-object.

2 Magnetic Anisotropy in Ultrathin Films

There are many ways to induce changes in the magnetic properties of materials. Among them, a well-known favorite is the modification of the exchange interaction (whose range is typically the interatomic distance) by alloying. This is of course important in basic studies, but although it affects the saturation magnetization in an oft-controlled way, applications generally also require other characteristics. The orientation of the magnetic field relative to a sample plane, the structure of magnetic domain walls, and more generally the magnetic-reversal properties under an applied external magnetic field – all these are often crucial. The availability of artificial magnetic thin-film arrays has opened up vast new possibilities in this area, and the use of ion beams is one of the valuable additions to the latter. A major novel feature brought by the introduction of metallic superlattice structures, besides the possibility of controlling the interlayer exchange interaction, is precisely a control over the magnetic anisotropy and the magnetic-reversal properties. The use of ion beams to literally design these properties [2] will be detailed here. Another example involving beam-controlled modifications of the so-called exchange-bias phenomenon has been reviewed elsewhere [3].

Magnetic anisotropy is due to spin-orbit coupling [4], hence it is directly related to a broken symmetry in the atomic environment. The origin of the asymmetry may be magnetocrystalline (lattice space group), magnetoelastic (uniaxial strain) or interfacial. The former contributions are proportional to the magnetic volume: (i) The magnetocrystalline anisotropy is due to the anisotropic crystal structure, and its energy functional reflects the crystal symmetries. For example, hcp cobalt exhibits a strong uniaxial anisotropy along the c-axis, whereas the anisotropy energy of the far more structurally isotropic fcc cobalt is ten times lower [5]. A given material may thus have very different anisotropies if it exists in different crystal structures, (ii) Magnetostriction can also be a major source of anisotropy [6]. Applying uniaxial stress to a magnetic system changes its magnetization. Since the resulting unit-cell deformation amplitude depends on the applied force's direction, the magnetization change also depends on the strain tensor. The origin of this magnetic anisotropy is magnetoelastic. For instance, a cobalt thin-film crystal whose Co–Co distances are different for out-of-plane bonds and for in-plane bonds exhibits uniaxial (magnetoelastic) anisotropy. The cobalt planes in tensile strain are magnetically harder than in unstrained cobalt, while the compressed crystal axis is an easier axis of magnetization. Large strain may occur in ultrathin films, especially when grown by epitaxy on a buffer with some lattice mismatch. Ion beam effects on strained magnetic materials are well documented [7]. The latter two contributions to magnetic anisotropy arise from bulk properties, in contrast to (iii) the interface magnetocrystalline anisotropy, which stems [8] from a symmetry breaking of the magnetic environment of the magnetic atom. It can be generated by an asymmetry in the number of bonds: e.g., in ultrathin cobalt films, there are more Co–Co bonds in the film plane than in the growth direction, due to the stacking character of the sample. The magnitude of the interfacial anisotropy depends very sensitively on the details of the interface state.

The easy axis of magnetization is determined by the minimum of the magnetic anisotropy total energy. In its simplest form, it is expressed as:

$$E_{\text{tot}} = \left(-2\pi M_{\text{S}}^2 + K_1\right) \sin^2\theta, \tag{1}$$

where K_1 is the first-order magnetocrystalline anisotropy parameter. θ is the angle between the magnetization and the sample surface normal. The quantity $-2\pi M_{\text{S}}^2$ is the macroscopic shape anisotropy term, that strongly favors inplane magnetization. When $K_1 < 2\pi M_S^2$, the sample exhibits an easy plane of magnetization, which is the sample plane. Conversely, when the magnetocrystalline anisotropy is stronger than the shape anisotropy, the thin film exhibits an easy axis of magnetization that is perpendicular to the film plane.

Adjusting the various anisotropy contributions has led to numerous perpendicular magnetic anisotropy (PMA) systems, a property of major interest for ultrahigh-density magnetic recording [9] in which the easy axis of magnetization is perpendicular to the thin film plane. This requires that the magnetic anisotropy overcome the dipole interaction, which – given the thin film's shape – should favor inplane magnetization. Among PMA structures, Co–Pt multilayers have attracted considerable interest because of a huge magneto-optical Kerr effect (MOKE) in the blue/UV spectrum [10]. When Co thicknesses are sufficiently small in these structures, interface effects are strong enough to overcome the macroscopic shape anisotropy and to induce an easy PMA axis. Studies of these particular artificial multilayer structures will be used as examples in the following.

In order to modify and control the anisotropy *magnitude*, one has to change the property of the interface alone, and not that of the crystal structure because that would affect the anisotropy energy functional *symmetry.* We show how this may be done via ion-beam modification, detailing the steps in the reasoning and the experiments in order to clarify how irradiation affects the microscopic origin of the changes in magnetic anisotropy, Curie temperature and magnetic-reversal properties.

3 Controlling Thin-Film Magnetic Anisotropy by Ion Irradiation

3.1 The Strategy

The magnetic anisotropy of these multilayers is dominated by the magnetocrystalline term. Symmetry breaking at interfaces induces PMA; conversely, the magnetic anisotropy of the layers – which is most sensitive to the asymmetry of the Co atoms' immediate environment – may be modified in a controlled way by changing the short-range order around the Co and Pt atoms at the interface, i.e., by roughening the latter. In the case of ultrathin films, any change in the interface structure actually modifies the local environment around a large fraction of the film's atom population. This may be done by using a very simple version of "ion-beam mixing", involving room-temperature collision-induced mixing by light (He) ions at energies typically ranging from 10 to 150 keV [11, 12]. In that energy range, the average energy transferred by the He ion in a collision is very low (~100–200 eV), so that Pt or Co recoils only travel 1–2 interatomic distances on average. The temperature is sufficiently low to avoid diffusion (but relaxation occurs). The displacement cross sections are low (typically 1% per atomic layer), so that the irradiation parameters (including beam flux and fluence) allow very precise control over the "mixing rate". This is also an ideal case for Monte Carlo simulations of the corresponding interface modifications. The experimental strategy for nanostructure patterning of Co/Pt ultrathin layers by light-ion beam-induced intermixing was described [2] in 1998.

3.2 Modeling Ballistic Recoil-Induced Structural Modifications

A simple, purely collisional mixing model predicts the system's structural evolution under keV light-ion irradiation. The distribution $d\sigma/dT$ of energy T transferred in motional degrees of freedom when a beam of 30-keV He^+ ions enters an infinitely thick Pt layer varies approximately as $1/T^2$, so that low transferred-energy collisions are by far the most probable events [13]. When T is below the displacement threshold T_d in metals (~25 eV), the energy is dissipated into phonons and the Pt atom displacement probability vanishes. About 92% of the collisions above T_d transfer less than 200 eV. Only 25% of the collisions above T_d exceed $3T_d$ and may induce secondary collisions [14, 15]. Collision cascades are thus absent and the structural modifications are confined to the immediate vicinity of the ion path in the metal. The projected range of a recoiling atom penetrating a host (assumed to be amorphous), may be calculated from the "stopping and ranges of ions in matter" (SRIM) Monte Carlo code [16] that assumes binary collisions. For the hosts discussed here, it is found that the interstitial-to-vacancy distance is almost always below 8 Å, so that this defect disappears via recombination of the vacancy with a neighboring atom. The irradiation process thus reduces exclusively to substitutions of atoms: in our case, either Co by Co, Pt by Pt, Co by Pt or Pt by Co.

For example, Fig. 2 shows the result of a SRIM simulation performed on a sequence Pt(41 Å)/Co(13 Å)/Pt(22 Å) with initially pure layers, in which the

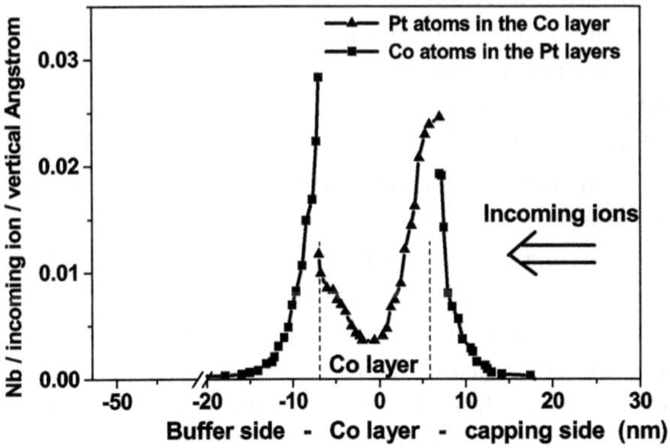

Fig. 2. Distribution of foreign atoms in an irradiated Pt/Co(13 Å)/Pt structure, i.e., cobalt atoms (*squares*) in the buffer and capping layers and Platinum atoms (*triangles*) in the nominal Co layer. SRIM simulation after 8×10^5 incoming ions

first stages of ion-beam mixing with 30-keV He ions leads to a foreign-atom distribution, i.e., Co atoms scattered into the Pt buffer or into the Pt cap layers, and Pt atoms scattered into the Co layer. There are 40% more Pt atoms knocked into the Co layer than Co atoms knocked into the Pt layers. Thus, immediately after irradiation (before any diffusion or recombination process) the atomic density of the different layers has changed, and differs on both sides of the Co film. The probability for a given atom to undergo two collisions with He ions, even at an incoming fluence of 10^{16} ions cm^{-2} is low (below about 10%); cumulative mixing (leading to metastable alloying) is only expected at fluences well above this value.

The calculated foreign-atom distribution provides (i) the characteristic lengths of the problem, (ii) the mixing rates, and (iii) an insight into sandwich roughness asymmetry. The mean transferred energy in a He–Co (resp. He–Pt) collision is 85 eV (resp. 68 eV). The corresponding path of 85-eV Co in a Pt layer is 3 Å, whereas that of 68-eV Pt in a Co layer is 5 Å. As seen in Fig. 2, the postcollision Co atom distributions fall off exponentially away from the initial interfaces with decay lengths, respectively, 2.7 Å and 1.5 Å. The Pt distribution is smoother: the typical decay lengths are ~4.5 Å on both the cap and buffer side, so that the Pt concentration is nonzero at the center of the Co layer. These numbers correspond to all events weighted by their respective probability, and are related to the most probable range of a recoil after the collision. Since Pt atoms penetrate deeper in the Co layer than do Co atoms in the Pt layers, the Pt concentration gradient inside the Co layer is less abrupt than that of Co in Pt. Because of momentum conservation, Co atoms are mainly driven into the underlying layer, whereas Pt cap atoms are driven into the upper part of the cobalt layer. Secondary collisions, in which a previously displaced atom can collide with a neighbor and may send some recoils towards the surface, are rare events in light-ion irradiation and have very low energies. Hence, we expect a significant asymmetry between the mixing rates above and below the cobalt layer.

The interface roughness is obviously affected by the fact that the number of foreign atoms is maximum at the interfaces. To evaluate alloying, we estimated the number of foreign atoms in each entire layer. To estimate interface roughness, we determine the proportion of foreign atoms located closer than 3 Å apart from a given interface. From Fig. 1, this proportion is 36% near the lower Co interface for a fluence of 10^{16} ions cm^{-2}, and is 1.4 times larger near the upper Co interface. From the discussion above, Co atoms recoiling inwards travel far enough for alloy formation to be assumed, whereas the Co atoms transferred towards the surface (via secondary collisions) travel lower distances, staying in direct contact with the Co layer, so that a roughness increase is triggered. A schematic of this scenario is displayed in Fig. 3.

The purpose of this somewhat pedestrian discussion was to show how a straightforward ballistic model predicts the different mixing rates, the mixing range in Pt and Co, and the roughness asymmetry of the Co layer, all of which

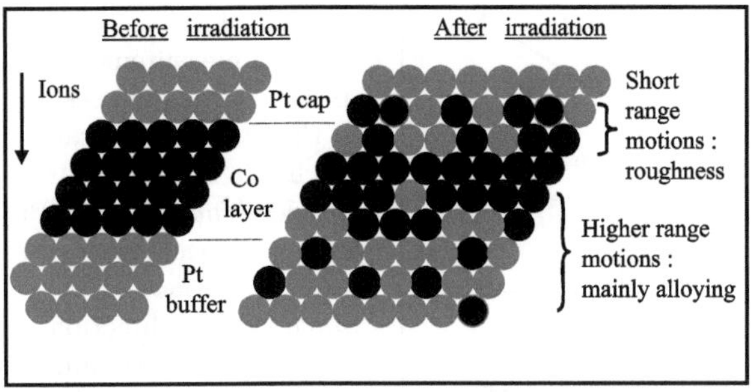

Fig. 3. Schematic of a sandwich structure before and after irradiation. Co (resp. Pt) atoms are in *black* (resp. *gray*). Irradiation towards the bottom of the figure. Co atoms moving in the ion direction travel more than one interatomic distance and contribute to Co–Pt alloying, whereas Co atom moving in the opposite direction typically travel only one interatomic distance and contribute to roughness

should determine the essential features of the magnetic anisotropy variations under irradiation (no corrections are made for any thermodynamic potential gradient contribution or for irradiation-induced crystal-lattice relaxation). We now compare with experiments.

3.3 Experimental Measurements of Structural Modifications

The interface mixing produced in Pt/Co/Pt films by a 30-keV He ion-irradiation was measured [17] via grazing X-ray reflectometry (GXR) [18–20]: modeling the spectra provides the composition modulation perpendicular to the interfaces. Consider the representative behavior of a Pt/Co 13 Å/Pt sandwich sputtered on a sapphire (Al_2O_3) crystal substrate. We can measure the fluence-dependent increase of interfacial roughnesses, and model it more precisely by distinguishing between short- and long-range mixing as suggested above. The pristine Co layer of a Pt/Co/Pt sandwich is symmetric (the initial roughness of the upper and lower interfaces are 3.2 Å and 3.4 Å, resp.). Irradiation breaks this symmetry: the *upper* Co interface roughness increases at a much higher rate than that of the lower interface, as predicted in the simulation (Fig. 4).

The figure shows that irradiation progressively roughens the upper Co/Pt interface, whereas the lower Pt/Co interface undergoes much less roughening but forms a Pt-rich alloy layer at a distance greater than 5 Å from the Pt/Co interface. Thus, even in these nanometer-thick sandwiches, irradiation affects the upper and lower interfaces of a given Co layer differently: the Co atoms crossing the upper interface undergo short-range (2 Å) motion generating roughness increase, whereas Co atoms crossing the lower interface undergo higher-range (>5 Å) motions resulting in alloy formation. This confirms the He irradiation-induced local compositional changes around the magnetic Co

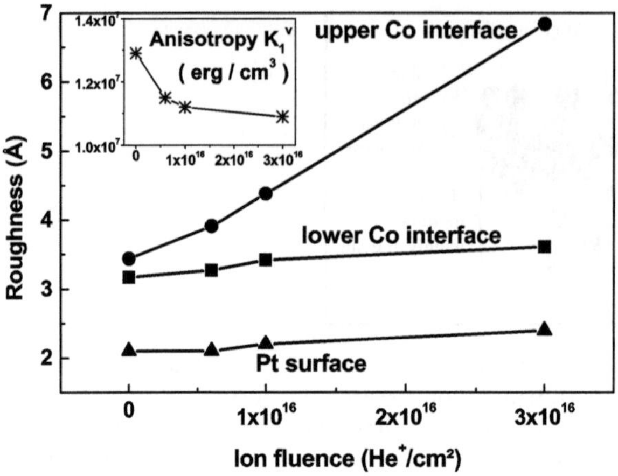

Fig. 4. Experimental (GXR) measurement of interface roughnesses for a Pt/Co(13 Å)/Pt sandwich: Pt buffer/Co interface (*squares*), Co/Pt cap interface (*circles*) and Pt surface (*triangles*), as a function of the 30-keV He-ion fluence. The inset displays the value of the first-order magnetocrystalline anisotropy parameter K_1

atoms expected from the SRIM Monte Carlo calculations (also as expected, evidence for the formation of a more concentrated alloy only appears at the highest fluence, around 10^{17} He ions cm^{-2}). However, although the roughness asymmetry is obviously important to understand the magnetic properties, the experimental correlation between variations in roughness and magnetic anisotropy is too weak to account entirely for the latter.

GXR measurements ignore both the atomic structure and crystallographic order – hence local structural symmetry – which in fine largely determine the magnetic anisotropy. A more direct study of the local environment around cobalt atoms, and its symmetry, was performed via extended X-ray absorption fine structure (EXAFS) measurements [21], which determine the local structure (elements and interatomic distances in near-neighbor shells) around, e.g., Co atoms, and provide quantitative insight into the microscopic origin of the magnetic anisotropy decrease after irradiation. In EXAFS, the electronic transition resulting from atomic excitation by an X-ray photon is related to the electric-dipole Hamiltonian, so the EXAFS signal intensity depends on the angle between the direction of the atomic bond and that of the X-ray electric vector **E**. Measurements carried out with **E** in the sample plane or perpendicular to it thus provide a measure of the local structural anisotropy [22]. EXAFS performed on Co/Pt multilayers [23] irradiated at low fluences first revealed a major effect that was ignored in the SRIM simulations. The perpendicular bond length of Co atoms was not significantly affected by irradiation, whereas the inplane lattice parameter varied from 2.6 Å to 2.51 Å: thus, the initial tensile stress was *entirely relaxed* by the first stages of irradiation, the unit-cell volume around cobalt atoms being reduced by 6%. The data also indicate less dispersion in the interatomic distance, consistent with a relaxed structure. Hence, even before any significant intermixing takes place, the initial stage of irradiation triggers a complete release of the tensile strain of the cobalt layers. This explains the initial drop in K_1 (Fig. 3, inset) and has a major impact on the magnetic anisotropy (Fig. 5).

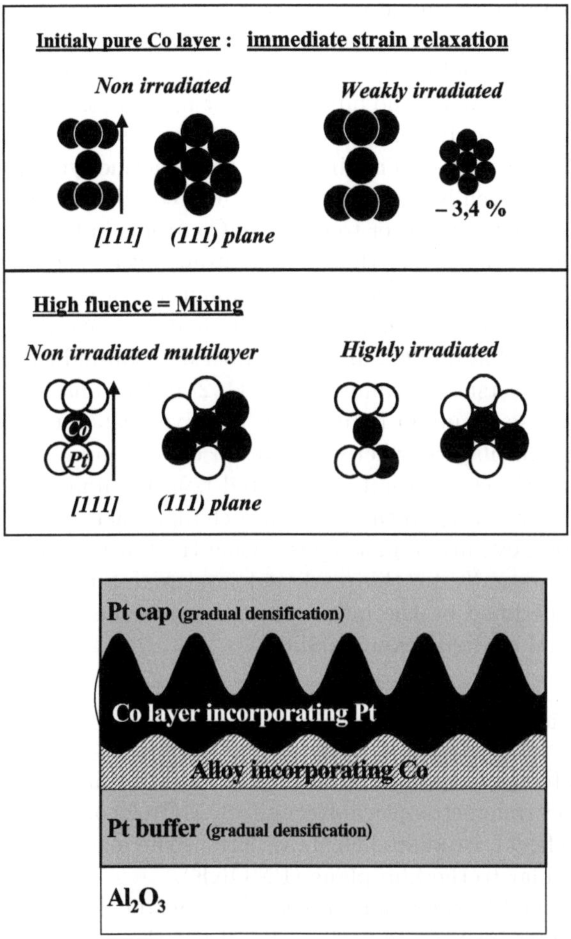

Fig. 5. Schematic of the two main structural effects of ion-irradiation. *Upper*: structural anisotropy comes from both the Co–Co bond lengths (modified by strain relaxation) and the repartition of the number of these bonds (due to ion-beam mixing). *Lower*: Mixing characteristics: asymmetric roughness increase and formation of an alloy layer

At higher irradiation fluences, intermixing (Fig. 5b) sets in progressively [2, 24, 25] as expected and the EXAFS results show that irradiation ultimately drives the Co environment towards a far more isotropic state (a structure resembling that of a $CoPt_3$ reference compound). Replacing Co neighbors by Pt neighbors decreases the structural anisotropy around Co atoms, increasing and uniformizing their interatomic distances, while preserving the layers' crystal structure (fcc and hcp, resp.). As shown below, the easy magnetization axis finally tilts and falls in the film plane, because the weakened magnetocrystalline anisotropy no longer overcomes the shape anisotropy. Experiments and calculations agree as regards the different behavior at the Pt underlayer and the Pt buffer interfaces, and the asymmetry in alloy concentrations over and under the Co layer is also found as anticipated. The mixing effect is thus mainly driven by the collisional mechanism simulated by SRIM. But a detailed examination reveals a bias in this process. Ballistic recoils alone do not account exactly for the asymmetry between the upper and lower Co/Pt interface roughnesses, whose experimental ratio is 2, versus the calculated value of 1.4. As for the mixing rate, it appears that Co atoms enter Pt as predicted by purely collisional mixing, whereas fewer Pt atoms enter the Co layer than expected. This indicates that thermodynamics cannot be entirely neglected even in these nonequilibrium conditions: the Co–Pt phase diagram shows[3] that it is easier to stabilize Co atoms in the Pt fcc matrix (favoring alloy formation) rather than the Pt atoms inside the Co hcp layer (segregation resulting in roughness). The combination of equilibrium and irradiation-induced effects was quantified in the linear regime corresponding to very light ion irradiation [28], and illustrates why ion-induced modifications are rather stable versus annealing at 200°C [2] in this miscible binary alloy system. The reverse occurs in immiscible (positive heat of mixing) systems such as Fe/Ag multilayers, where irradiation or annealing can even lead to more abrupt interfaces than initially existed [29]. In summary, ion irradiation contributes in two ways to the structural changes that determine the magnetic anisotropy's evolution: first, by releasing the tensile strain (a large, ∼30% effect), and secondly by the predicted intermixing effect – the latter being somewhat modified by the influence of thermodynamics in determining the postcollisional configuration stabilities.

3.4 Experimental Variation of the Magnetic Anisotropy

Figure 6 displays the magneto-optical ellipticity of Pt/Co/Pt ultrathin sandwiches on reflecting substrates (magneto-optical Kerr effect, MOKE) or transparent substrates (Faraday effect), irradiated as above, versus the magnitude of an applied field perpendicular to the film plane (PMOKE).

The ellipticity is proportional to the magnetization's perpendicular component $M_\perp$, the proportionality constant depending slightly on the structure. We thus obtain experimental magnetic hysteresis cycles as a function of the irradiation fluence (in ions cm^{-2}) measured at 300 K. Before irradiation, the easy axis lies along the film normal, signaling strong uniaxial anisotropy along this direction. Several features are apparent, which will be discussed in the following sections: (1) the easy magnetization axis remains perpendicular to the film plane, the hysteresis cycles remain square, and both the saturation and remanent magnetizations remain approximately constant up to quite high

[3] At 300°C, the maximum stable concentration of Pt in hcp Co is ∼15%, whereas the maximum amount of Co in fcc Pt is more than 60% (no data available at room temperature because of kinetic limitations), from [26, 27].

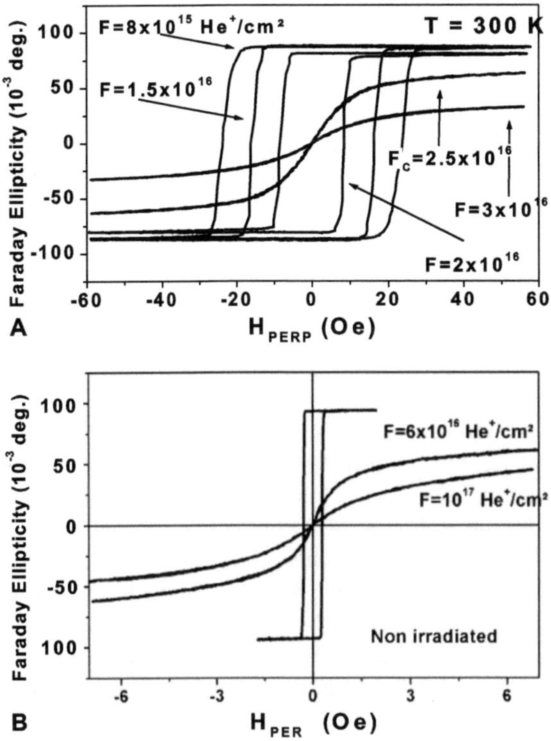

Fig. 6. Polar magneto-optical Faraday effect hysteresis loops of Pt/Co(5 Å)/Pt/Al_2O_3 films. The loops are measured with strictly identical sweeping rates. (**A**): After medium irradiation fluences. (**B**): Before irradiation and after high fluences

fluences, while (2) the coercive field is progressively reduced as the fluence is increased; (3) above a critical fluence F_C, the easy magnetization axis is progressively tilted into the film plane. It is already clear that ion irradiation can tailor interesting magnetic properties. Let us see how.

3.5 Relation Between Structural and Magnetic Anisotropies

We consider in turn the various contributions. (1) We saw that the pristine Pt/Co(13 Å)/Pt sandwich exhibits significant strain, whose release at low irradiation fluences leads to a large reduction in magnetic anisotropy (when the initial tensile strain is low, no change in magnetocrystalline anisotropy is observed in the early stages of ion mixing). Due to the lattice mismatch between Pt and Co, the as-deposited Co layer is in tensile stress. The magnetoelastic anisotropy of Co being negative, it is lowered by any reduction in the Co layer tensile stress. The magnetoelastic anisotropy energy per unit volume [6] being $(3/2)\times\lambda\times\varepsilon\times E_y$ where $l = -5\times10^{-5}$ is the magnetostriction coefficient of Co dense planes, $E_y = 2.1\times10^{12}\,\mathrm{dyn\,cm^{-2}}$ their Young's modulus, and ε is the in-plane stress in the magnetic layer. As $\lambda \leq 0$, a cobalt layer under *tensile* strain ($\varepsilon < 0$) has an increased anisotropy. Relaxing this strain reduces the magnetoelastic contribution to the total magnetic anisotropy energy. Strict application of this model to the above case with $\varepsilon = 3.4\%$ would overestimate the magnetoelastic contribution's reduction considerably, since the calculation assumes that the strain relaxation occurs at constant unit-cell volume, but the qualitative variation is correct. (2) The structural anisotropy reduced by irradiation (inset in Fig. 4) leads to a decrease in MA and coercivity, then to a reorientation of the easy magnetization axis. From (1), in the absence of the interface anisotropy the easy axis of magnetization would be in-plane for all layer thicknesses and irradiation fluences. Also [30], the more abrupt the interfaces, the larger the interface anisotropy. Hence, irradiation-induced interface mixing can only lower the interface anisotropy. Such a decrease should scale with the roughness evolution, which was assessed to be linear with regards to the fluence in the previous section. Note, however, that only the upper interface roughness is significantly affected by irradiation, so that the contribution of each interface should not be altered in the same way; schematically, the interface contribution to the total magnetic anisotropy can not be reduced by more than a factor of two. In the case of the thinnest, initially relaxed (5 Å Co) Pt/Co sandwiches, the irradiation-induced decrease of anisotropy is linear for low irradiation fluences. In such ultrathin Co layers all Co atoms are at interfaces, so that the anisotropy decrease may be ascribed exclusively to interface roughening. For sandwiches with larger cobalt thickness, after the abrupt anisotropy decrease at low fluences due to strain relaxation, the mixing process dominates at intermediate and high fluences. Upon room-temperature irradiation of Pt/Co systems, the random incorporation of Pt into bulk Co leads to a decrease of the magnetocrystalline contribution from $4.5 \times 10^6\,\mathrm{erg\,cm^{-3}}$ (pure hcp Co) to ~ 0 (random fcc Co–Pt alloy). Since EXAFS results rule out any change in the local lattice space group, the symmetry of this contribution to the total magnetic anisotropy is not altered by irradiation. The magnetocrystalline anisotropy of chemically disordered Co–Pt alloys is much smaller than the interfacial contributions and may be neglected.

The evolution of the Co sandwich anisotropy is thus accounted for. K_1 first decreases because of combined strain relaxation in the Co layer and upper interface roughening. Then, when the stress is relaxed, roughening of the upper Co/Pt interface leads to a decrease in the magnetic anisotropy. This interfacial magnetocrystalline anisotropy contribution simply arises from an unbalanced number of Co–Co bonds in the growth direction and in the plane of the sample. Its evolution versus ion fluence can be estimated using the Néel model [8]. The magnetocrystalline anisotropy energy w_i of a cobalt atom and its neighbor i is:

$$w_i = (\ell_i + m_i \delta r_i)\left[\cos^2\theta_i - 1/3\right],$$

where θ_i is the angle between bond Co-i and the magnetization, δr_i is the experimental bond length minus the equilibrium bond length r_0. ℓ_i reflects the nature of the bond and m_i reflects the bond magneto-elasticity. Using known values from the literature [31] and neglecting possible lattice parameter gradients, the magnetocrystalline anisotropy per Co atom K_{mc} is [23]:

$$K_{\mathrm{mc}} = (1/4)(b_{\mathrm{Co}//} - b_{\mathrm{Co}\perp})(2\ell_{\mathrm{Co\text{–}Pt}} - \ell_{\mathrm{Co\text{–}Co}}),$$

where the b parameters are the numbers of Co–Co bonds in-plane or perpendicular to the film.

At this stage of irradiation, the magnetic anisotropy thus scales with the imbalance between the number of Co–Co bonds in the growth direction and in the sample plane: using Néel's model, the EXAFS data provide the interfacial magnetic anisotropy contribution.

In summary, the magnetoelastic contribution due to the tensile strain of the cobalt-rich dense planes dominates in the early stages of irradiation (fluences up to $2 \times 10^{15}\,\mathrm{He\,cm^{-2}}$), and the irradiation-induced change in the ratio of in-plane versus out-of-plane cobalt bonds accounts for the interfacial magnetic anisotropy decrease.

3.6 Magnetic Reversal Properties Under Irradiation

In magnetic media with strong uniaxial perpendicular anisotropy, magnetization reversal proceeds through nucleation of reversed magnetized domains followed by possible growth of these domains through domain-wall (DW) motion. If the magnetic sample is structurally inhomogeneous, DW motion may be perturbed. If it is totally inhibited, the entire sample can only reverse its magnetization through nucleation and growth of reversed domains. Nucleation is then a thermally activated process occurring at major structural extrinsic defects, with a distribution of local nucleation fields, leading to a significant coercivity slope in the corresponding hysteresis loop (Fig. 7).

On the other hand, magnetization reversal in high-quality ultrathin-films with large perpendicular magnetic anisotropy occurs through easy DW propagation following very rare (ideally, a single) nucleation events located at some extrinsic structural defect(s). The DW sees a homogeneous energy landscape and may thus move without deformation. Far from the initial nucleation site, the DW then has a very large curvature radius. The corresponding hysteresis loop is essentially square, its coercivity being the nucleation threshold of the entire sample. Simple models [32] determine the volume fraction in which magnetization reverses through nucleation with respect to the total magnetic area. When this fraction is low, the energy barriers for nucleation are higher than those of reversed domain expansion by domain-wall motion: the nucleation field H_{N} is greater than the DW propagation field H_{P}. Magnetic aftereffect experiments reveal that the magnetization reversal process is invariant by a scaling factor $[H/\ell n(t)]$, meaning that all reversal steps are thermally activated. The DW velocity follows [33] an Arrhenius law:

$$v = v_0 . \exp\left[-\left(E_{\mathrm{p}} - 2M_{\mathrm{S}} \times t_{\mathrm{Co}} \times \ell_{\mathrm{B}}^2 \times H\right)/k_{\mathrm{B}}T\right],$$

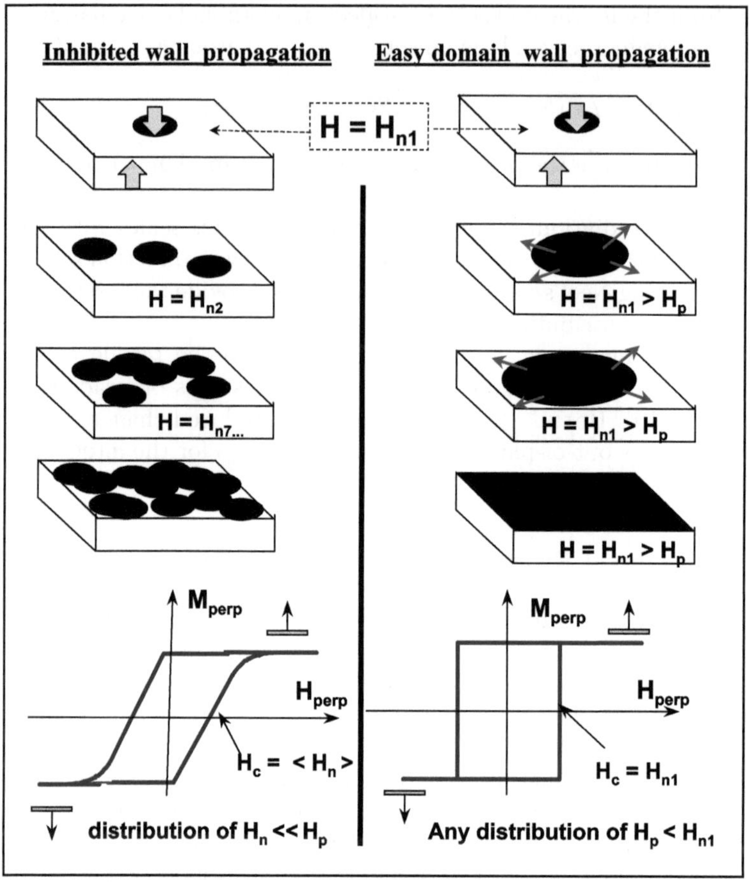

Fig. 7. Schematics of magnetization reversal modes in a perpendicularly magnetized high-anisotropy ultrathin-film

where E_p is an energy barrier, ℓ_B is the so-called Barkhausen length, i.e., the typical elementary jump distance of a propagating DW, M_S the saturation magnetization and t_Co the Co film thickness. It involves a convolution of the wall width and of the typical distance between the domain-wall friction centers: the higher ℓ_B, the less numerous the defects affecting, domain-wall velocity – it is thus a measure of a magnetic sample's lateral homogeneity.

How does changing the defect size or population affect DW motion? Thermal annealing of Pt/Co/Pt ultrathin sandwich samples, e.g., for 30 mn at 160°C, drastically increased the coercivity and strongly reduced the magneto-optical signal, while the loop rounded off, changing the magnetization-reversal mechanism [34] from easy DW propagation to DW pinning and progressive nucleation. This contrasts with the effect of He irradiation (Fig. 5), which reduced the coercivity while retaining the loop's square shape, demonstrating [35] that the magnetization-reversal mechanism occurs through very few nucleation events followed by easy DW propagation. Coercive fields as low as a few Oe were reproducibly obtained by these He-ion irradiations. This evolution is due [36] to a reduction of the magnetocrystalline anisotropy parameters. For example, in the case of Pt/Co(5 Å)/Pt, K_1 drops continuously from 13×10^6 erg cm^{-3} to 7×10^6 erg cm^{-3} as the fluence is increased from zero to 2×10^{16} He cm^{-2}. Above a critical ion fluence $F_C = 2.5\times10^{16}$ He$^+$ cm^{-2}, the remanent magnetization vanishes and the loop becomes fully reversible (Fig. 6b) at room temperature. Its Langevin function shape indicates a paramagnetic state with large magnetic susceptibility. The Curie point (T_C) is then found to be just below room temperature. The susceptibility diminishes further when $F > F_C$, indicating that T_C also decreases. Irradiation thus reduces both the anisotropy and the Curie temperature. Its detailed influence also depends on the cobalt film thickness [37]. The effect of irradiation on the anisotropy of magnetic multilayers is similar, except that the initial Curie temperature being significantly higher, irradiation never reduces it enough to trigger a paramagnetic transition at room temperature; but the anisotropy reduction does trigger an easy magnetization-axis reorientation into the sample plane.

Hence, the effect of irradiation on Pt/Co sandwiches or multilayers is first to reduce the magnetocrystalline anisotropy (hence the coercive force H_C), and then reorient the easy magnetization axis towards the in-plane orientation when the magnetocrystalline anisotropy becomes smaller than the shape anisotropy. Throughout, DW propagation dominates (square hysteresis loops) with a remarkable reduction in DW nucleation. The volume in which magnetization reverses through nucleation events is always a negligible fraction ($<10^{-4}$) of the total magnetic area. High-resolution magnetic imaging studies [38] suggested that typically less than one nucleation event per mm^2 occurred. They also showed that DW propagation occurred without any change of DW curvature from micrometer scales to at least millimeter scales, and hence that the dispersion in DW motion energy barriers is very low. The energy barriers for nucleation are higher than that of reversed domain expansion by DW motion, so that $H_N > H_P$. Even after irradiation, the propagation field H_P does not vary significantly over the sample. In Pt/Co(5 Å)/Pt/Al_2O_3 films, the Barkhausen length ℓ_B is 38 nm before irradiation and increases to 61 nm after an irradiation fluence of 10^{16} He cm^{-2}. Thus, irradiation actually *reduces* the density of effective DW pinning sites and enhances the lateral homogeneity of magnetic properties related to DW motion and domain nucleation. Since the number of irradiation-induced point defects certainly increases, this is counterintuitive and requires an explanation.

The pinning centers may be due to any structural inhomogeneity in the magnetic film: grain boundaries, cobalt terrace steps, roughness, short- or long-range fluctuations of cobalt thickness or film strain. The grain size is not affected by helium irradiation. Most atomic displacements involved in the irradiation-induced mixing process are very small, typically 2 to 5 Å. As a result, no extended defect such as a grain boundary or long-range cobalt thickness fluctuations can disappear in the irradiation process. Such defects must still pin the DW in the irradiated samples. In contrast, the cobalt terrace steps, or any initial abrupt variation of local structure will be strongly modified by irradiation. A fluence of 10^{16} ions cm^{-2} induces a Co/Pt interface intermixing that corresponds to a roughness slightly larger than the height of a terrace step (2 Å), and larger than many other possible abrupt local structure variations. This gradual blurring of the terrace steps and local abrupt structural variation as the ion fluence increases may enhance the mean distance between strong pinning centers and thus account for the irradiation-induced increase of the Barkhausen length. Quantitatively, in perpendicularly magnetized ultrathin-films, the DWs are Bloch walls of width $\Delta_{\mathrm{Bloch}} = 2\sqrt{A/K}$, where A is the exchange stiffness, $A \approx 10^{-6}\,\mathrm{erg\,cm}^{-1}$, and K is the effective total anisotropy energy as determined above. Thus, Δ_{Bloch} is ~6 nm before irradiation and ~15 nm just below F_{C}. It is well known [39, 40] that DWs are mainly affected by spatial variations approximating their width Δ_{Bloch}: an isolated defect of size Δ_{Bloch} is a pinning center but a dense assembly of similar defects will not pin the domain-wall, provided that the defect density is greater than several units per $\Delta^2_{\mathrm{Bloch}}$ surface. In magnetic multilayers and sandwiches, the typical irradiation-induced point defect is the substitution of a magnetic atom by a nonmagnetic atom. Because atomic displacements have a typical range of 2–5 Å, all cobalt atoms of a Pt/Co(5 Å)/Pt film may create such point defects. To estimate the density of point defects and its fluctuation, as introduced by irradiation, we consider N random atomic substitutions, on a macroscopic surface S, with mean surface density N/S. Since a 30-keV He ion typically induces 0.02 displacements per Å of matter along its track, a fluence of 10^{16} ions cm^{-2} induces at least $N/S \sim 10$ substitutions nm^{-2} in a 5-Å thick magnetic layer. The probability $P(n)$ of having n atomic substitutions on a surface $\sigma \ll S$ is

$$P(n) = \binom{n}{N} p^n (1-p)^{N-n},$$

if $p = \sigma/S$. This leads to a mean value $\langle n \rangle = Np$, i.e., the surface area multiplied by the surface density of point defects. The variance of n is $\Delta n^2 = Np(1-p)$, which reduces to $\Delta n^2 = Np$ in the limit of small surfaces $\sigma \ll$ S. The relative defect density fluctuation is $\Delta n / \langle n \rangle = 1/\sqrt{\langle n \rangle}$. Thus, for a $10 \times 10\,\mathrm{nm}^2$ surface, the defect density fluctuation is only 3% for a typical irradiation fluence of $10^{16}\,\mathrm{He}^+\,\mathrm{cm}^{-2}$. Hence, for surfaces whose size is relevant for DW propagation, e.g., $\sigma \sim \Delta^2_{\mathrm{Bloch}}$ and $\sigma \sim \ell^2_{\mathrm{B}}$, the irradiation-induced defect density does not exhibit significant fluctuations. The fact that irradiation induces many atomic substitutions ("defects") leads to such a large density of the latter that a DW covers many of them at the same time inside its width: the structural disorder introduced by the irradiation process actually eases DW propagation at room temperature. This conclusion is obviously of interest well beyond the samples and specific conditions from which it was drawn. A rather spectacular confirmation is provided by experiments [41] in which DW propagation in Pt/Co/Pt layers, as a function of applied field, was enhanced by a factor $\sim 10^4$ by prior keV He-ion irradiation. These irradiation effects on DW motion have produced excellent model systems to study [42] general properties of interface motion at varying speeds (creep, viscous motion, thermally assisted motion, etc.). They are also crucial to study mechanisms that may enhance magnetization reversal in recording media.

3.7 A Magnetic Anisotropy Phase Diagram

Let us summarize the fluence-dependent and thermal variations of the hysteresis properties. As before, we consider results relative to the Pt/Co(5 Å)/ Pt/Al_2O_3 sandwich. At all temperatures from 2 to 300 K, the easy axis of the unirradiated Pt/Co(5 Å)/Pt/Al_2O_3 sandwich is perpendicular to the sample plane, the hysteresis loop of the nonirradiated sample is square and the coercivity is determined by the lowest nucleation field H_N (Fig. 8). After irradiation, the remanent ratio $M_{H=0}/M_S$ diminishes upon cooling, even in those samples that have full remanent magnetization at 300 K. Below $\sim$100 K, M_S still increases, but the remanent magnetization does not. The cooling-induced magnetization increase enhances the shape anisotropy, which finally overwhelms the magnetocrystalline anisotropy, triggering a tilt of the easy magnetization axis. Also, there is a pronounced rounding of the hysteresis loop near H_C at low temperatures $T \leq 40$ K, indicating [44] that magnetization reversal no longer occurs by easy DW propagation. Low temperatures inhibit DW propagation and the propagation field H_P becomes larger than the nucleation fields H_N which now have multiple values.

At and above 300 K, typical hysteresis loops show that an irradiated, perpendicularly magnetized Pt/Co(5 Å)/Pt/Al_2O_3 sandwich experiences a ferromagnetic-to-paramagnetic transition at increasing fluences as the temperature is raised. The phase transition takes place in a very narrow temperature interval over the entire cm-sized sample, confirming the high lateral homogeneity of the irradiation's influence, at least as regards the Curie temperature. The critical magnetization $M(T \leq T_C)$ behavior shows universal scaling, with a rather small critical exponent b around 1/8. The magnetic susceptibility χ scales as $\chi \sim (1 - T/T_C)^{-\gamma}$ with $T_C = 340 \pm 1$ K and $\gamma \cong 7/4$. Scaling theories [45, 46] indicate that such values of the β and γ critical exponents are those of quasiperfect 2D Ising systems. The Curie temperature decrease upon irradiation signals that the system has 2D Ising spins when perpendicularly magnetized, and then has a 2D XY spin lattice when in-plane magnetized. This reduction in the order parameter dimensionality reduces the Curie temperature. For a complete discussion of the phase diagram in Fig. 9, see [43].

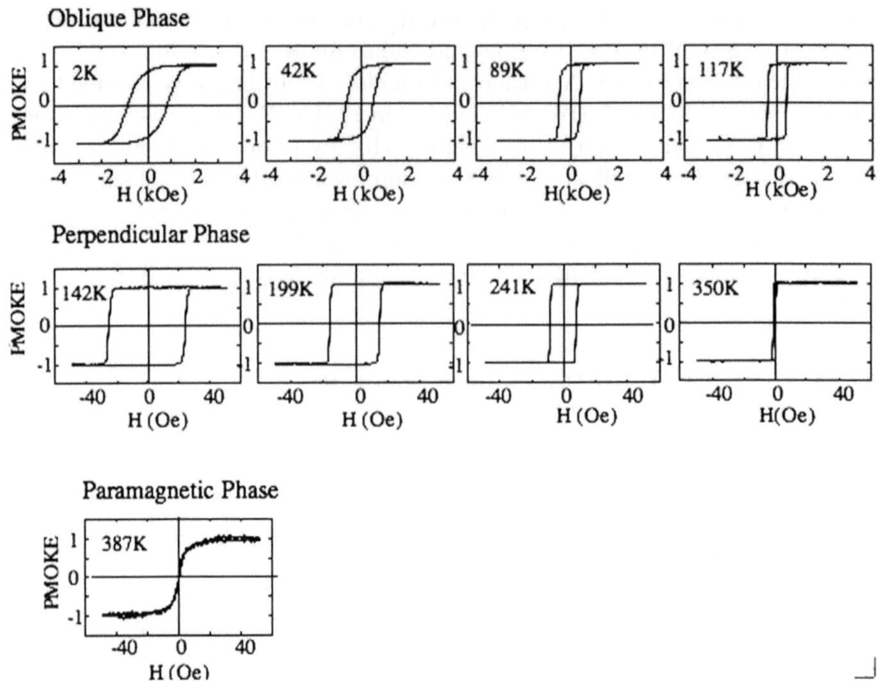

Fig. 8. Temperature dependence of the renormalized PMOKE hysteresis loops for the Pt/Co(5 Å)/Pt films irradiated to a fluence $F = 1.5 \times 10^{15}\,\mathrm{He\,cm^{-2}}$, in the oblique (low-temperature), perpendicular and paramagnetic phases. From [43]

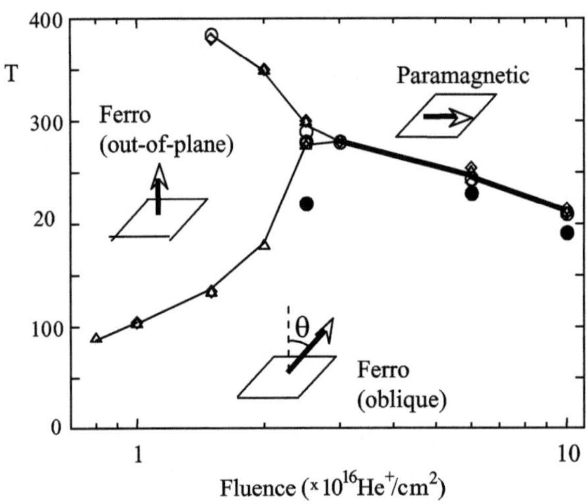

Fig. 9. Temperature-Fluence phase diagram of a Pt/Co(5 Å)/Pt film. The T_{C} (Curie temperature) and T_{D} (spin-deviation temperature) values were deduced from MOKE (*open* and *full triangles*), coercive fields (*diamonds*), susceptibility (*open* and *full dots*). From [43]

3.8 Summary

The structural modifications induced by keV light-ion irradiation alter the magnetoelastic and magnetocrystalline anisotropies, and hence the coercivity of films with perpendicular magnetization such as Pt/Co(5 Å)/Pt. In the latter, for example, the magnetocrystalline anisotropy reduction induces a paramagnetic transition near room temperature at a critical fluence (typically, $F \geq F_C = 2.5 \times 10^{16}\,\mathrm{He\,cm^{-2}}$). Ion irradiation may thus allow fabrication of perpendicularly magnetized samples with an adjustable Curie point around room temperature, as well as samples with a controlled orientation of the magnetization orientation and amplitude. As regards the paramagnetic phase transition, we conclude that the films irradiated at fluences strictly below F_C exhibit the characteristics of high-quality two-dimensional systems whose exchange interaction is described by an Ising Hamiltonian. At higher irradiation fluences, the easy magnetization axis abruptly tilts towards an easy cone, while the correlated change in the order-parameter dimensionality abruptly lowers the Curie temperature. Such films provide an interesting medium to study these phase transitions. For magnetic multilayers, the initial Curie temperature is far above room temperature, and irradiation never triggers a ferromagnetic to paramagnetic transition. As in the case of ultrathin sandwiches, it first reduces the coercive force, but then it induces a reorientation of the perpendicular easy axis of magnetization towards an easy plane, which is the sample plane. Below the critical fluence F_C, the magnetization reversal at room temperature occurs through a very limited number of nucleation events followed by easy DW propagation sweeping the whole sample surface. The density of pinning sites is reduced by the irradiation, because the defects created by irradiation are too densely packed to alter the DW propagation mode. The ability to adjust the technical magnetic properties via irradiation suggests that light-ion irradiation of magnetic thin-films could be of great interest for applications as well as a means of building model systems for various magnetic and phase-transition studies.

In the foregoing discussion, we have concentrated on *light*-ion irradiations at keV energies, in view of their conceptual and technical advantages for controlling the corresponding modifications. Why not use *heavy*-ion irradiations at energies ranging from a few tens to a hundred keV or more? Besides some archival interest, a prime reason to study this question is the potential advantage of using a focused ion-beam (FIB), usually of Ga, possibly of Si, Co or Au, to design chosen arrays of nanosized structural modifications. It has been shown [47, 48] that, as expected, the magnetic anisotropy changes contributed by atomic displacements scale roughly with the heavy ions' deposited-energy density. The nonlinearities due to damage cascades influence this scaling for ions heavier than, say, mass 20, but although more difficult, it is still possible to control magnetic anisotropy modifications at the sub-100 nm scale, and thus support applications with the 30-keV Ga beam of a FIB [49–51].

4 Magnetization Reversal in Irradiation-Fabricated Nano-Structures

At its present areal density growth rate, conventional magnetic recording on disks will very soon reach two major barriers [52]. The first is the so-called "contact recording limit", when the head-to-media clearance falls below 10 nm [53–56], requiring controlled ultra-smooth surfaces. In the case of heat-assisted magnetic recording, with heating provided by a near-field optical spot, abrupt spatial variations of the optical index will have to be avoided to ease control of thermal gradients around the laser spot. The second barrier is the thermal stability of written bits, the so-called "superparamagnetic limit". For particulate in-plane magnetization media, this limited the recording areal density to $\sim$200 bits μm^{-2}. Among the alternatives could be patterned magnetic media, where bit borders are geometrically defined using high-resolution large-area lithography techniques [57–60]. Bit sizes as small as 25 nm have indeed been demonstrated in this way. Combined with nm-scale masking techniques, light-ion irradiation would allow magnetic patterning (through a lithographically defined mask) of a continuous magnetic film without significant modification of the surface roughness or of the film's optical indices [2]. This potential application is briefly reviewed here, with some examples on Co/Pt multilayers.

Magnetization reversal is affected by size reduction in a very specific way, and ion-beam magnetic patterning allows fabrication of a new type of planar, optical contrast-free, magnetic storage media. The use of light-ion irradiation, with a low displacement density, is required for precise tuning of the magnetic properties; it is also needed to avoid degrading the surface planarity (via surface sputtering or implantation-induced swelling [61, 62]) after mask removal. The lithographed masks can be either contact masks, fabricated directly on the sample surface, or noncontact "stencil masks" with drilled holes, suspended above the surface [61, 62]. Among the latter's crucial advantages for magnetic storage is easy preservation of the initial sample topography. Another, crucial advantage of the technique is to ensure the existence of low-field nucleation centers at the borders between magnetically soft (irradiated) and hard (nonirradiated) areas that strongly influence the magnetization reversal processes in submicrometer structures. This is the result of so-called "collateral damage", as schematized in Fig. 10: a proportion of beam-induced recoils is scattered out of the mask and induces additional atomic displacements in the magnetic film, near the mask edge. The total amount and lateral extent of the added displacement density depends on the mask material – a typical evaluation from modified SRIM simulations [63–65] is presented in the figure for a Pt/Co/Pt film under an ultranarrow silica line mask (the corresponding overshoot value for a W mask would be 11%). Note that this effect must be carefully taken into account when designing arrays of close-lying lines or dots where irradiated zones may overlap.

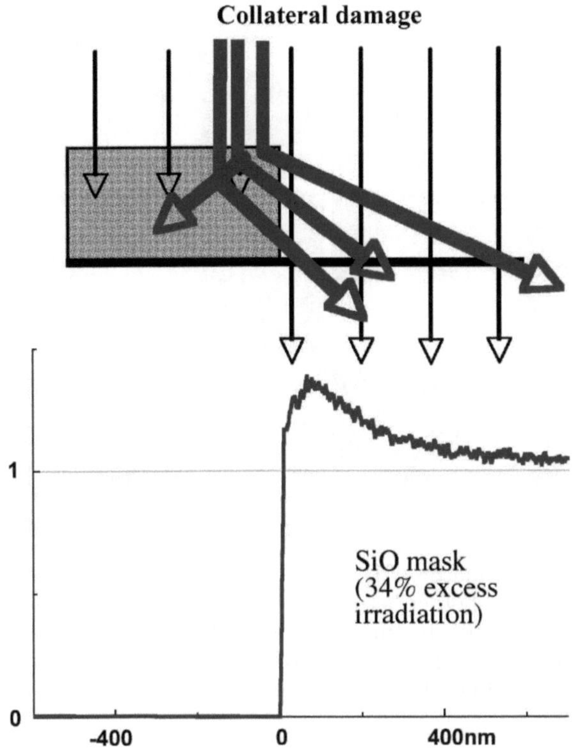

Fig. 10. Incoming ions hitting a mask near an edge have a significant probability of exiting from the mask after a single scattering event, thus "overirradiating" the sample (leading to a reduced magnetic anisotropy) in the mask vicinity. The amplitude and lateral spread of the excess irradiation depend on the mask material, and may be estimated from modified SRIM simulations [64]

Because of the displacement density excess, this collateral damage effect creates a line of *low field nucleation centers* in the immediate vicinity of the mask edge (or in the entire line or dot if it is nm-sized). These initiate magnetization reversal as shown in Fig. 11, in which the upper and lower middle images clearly show that nucleation of reverse domains occurs in the border's immediate vicinity, the latter being a preferred DW propagation path due to the collateral damage effect. The same behavior was found in isolated lines down to sizes 50 nm or less [63], and was also found [65] in highly irradiated structures with in-plane magnetization.

The above example is relevant to magnetically soft lines or dots produced in a magnetically hard (unirradiated) environment, which could provide a novel type of magnetic media in which writing (reversing the magnetization of the soft area) and erasing a magnetic bit would both occur through DW motion at the same field H_P. An example (100 nm dots) is detailed elsewhere [66]. The same effect plays a significant role when producing the reverse situation (hard dots or lines in a soft host). Hard line reversal occurs at much higher applied fields, but – as opposed to cases where they are produced via ion milling or standard lithography – the border effects due to ion irradiation produce coercivities whose distribution *narrows* as the elements are down-

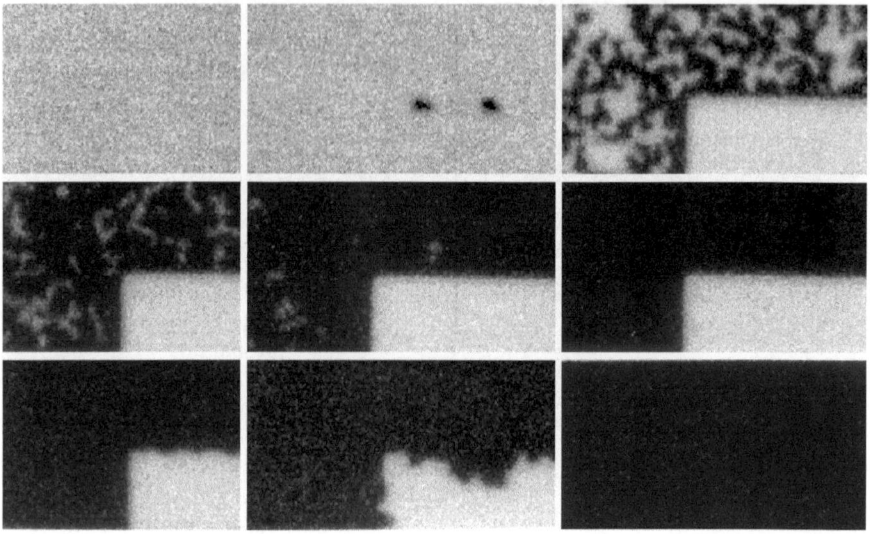

Fig. 11. $11 \times 27\,\mu m^2$ magneto-optical images of DW propagation in a [Co 3 nm/Pt 0.6 nm]$_6$ multilayer, taken close to the border between an unirradiated area (*lower right*) and an area irradiated with 2×10^{15} He cm^{-2} at 30 keV. Reverse magnetized areas appear in *black*. Images are taken after field pulses (duration 10 s) of increasing and amplitude from 46 Oe (**A**) to 742 Oe (**I**)

sized, i.e., hard area reversal occurs through DW motion out of the softer irradiated areas. This is of great potential interest in patterning applications to information storage.

5 Ion Beam-Induced Ordering of Intermetallic Alloys

Efforts to produce high-density recording media lead to the search for alloys that maximize magnetic anisotropy (and now preferentially perpendicular magnetic anisotropy for reasons related to size and sensitivity). The tetragonal $L1_0$ phase of FePt is the best known candidate in this regard. Potentially terabit-density FePt (2–5 nm diameter) nanocluster-ordered arrays are produced by solution-phase syntheses [67], but the method synthesizes the disordered face-centered cubic phase. The latter may be transformed to the $L1_0$ phase (Fig. 12) by annealing above 900 K, an expensive high-temperature process that induces sintering and coalescence of the initial nanoclusters.

How do we obtain low-temperature ordering? Irradiating with He as before, but at temperatures around or below 550 K (favoring limited thermal mobility and relaxation), the cubic-to-$L1_0$ phase (chemical ordering) transformation was promoted in intermetallic FePt and FePd thin-films [68]. Above, we saw that a 300 K irradiation led to interface disordering because atomic

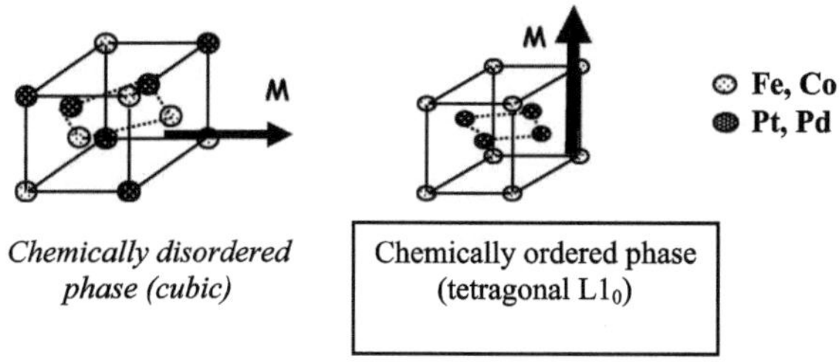

Fig. 12. Structure and magnetic anisotropy orientation of cubic versus L1$_0$ phase of equiatomic Fe (or Co) Pt (or Pd) alloys. The cubic phase's magnetic anisotropy is weak and inplane; because of symmetry breaking in the L1$_0$ structure, its magnetic anisotropy is perpendicular to the (100) planes. For FePt, this is the largest known magnetic anisotropy for transition-metal alloys

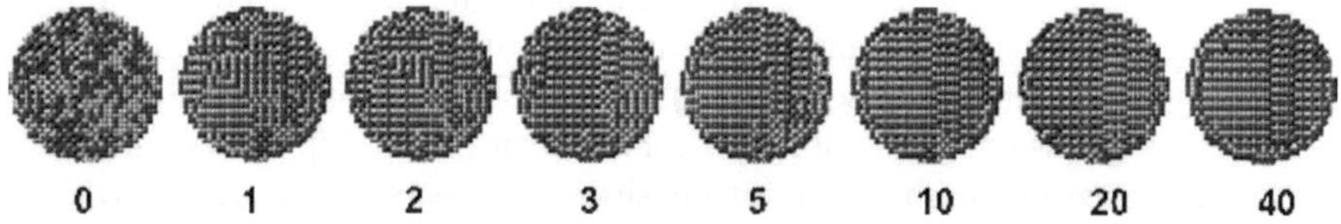

Fig. 13. Kinetic lattice Monte Carlo simulation of chemical ordering in a FePd nanocluster. Vacancies are introduced one by one (experimentally, via low-energy He irradiation), forcing successive pairwise exchanges. In a film, a single *c*-axis variant is selected by inducing directional short-range order during deposition [68]. A challenge for applications is to do the same for all nanoclusters in an array (figure courtesy of K.H. Heinig, FZ-Dresden)

site exchange processes are suppressed by high energy barriers. Introducing mobile vacancies by irradiation at sufficiently high temperatures (typically 450–550 K) allows successive pairwise exchanges that are determined by the vacancy-jump probability that, in turn, depends exponentially on the ratio of Fe–Fe, Fe–M and M–M binding energies to kT (where M = Pt or Pd). The system thus explores the nonequilibrium paths towards those low-energy configurations corresponding to chemical ordering. Kinetic lattice Monte Carlo simulations (see chapters "Fundamental Concepts of Ion-Beam Processing" by Averback and Bellon and "Precipitate and Microstructural Stability in Alloys Subjected to Sustained Irradiation" by Bellon) show that chemical order propagates along the directions that correspond to the lower energy barriers, thus favoring Fe–Fe and Pt–Pt plane formation. Note that applications (notably to magnetic recording) require a unique, well-defined magnetic anisotropy orientation, whereas the L1$_0$ phase has three variants along the three equivalent (100) directions. The introduction [68] of directional short-range order in the film plane, via layer-by-layer growth during initial film deposition selects the single variant perpendicular to the film plane (Fig. 13). Thus, ion irradiation may be combined with other features of the process (here, heat and adequate initial film-growth conditions) to control the ultimate physical properties. Major challenges remain in this area, related to the production of FePt nanocluster arrays and to the control of an entire population's magnetic anisotropy.

6 A Word on Control of Exchange-Bias Systems via Ion Irradiation

The exchange-bias phenomenon [69–71] occurs in samples involving a ferromagnetic (FM) layer in contact with an antiferromagnetic (AFM) layer, when they have been prepared (or when the AFM has been cooled below the Néel temperature) in an applied field. The uncompensated spin density then acts as an effective field, the FM hysteresis loop being shifted by the so-called exchange-bias field, while the coercivity is enhanced. Although the effect is being implemented in a number of applications (spin-valves, magnetic tunnel junctions, sensors...), it has been the subject of conflicting interpretations ever since its discovery. Recently [72, 73], it was ascribed to the existence of nonmagnetic atom sites or defects in the AFM, which are partial DW pinning sites (defects reduce the DW energy, hence induce DW pinning and stabilization of the domains). Because their populations are statistically unequal in the AFM sublattices, such defects could induce a residual magnetization. Changing the overall population of nonmagnetic sites in the AFM was indeed shown to affect both the exchange-bias field amplitude and its orientation. Ion irradiation was shown [74] to allow such control in a FM permalloy/AFM (e.g., FeMn) bilayer. Irradiation experiments and simulations [75, 76] show that the observed changes, and pinning, involve local changes (inside the AFM) in the uniaxial magnetic anisotropy. The latter being determined over a scale significantly larger than a single atomic site, irradiation defects must act on the local symmetry in the AFM via both strain relaxation and Fe–Mn pairwise exchanges, leading to a local adjustment of the magnetic anisotropy amplitude and direction. This would account for the fact that the comparatively weak magnetic interactions override all others (including those involved in defect creation) in determining exchange-bias properties. Other consequences are discussed in [3], which also provides references to experiments in which ion-irradiation or implantation were implemented to modify (or synthesize) magnetic materials often in more empirical ways.

References

1. N. Mathur, P. Littlewood, Nat. Mater. **3**, 207 (2005) 227, 228
2. C. Chappert, H. Bernas, J. Ferré, V. Kottler, J.P. Jamet, Y. Chen, E. Cambril, T. Devolder, F. Rousseaux, V. Mathet, H. Launois, Science **280**, 1919 (1998) 229, 231, 235, 236, 246

3. J. Fassbender, D. Ravelosona, Y. Samson, J. Phys. D, Appl. Phys. **37**, R179 (2004) 229, 250
4. P. Bruno, in *Magnetismus von Festkörper und Grenzflächen*, Ferienkurse des Forschungszentrums Jülich, Germany, 1993, Chap. 24 229
5. *Cobalt Monograph* (Centre d'information du cobalt, Brussels, 1960) 229
6. W.J.M. de Jonge, P.H.J. Bloemen, F.J. A den Broeder, in *Ultrathin Magnetic Structures*, ed. by B. Heinrich, J.A.C. Bland, vol. I (Springer, Berlin, 1993) 229, 237
7. G.A. Müller, E. Carpene, R. Gupta, P. Schaaf, K. Zhang, K.P. Lieb, Eur. Phys. J. B **48**, 449 (2005) 229
8. L. Néel, J. Phys. Rad. **15**, 376 (1954) 229, 238
9. M. Mansuripur, in *The Physical Principles of Magneto-optical Recording* (Cambridge University Press, Cambridge, 1995) 230
10. D. Weller, H. Brändle, G. Norman, C.-J. Lin, H. Notarys, Appl. Phys. Lett. **61**, 2726 (1992) 230
11. M.G. Le Boité, A. Traverse, H. Bernas, C. Janot, J. Chevrier, J. Mater. Lett. **6**, 173 (1988) 230
12. M.G. Le Boité, A. Traverse, L. Névot, B. Pardo, J. Corno, J. Mater. Res. **3**, 1089 (1988) 230
13. J. Lindhard, M. Scharff, H.E. Schiott, Mat. Fys. Medd. Vid. Selsk **33**(14) (1963) 231
14. P. Sigmund, Rev. Roum. Phys. **17**, 823 (1972) 231
15. P. Sigmund, Rev. Roum. Phys. **17**, 969 (1972) 231
16. J. Ziegler, J. Biersack, U. Littmark, *The Stopping and Range of Ions in Solids* (Pergamon, New York, 1985). http://www.srim.org 231
17. T. Devolder, Phys. Rev. B **62**, 5794 (2000) 233
18. H. Kiessig, Ann. Phys. **10**, 715 (1931) 233
19. L. Névot, Acta Electron. **24**, 255 (1981–1982) 233
20. X.L. Zhou, S.H. Chen, Phys. Rep. **257**, 223 (1995) 233
21. P.A. Lee, J.B. Pendry, Phys. Rev. B **11**, 2795 (1975) 234
22. C. Meneghini, M. Maret, V. Parasote, M.C. Cadeville, J.L. Hazemann, R. Cortes, S. Colonna, Eur. Phys. J. B **7**, 347 (1999) 234
23. T. Devolder, S. Pizzini, J. Vogel, H. Bernas, C. Chappert, V. Mathet, M. Borowski, Eur. Phys J. B **22**, 193 (2001) 234, 239
24. M. Cai, T. Veres, S. Roorda, R. Cochrane, R. Abdouche, M. Sutton, J. Appl. Phys. **81**, 5200 (1997) 235
25. D. Weller et al., J. Appl. Phys. **87**, 5768 (2000) 235
26. T.B. Massalski, *Binary Alloy Phase Diagrams*, 2nd edn. (Metal Information Society, Metals Park, 1990) 236
27. J.M. Sanchez, J.L. Moran-Lopez, C. Leroux, M.C. Cadeville, J. Phys. C, Solid State Phys. **21**, 1091 (1988) 236
28. A. Traverse, M.G. Le Boité, G. Martin, Europhys. Lett. **8**, 633 (1989) 236
29. D. Kurowski, K. Brand, J. Pelzl, Nucl. Instrum. Methods Phys. Res. B **148**, 936 (1999) 236

30. G.A. Bertero, R. Sinclair, C.-H. Park, Z.X. Shen, J. Appl. Phys. **77**, 3953 (1995) 238
31. A. Moser, D. Weller, IEEE Trans Magn. **35**, 2808 (1999) 239
32. A. Kirilyuk, J. Ferré, V. Grolier, J.-P. Jamet, D. Renard, J. Magn. Magn. Mater. **171**, 45 (1997) 239
33. J. Ferré, in *Spin Dynamics in Confined Magnetic Structures I*, ed. by B. Hillebrand, K. Ounadjela, Topics Appl. Phys., vol. 83 (2002), pp. 128–165 239
34. J. Pommier, P. Meyer, G. Penissard, J. Ferré, P. Bruno, D. Renard, Phys. Rev. Lett. **65**, 2054 (1990) 240
35. J. Ferré, C. Chappert, H. Bernas, J.-P. Jamet, P. Meyer, O. Kaitanov, S. Lemerle, V. Mathet, F. Rousseaux, H. Launois, J. Magn. Magn. Mater. **198–199**, 191 (1999) 241
36. T. Devolder, H. Bernas, D. Ravelosona, C. Chappert, S. Pizzini, J. Vogel, J. Ferré, J.P. Jamet, Y. Chen, V. Mathet, Nucl. Instrum. Methods Phys. Res. B **175**, 375–381 (2001) 241
37. T. Devolder, J. Ferré, C. Chappert, H. Bernas, J.-P. Jamet, V. Mathet, Phys. Rev. B **64**(1–7), 064415 (2001) 241
38. J. Ferré, J.-P. Jamet, P. Meyer, Phys. Status Solidi **175**, 213 (1999) 241
39. R. Becker, W. Döring, *Ferromagnetismus* (Springer, Berlin, 1939) 242
40. L. Néel, Physica **15**, 225 (1949) 242
41. V. Repain, M. Bauer, J.-P. Jamet, J. Ferré, A. Mougin, C. Chappert, H. Bernas, Europhys. Lett. **68**, 460 (2004) 243
42. M. Bauer, A. Mougin, J.-P. Jamet, V. Repain, J. Ferré, R.L. Stamps, H. Bernas, C. Chappert, Phys. Rev. Lett. **94**, 207211 (2005) 243
43. J. Ferré, T. Devolder, H. Bernas, J.P. Jamet, V. Repain, M. Bauer, N. Vernier, C. Chappert, J. Phys. D, Appl. Phys. **36**, 1 (2003) 243, 244
44. A. Herpin, *Théorie du magnétisme* (Presses Universitaires de France, Paris, 1964) 243
45. G. Toulouse, P. Pfeuty, *Introduction to the Renormalisation Group and to Critical Phenomena* (Wiley, New York, 1977) 243
46. F.J. Himpsel, J.E. Ortega, G.J. Mankey, R.F. Willis, Adv. Phys. **47**, 511 (1998) 243
47. C.T. Rettner, S. Anders, J.E.E. Baglin, T. Thomson, B.D. Terris, Appl. Phys. Lett. **80**, 279 (2001) 245
48. B.D. Terris et al., Appl. Phys. Lett. **75**, 403 (1999) 245
49. J. Gierak et al., Appl. Phys. A **80**, 187 (2004) 245
50. V. Repain, J.P. Jamet, N. Vernier, M. Bauer, J. Ferré, C. Chappert, J. Gierak, D. Mailly, J. Appl. Phys. **95**, 2614 (2004) 245
51. C. Vieu et al., J. Appl. Phys. **91**, 3103 (2002) 245
52. E. Grochowski, D.A. Thompson, IEEE Trans. Magn. **30**, 3797 (1994) 246
53. E. Betzig et al., Appl. Phys. Lett. **61**, 142 (1992) 246
54. B.D. Terris, H.J. Mamin, D. Rugar, W.R. Studenmund, G.S. Kino, Appl. Phys. Lett. **65**, 388 (1994) 246

55. T.J. Silva, S. Schultz, D. Weller, Appl. Phys. Lett. **65**, 658 (1994) 246
56. V. Kottler, C. Chappert, N. Essaidi, Y. Chen, IEEE Trans. Magn. **34**, 2012 (1998) 246
57. S.Y. Chou, M.S. Wei, P.R. Krauss, P. Fischer, J. Appl. Phys. **76**, 6673 (1994) 246
58. S.Y. Chou, P.R. Krauss, P.J. Renstrom, Science **272**, 85 (1996) 246
59. Y. Chen et al., J. Vac. Sci. Technol. B **12**, 3959 (1994) 246
60. A. Fernandez, P.J. Bedrossian, S.L. Baker, S.P. Vernon, D.R. Kania, IEEE Trans. Magn. **32**, 4472 (1996) 246
61. B. Terris, L. Folks, D. Weller, J. Baglin, A. Kellock, H. Rothuizen, P. Vettiger, Appl. Phys. Lett. **75**, 403 (1999) 246
62. B.D. Terris, D. Weller, L. Folks, J.E.E. Baglin, A.J. Kellock, J. Appl. Phys. **87**(9), 7004 (2000) 246
63. T. Devolder, Y. Chen, H. Bernas, C. Chappert, J.P. Jamet, J. Ferré, E. Cambril, Appl. Phys. Lett. **74**, 3383 (1999) 246, 247
64. T. Devolder, C. Chappert, V. Mathet, H. Bernas, Y. Chen, J.P. Jamet, J. Ferré, J. Appl. Phys. **87**, 8671 (2000) 246, 247
65. G.J. Kusinski, K.M. Krishnan, G. Denbeaux, G. Thomas, B.D. Terris, D. Weller, Appl. Phys. Lett. **79**, 2211 (2001) 246, 247
66. T. Devolder, M. Belmeguenai, C. Chappert, H. Bernas, Y. Suzuki, Mater. Res. Soc. Symp. Proc. **777**, T6.4 (2003) 247
67. S. Sun, C.B. Murray, D. Weller, L. Folks, A. Moser, Science **287**, 1989 (2000) 248
68. H. Bernas, J.-Ph. Attané, K.-H. Heinig, D. Halley, D. Ravelosona, A. Marty, P. Auric, C. Chappert, Y. Samson, Phys. Rev. Lett. **91**, 077203 (2003) 248, 249
69. W. Meiklejohn, C.P. Bean, Phys. Rev. **105**, 904 (1957) 250
70. J. Noguès, I.K. Schuller, J. Magn. Magn. Mater. **192**, 203 (1999) 250
71. I.K. Schuller, Mater. Res. Bull. **29**, 642 (2004) and refs therein 250
72. P. Miltenyi, M. Gierlings, J. Keller, B. Beschoten, G. Güntherodt, U. Nowak, K.D. Usadel, Phys. Rev. Lett. **84**, 4224 (2000) 250
73. U. Nowak, K.D. Usadel, J. Keller, P. Miltényi, B. Beschoten, G. Güntherodt, Phys. Rev. B **66**, 014430 (2002) 250
74. A. Mougin, T. Mewes, M. Jung, D. Engel, A. Ehresmann, H. Schmoranzer, J. Fassbender, B. Hillebrands, Phys. Rev. B **63**, 060409 (2001) 250
75. S. Poppe, J. Fassbender, B. Hillebrands, Europhys. Lett. **66**, 430 (2004) 250
76. J.V. Kim, R.L. Stamps, Appl. Phys. Lett. **79**, 2785 (2001) 250

Index

Structure and Properties of Nanoparticles Formed by Ion Implantation

A. Meldrum[1], R. Lopez[2], R.H. Magruder[3], L.A. Boatner[4], and C.W. White[4]

[1] Dept. of Physics, University of Alberta, Edmonton, AB, T6G2G7, Canada, e-mail: ameldrum@ualberta.ca
[2] Dept. of Physics and Astronomy, University of North Carolina, Chapel Hill, NC, 27599, USA
[3] Department of Physics, Belmont University, Nashville, TN, 37212, USA
[4] Materials Science and Technology Division, Oak Ridge National Laboratory, Oak Ridge, TN, 37831, USA

Abstract. This chapter broadly describes the formation, basic microstructure, and fundamental optoelectronic properties of nanocomposites synthesized by ion implantation. It is not meant as a complete literature survey and by no means includes all references on a subject that has seen a considerable amount of research effort in the past 15 years. However, it should be a good starting point for those new to the field and in a concise way summarize the main lines of research by discussing the optical, magnetic, and "smart" properties of these nanoparticles and the dependence of these properties on the overall microstructure. The chapter concludes with an outlook for the future.

1 Introduction

The emphasis on nanotechnology-related research has now been sustained for several years. Strong research efforts focusing on nanotechnology are common and crosscut the disciplines of chemistry, physics, biochemistry, medicine and engineering. Much of the interest is stimulated by the fact that many properties emerge in nanophase systems that do not occur in bulk materials. The practical motivation for this intense research effort derives both from the fundamental characteristics of small particles as well as their numerous potential applications, particularly in the areas of optical devices, micromechanical devices, and information storage.

The unique properties of nanophase precipitates arise from two principle factors. First, the large surface-to-volume ratio can have profound physical and thermodynamic effects that lead to significant modifications of melting temperatures and other phase transitions. Second, the three-dimensional spatial confinement of electrons creates a variety of novel optical and electronic effects. Surface effects and electron confinement together combine to produce new properties that can be manifested in a wide range of effects, such as a large nonlinear optical susceptibility, intense photoluminescence, altered band structures, and superparamagnetism, to name just a few.

H. Bernas (Ed.): Materials Science with Ion Beams,
Topics Appl. Physics **116**, 255–285 (2010)
DOI: 10.1007/978-3-540-88789-8_9,

As a result of the many interesting properties and potential applications of nanophase materials, many experimental techniques have been developed to produce and synthesize different types of nanocomposites. Depending on the application or property of interest, the synthesis technique must meet certain requirements. For example, since the critical properties are often directly dependent on particle size, it can be vital to obtain narrow size distributions. The ability to create core-shell structures or otherwise control the chemistry of the particles can greatly enhance certain optoelectronic effects. Solution chemistry has yielded some of the best results in these specific areas, although there has been some difficulty in producing materials that can be readily utilized in solid-state devices.

Here, we will review the experimental methods, formation, and properties associated with nanocrystalline materials produced by ion implantation. Nanocomposites produced by ion implantation have a number of attractive characteristics. The technique can produce a wide variety of single-element or compound nanocrystals embedded below the surface of virtually any solid host material with well-controlled depth and concentration. Because the particles are formed below the surface and are embedded within the host, the resulting composites are durable and the nanoparticles are protected from the environment. The physical properties of the nanocomposite can be optimized and tailored by controlling the concentration, crystal structure, orientation, and average size of the precipitates. By adjusting the experimental parameters, useful properties of two or more precipitated phases can be combined into a single integrated structure. Finally, ion implantation is widely employed in the semiconductor industry for doping silicon wafers, and therefore, it constitutes a mature materials technology that is established in the commercial synthesis and processing of materials with microscopic precision and control. The main disadvantages associated with the ion-implantation technique are the inability – so far – to produce suitably narrow size distributions and to control the lateral spacing of the precipitates. Additional difficulties can be caused by the effects of ion-beam damage, and high-temperature thermal processing may be necessary to remove matrix defects created in the implantation steps or control size distribution of the nanoparticles.

In this chapter, the main methods for using ion beams to produce nanocomposite materials will be discussed. The factors controlling the microstructure of the nanocomposites will be summarized. The microstructure has a major effect on the optoelectronic properties of the composite, and in this respect ion implantation offers a number of special advantages, as well as presenting some major difficulties. Next, some of the most attractive optical and electronic properties of nanocomposites produced by ion implantation will be briefly surveyed. Some aspects of the following will be based on a recent comprehensive review paper on nanocrystals produced by ion implantation, which we refer the reader to as the main citable reference [1]. The physics responsible for the optoelectronic properties of nanocrystals produced by ion implantation will also be touched on in this chapter, and is derived in more detail in subsequent chapters. Since there are thousands of papers on nanoclusters produces by ion implantation, a complete literature survey is not the intent of this short work; an attempt was made to keep the number of citations to a nonoverwhelming level.

2 Nanoparticle Synthesis

Ion beams have been used in basically two different ways to create nanocomposite materials. Some of the early work used ion irradiation of doped glasses to produce various types of nanoclusters. In this technique, a glass containing selected impurities – usually metal oxides – is irradiated to high fluences with H, He or Ar ions (Fig. 1a). Electronic energy-loss mechanisms induce nucleation and growth through a complex radiolytic process involving interactions between the metal ions and irradiation-induced defects in the glass. More recently, heavy-ion irradiation (up to $Z = 35$) was used to nucleate metal clusters in a glass [2]. Precipitate nucleation was driven by electronic energy-loss processes, similar to the case for light ions, but there is clear evidence for a threshold electronic energy loss, below which nucleation does not occur.

One advantage of the irradiation-induced nucleation technique is that the nucleation and growth stages can be temporally separated. Seed nuclei can be formed during the irradiation procedure, which then ripen during subsequent thermal processing. This provides a means for controlling the volume density and narrowing the size distribution of the precipitates. Generally, however, the concentration of particles produced is low (typical of thermal nucleation in a glass), the range of possible particle–host combinations is small compared to ion-implantation techniques, and there are none of the advantages associated with the production of nanocrystals embedded in a crystalline matrix.

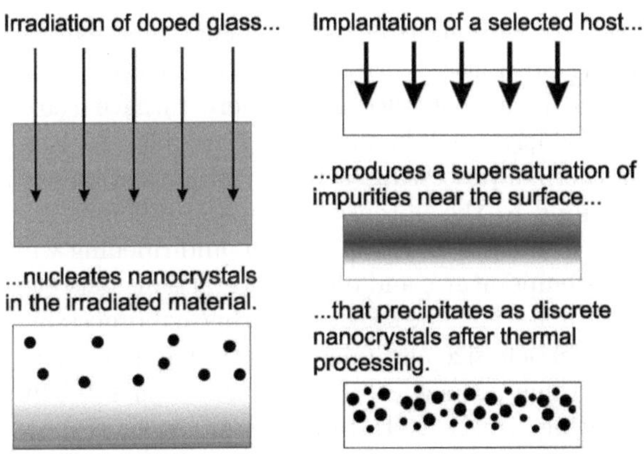

Fig. 1. Schematics of the ion-beam-induced nucleation and ion-implantation processes

The second main technique, now more commonly employed, uses ion implantation to inject impurities that subsequently nucleate as nanoparticles (Fig. 1b). In a method borrowed directly from the semiconductor industry, a high concentration of impurity ions is implanted into the near-surface region of a selected host material. This produces an impurity supersaturation within a few hundred nanometers of the surface, depending on the implantation energy and ion mass. Particularly for glass substrates it provides a method of achieving nonequilibrium concentrations of dopant ions without the need for glass modifiers to avoid phase separation in the glass. These glass modifiers can lead to increased absorption and degradation of the linear and nonlinear optical properties. Ion implantation can yield high-purity materials with large nanoparticle volume fractions and with accurately determined dopant concentration. In practice this permits the use of very high purity silica and other important optoelectronic materials without unwanted dopants that can degrade performance. Fluences can range up to 10^{17} ions cm^{-2} or higher, depending on the average size and particle concentration desired. One or more elements can be sequentially implanted in order to produce compound or chemically doped nanocrystals. The freeware computer program SRIM [3] is generally used with good results to ascertain the range and straggling in order to ensure overlapping concentration profiles. If the implants are done “hot” (i.e., above room temperature), particles may nucleate and grow during the implantation stage. On the other hand, if the specimen is cooled to liquid nitrogen temperatures, the implantation process can generally be completed without the nucleation of nanoparticles in the implanted layer, depending on the interactions of the implanted ions with the host material (e.g., diffusion rates). Care must be taken, however, because many crystalline host materials can become amorphous during implantation at low temperature, potentially leading to zone refining and other issues discussed below. The total current on the sample is a delicate balance between implantation time and the potentially negative consequences of specimen heating and adverse dose-rate effects. In general, beam current densities less than $\sim 20\ \mu A\,cm^{-2}$ are recommended.

Once the implantation process is completed, subsequent steps can be taken to nucleate and ripen the precipitates. This is usually done by high-temperature thermal processing in a controlled-atmosphere furnace (note that the term “annealing” is widely used in metallurgy to refer to heat treatment of metals). During thermal processing, a variety of nucleation and growth mechanisms occur, leading to the potential for many different microstructures and particle sizes. Diffusion-driven nucleation and ripening will depend on processing time and temperature, and on the interactions between the implanted ions and the host material. The selection of thermal processing atmosphere can also affect the particle size and microstructure and can have a major effect on the optoelectronic properties, due to surface passivation. Most anneals are done in reducing (ArH_2, N_2H_2) or neutral (Ar or N_2) atmospheres in order to prevent unwanted oxidation of the particles or the host. However, in the case of oxide substrates, annealing in oxygen can promote diffusion of the implanted ions and reduce unwanted defects introduced by the implantation process, in particular for some of the noble metals, or it can create oxide nanoparticles instead of metals. In either case the annealing atmosphere is an important parameter in determining the final size distribution. The cooling rate can also be a critical parameter – if the specimen is quenched, high-temperature metastable phases may be formed. So-called "zone refining" can be a problem if the specimen becomes amorphous during implantation. During the thermal processing step, a crystalline–amorphous boundary can progress toward the surface as the host material recrystallizes epitaxially, thereby "refining", or concentrating the implanted material nearer to the specimen surface. The affects of atmosphere, temperature, time, cooling rate, zone refining, etc. must be considered at the thermal processing stage.

The application of lasers for nucleating metal nanoclusters using frequencies on either side of the surface plasmon resonance (SPR) represents a powerful new tool for controlling the size and size distribution of the nanocrystals [4]. Various groups began investigating the use of laser annealing for creating metal nanocrystals with and without additional thermal treatments in attempts to narrow the size distribution of the metal nanocrystals. The effects on the metal nanocrystals were shown to depend upon the wavelength, power and pulse duration of the laser. Until very recently there were two processes used. In the first case, the substrate is transparent to the laser wavelengths and the radiation is absorbed by the metal nanocrystals. In the second case, the laser radiation is absorbed by the substrate in the UV region. At present, the fundamental understanding of the physics of either process on the changes of the metal nanoclusters is still an open question and the subject of continuing research.

More recently, the modification of the size of metal nanoclusters using a free-electron laser (FEL) with wavelengths in the range of 2–10 micrometers has been investigated [5]. In this scenario, the wavelengths corresponding to the vibration modes of the substrate are targeted. The excitation of the vibration modes can result in rapid heating and cooling of the composite with changes effectively occurring during the duration of the pulse. This technique was used in an effort to avoid difficulties produced with a high-temperature thermal processing discussed above. Initial efforts involving Au and Ag nanocrystals formed by ion implantation in type-III silica produced pronounced effects on the linear optical properties of the composite as well as on nanocrystal size. While conventional thermal treatments have been demonstrated to be sensitive to the annealing atmosphere, the FEL process showed no dependence on atmosphere.

Ion implantation will often produce a near-Gaussian concentration gradient of the implanted material, which can lead to wide size distributions if the larger clusters form at the peak of the implanted ion concentration. A commonly employed method to narrow this size distribution uses multiple implantation energies to obtain a more uniform concentration gradient in the sample. In order to narrow the size distribution, some workers avoid the thermal processing steps altogether – instead preferring to use a subsequent high-energy ion irradiation to nucleate precipitates in the implanted layer in a process similar to that for doped glasses, with some of the same advantages. In theory, at least, it should be possible to obtain narrower size distributions – although, unlike for doped glass, the concentration variation across the implant profile can remain problematic. At this point, irradiation of preimplanted layers has not yet produced any striking benefits over thermal processing.

Other methods for narrowing the size distributions have produced some success. One is the use of low-energy implants (e.g., <20 keV), which limits the implanted-ion distribution to very shallow depths. A second is the use of a sequential ion-implantation process to modify the substrate with dopants that are incorporated into the substrate and do not precipitate out as nanoparticles. These modifications of the substrate before the nanocrystals are formed can result in changes in the defect zone around the inclusion due to the distortion of the local matrix and yield greater control over nanocrystal microstructure by providing more nucleation sites in the implanted zone [6]. Both of these techniques have been shown to help narrow size distribution, but have not provided significant advancement to solving the lateral-spacing control issues.

3 Microstructures

Particle-size distributions are generally wide, as compared to those obtained by solution chemistry and thin-film methods (chemical and physical vapor deposition), and bimodal size distributions are not uncommon. Figure 2 is a

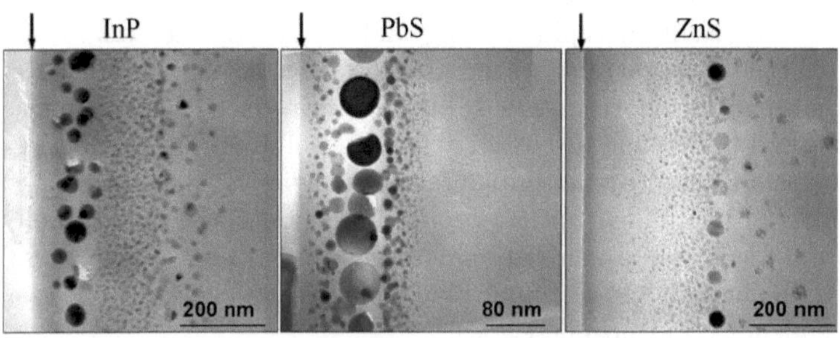

Fig. 2. Cross-sectional TEM images of InP, PbS, or ZnS nanocrystals in SiO_2. See [7] for implantation and thermal-processing conditions. Note the location of the band of larger particles in each of the specimens

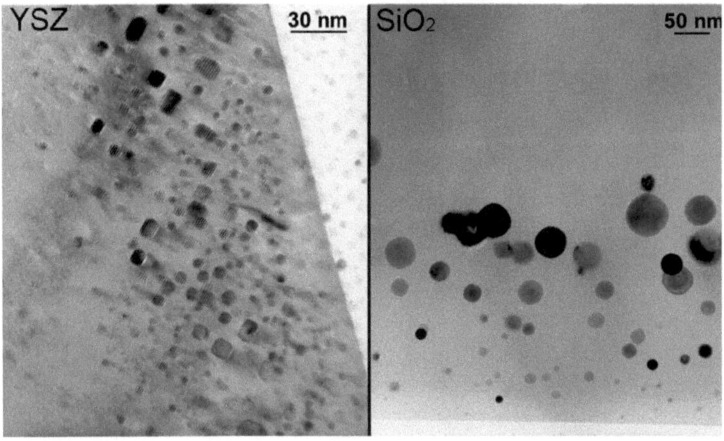

Fig. 3. Cross-sectional images showing Fe nanoparticles in a [110]-oriented YSZ wafer (*left*) and in a fused silica substrate (*right*). In YSZ the particles are well-faceted cubes, whereas in fused silica they show only a hint of faceting

set of cross-sectional TEM micrographs showing bimodal size distributions in compound semiconductor nanocrystals. Particularly interesting is the location of the layer of large particles. Pure Ostwald ripening would imply that the largest particles should form in the region of highest implanted-ion concentration (i.e., at the center of the implant profile), as is the case of PbS nanocrystals in SiO_2. However, in practice, a layer of larger precipitates can form near the surface (e.g., InP nanocrystals) or at the back of the implanted layer (ZnS nanocrystals) [7]. In the case of ZnS, the layer of large particles corresponds to the boundary between SiO_2 that was damaged by the implantation process and the pristine SiO_2 at greater depths. A similar effect occurs for metallic Zn nanocrystals in SiO_2. Zn rapidly diffuses into the substrate during thermal processing, probably at a faster rate in the ion-beam damaged surface layer. This leads to a "piling up" of implanted material at the boundary, creating the layer of larger nanocrystals. Other materials have a preference to migrate toward the surface during annealing – potentially leading to the type of microstructure characteristic of InP nanocrystals in fused silica. Attempts to narrow the size distributions by using lower-temperature thermal processing or irradiation-induced nucleation and growth have met with some success. Laser annealing has added to this success in controlling the size of the nanoparticles and may be more compatible with device-fabrication processes.

Precipitate faceting and crystallographic orientation can be controlled by the selection of host material and by the thermal processing conditions. Nanocrystals that form in a crystalline matrix have a strong tendency to be crystallographically aligned with the host. Figure 3 shows cross-sectional TEM micrographs of Fe nanocrystals in yttrium-stabilized zirconia (YSZ)

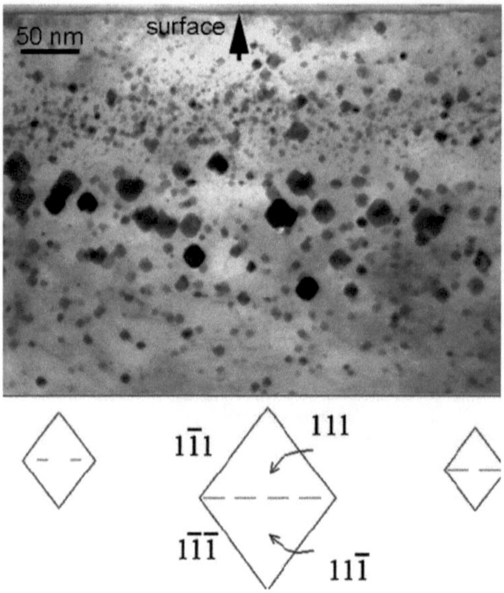

Fig. 4. Cross-sectional TEM micrograph showing Zn precipitates in silicon. The particles have an octahedral form in the silicon, despite the fact that zinc has a hexagonal crystal structure. After [9]

and fused silica hosts. In the glass, the Fe nanocrystals are spherical and randomly oriented; whereas in the YSZ, they have a well-faceted cubic shape and are crystallographically aligned [8]. The actual orientation relationship appears to be similar to the stress minimization that occurs in thin films – crystallographic directions that have similar spacings or multiples of spacings tend to align parallel to one another. In many cases, more than one specific orientation is possible due to the equivalence of crystallographic directions.

On the other hand, particle faceting is strongly controlled by surface-energy considerations. This is because the host material must "open up" to accommodate the particle. For example, nanocrystals embedded in pure silicon tend to form octahedra with apices pointing in the three equivalent [100] directions of the host. Figure 4 shows nanoscale octahedra of Zn embedded in crystalline Si [9]. The faces of the octahedra are parallel to silicon [111] planes – which have the lowest surface energy. The fact that zinc is not cubic and does not normally form octahedral crystals is clear evidence of the importance of the host material in governing precipitate morphology. For similar reasons, nanocrystals formed in YSZ have a cubic form, while those in sapphire frequently appear triangular or hexagonal when viewed down the [0001] axis of the Al_2O_3.

Thermal processing also affects the particle size and size distribution, and the defect concentration in the host material. Obviously, a higher processing temperature or a longer time tends to produce larger particles and/or fewer defects. This can be very important, for example in the case of silicon nanocrystals in which both the size and interface defects (dangling bonds) play major roles in the resulting luminescence properties [10]. Recently, a detailed investigation explained the physical origin of the lognormal size distributions characteristic of nanocrystals formed by ion-implantation and thermal-processing methods [11].

Clearly, the main factors controlling size distributions and precipitate microstructures are: ion species, implant fluence, implant temperature, thermal-processing time, temperature, atmosphere and cooling rate, and the selection of host material. Variation of one or more of these parameters provides experimental control over the average size, orientation, faceting, and composition of the resulting nanocrystals. The relatively wide size distributions and lack of lateral spatial control obtained by these methods remain some of the most significant problems associated with the ion-implantation technique. New methods, such as lithographic masking and focused ion beams offer hope for obtaining better size distributions in the near future, as discussed in Sect. 5.

4 Optoelectronic Properties

The versatility of the ion-implantation method has permitted the synthesis of a wide variety of nanoparticle-host combinations. The main goal of this research is in the creation of new materials with well-controlled optical, electronic, or magnetic properties. The following sections discuss the main physical effects and outline the optoelectronic properties of nanocomposites produced by ion implantation. A considerable portion of the work has concentrated on nonlinear optical effects, light-emitting applications, and magnetic properties of implanted nanocomposites, so these three topics will be given separate headings.

4.1 Nonlinear Optical Materials

Nanocomposites produced by ion implantation have many attractive characteristics with respect to nonlinear optics [1, 12, 13]. Both metal and semiconductor nanocrystals can exhibit pronounced third-order optical susceptibility, although for somewhat different reasons. Most of the work on ion-implanted materials has been done on metallic nanoclusters, due to the large local-field enhancement produced by the real part of the metal dielectric constant. Metal–nanocluster composites formed by ion implantation exhibit distinct optical effects including: (1) absorption due to surface-plasmon resonance, and (2) strong third-order nonlinear optical susceptibility.

Both classical electronic and quantum processes are responsible for these effects. The contribution of each process to the total optical response is a strong function of size, shape and ion species. The spatial confinement of the metallic electrons by the insulating host produces an enhanced electromagnetic field due to the large dipole moment induced by the optical field. In effect, the conduction electrons oscillate as a group with a characteristic frequency that depends on the electronic properties of the particles and of the host. In addition, the confinement of the electronic wavefunctions to a volume much smaller than the bulk mean free path produces an additional contribution to the electric susceptibility.

Assuming noninteracting particles, size effects can be divided into three regions for small spherical nanoparticles [14]. The first is for diameters <5 nm where quantum-size effects can become important and dominate the nonlinear optical properties. The second region extends from approximately 5 nm to 50 nm in diameter. For such particles the dielectric functions of the metal are expected to approach that of the bulk metal. In this region the extinction is dominated by the dipole absorption term for particles less than ~25 nm in diameter with scattering becoming more important as the diameter increases. The absorption and scattering cross section contributions as a function of particle size are also a function of ion species. The third region is for nanoparticles >50 nm in diameter. In this third region higher-order multipole terms are needed to describe the extinction both in absorption and scattering, again with higher-order scattering terms becoming more important with increasing diameter. The additional higher-order multipole terms lead to additional absorption peaks due to the higher-order multipole resonances. It is noteworthy that additional important effects can arise for the linear and nonlinear optical properties for nonspherical nanoparticles as discussed in [14, 15]. Other effects can also modify the optical absorption for nanocrystals. Particle–particle interactions can occur resulting in the appearance of a second peak in the absorption spectra that splits from the electric dipole SPR peak and redshifts away from the dipole peak with increasing degree of interaction.

For clusters of size in region 2, the optical field is nearly constant across the particles and the absorption is described by the electric dipole approximation derived by Mie and Maxwell-Garnett, and refined by Fröhlich in the first half of the twentieth century:

$$\alpha = p\frac{18\pi n_{\mathrm{d}}^3}{\lambda}\frac{\varepsilon_2}{(\varepsilon_1 + 2n_{\mathrm{d}}^2)^2 + \varepsilon_2^2}, \tag{1}$$

where α is the absorption coefficient, p is the volume fraction of the precipitates, n_{d} is the refractive index of the host, and ε_1 and ε_2 are the real and imaginary parts of the wavelength-dependent metal dielectric constant. The peak absorption occurs when $\varepsilon_1 + 2n_{\mathrm{d}}^2 = 0$, at the surface-plasmon resonance frequency. Further effects occur as particle sizes become smaller than the electron mean-free path, when free-electron contributions to the metal dielectric function are strongly affected by the particle surface. The surface-plasmon resonance absorption both weakens and broadens due to a resulting linear increase in the imaginary part of the dielectric constant (ε_2) with decreasing particle size, and there is also a weak blueshift of the resonant energy. One of the problems with the size distribution observed in nanoclusters formed by ion implantation is that surface-plasmon resonance peak position, magnitude, and full width at half-maximum are due to the distributions of particle sizes in most samples. The broader size distribution leads to greater absorption for all wavelengths, because larger nanocrystals have larger absorption cross sections than the smaller nanocrystals, as well as increasing the possibility of scattering. Larger particles can dominate the optical properties of the composite provided that the numbers of large and small particles is comparable. This increase in absorption can lead to some degradation of the figure of merit, as discussed below for these materials.

The effective third-order optical susceptibility ($\chi^{(3)}_{\text{eff}}$) is strongly enhanced by local electric fields near the surface-plasmon resonance frequency. This enhancement is considerably larger for metals than for semiconductors. In the dielectric approximation, the intensity of the nonlinear susceptibility depends on several of the same parameters as the surface plasmon absorption [16]:

$$\chi^{(3)}_{\text{eff}} = p\left[\frac{3n_{\text{d}}}{\varepsilon_1 + \varepsilon_2 + 2n_{\text{d}}^2}\right]^2 \cdot \chi^{(3)}_{\text{met}}, \tag{2}$$

where $\chi^{(3)}_{\text{met}}$ is the third-order susceptibility of "bare" metal clusters. The $\chi^{(3)}_{\text{met}}$ term varies with frequency due to variations of the energy and coherence relaxation times that are themselves dependent on the electronic properties of the specific metal. For cluster sizes smaller than the mean-free path of the conduction electrons, $\chi^{(3)}_{\text{met}}$ is proportional to $1/r^3$, where r is the radius of the nanoparticle [17]. Therefore, decreasing the particle size has the double benefit of enhancing the third-order susceptibility due to confined conduction-band transitions, and simultaneously decreasing the optical density due to an increase in ε_2. Note that the increase in ε_2 also decreases $\chi^{(3)}_{\text{eff}}$; however, this decrease is overwhelmed by the larger increase in $\chi^{(3)}_{\text{met}}$. Thermal effects can also enhance $\chi^{(3)}_{\text{eff}}$ through the thermo-optic coefficient of the composite, although the thermal relaxation lifetimes must be short for device applications. The effective medium theory used to model the nonlinear optical properties has recently been extended to include nanoparticles in a nonlinear medium and core-shell effects. In the latter case the core and shell can both be metallic, or one metallic and the other a semiconductor. This additional work has indicated that even greater enhancements of $\chi^{(3)}_{\text{eff}}$ of the nanoparticles is possible as the nonlinearity of the host matrix increases or for core-shell nanoparticles leading to the possibility of several orders of magnitude increase in the nonlinearity of the composite [18].

Experimental techniques can readily be used to measure the nonlinear refractive index, nonlinear absorption, obtain the decay rates, and deconvolute the thermal and electronic contributions to the optical susceptibility. Both four-wave mixing and Z-scan techniques have been used to measure the nonlinear properties of ion-implanted specimens, and each technique has its own specific advantages and limitations. In cases where absolute values of $\chi^{(3)}_{\text{eff}}$ in ion-implanted nanocomposites have been measured (e.g., Au nanocrystals in SiO_2 and Al_2O_3, and Cu clusters in SiO_2), the nonlinear refractive index exceeded by as much as several thousand times that of quenched nanocluster glass composites, with picosecond response times for the electronic component of the nonlinear index. Particle-size-induced enhancement in the third-order susceptibility was clearly observed. Short switching times and the large values of n_2 and $\chi^{(3)}_{\text{eff}}$ have led many authors to suggest potential applications for these materials in optoelectronics.

Despite these attractive characteristics of ion-implanted nanocomposites, difficulties remain to be addressed. A figure of merit for absorbing nonlinear optical materials is $\chi^{(3)}/\alpha\tau$ [19], where τ is a relaxation time. Metal nanocluster composites produced by ion implantation have high values of $\chi^{(3)}$; however, both $\chi^{(3)}$ and α have a peak near the same wavelength (when $\varepsilon_1 + 2n_{\text{d}} = 0$). The high value of α and the corresponding long thermal relaxation time decrease the figure of merit for these materials, particularly for laser excitation lasting longer than a few ps. For high-pulse-repetition-rate applications, reducing the overall thermal relaxation time is also critical.

Recent work has aimed at finding methods for reducing the optical density due to the surface-plasmon resonance, while maximizing $\chi^{(3)}_{\text{eff}}$. This can, in theory, be accomplished if one can synthesize a uniform size distribution of nanoparticles whose radius is smaller than the electron mean-free path. In the quantum-confinement regime, $\chi^{(3)}_{\text{eff}}$ varies as $1/r^3$, and α is decreased due to the weakening of the plasmon resonance. So far, however, it has not been possible to produce suitably narrow size distributions by ion implantation, so research has been instead focusing on other means to minimize α and maximize $\chi^{(3)}_{\text{eff}}$. For example, in one experiment both Au and Ti were implanted into silica glass [20]. The implanted Au precipitated as metallic colloids but the Ti remained dissolved in the glass. The polarizable Ti ions increased $\chi^{(3)}_{\text{eff}}$ by a factor of approximately 2. This type of multi-implantation protocol can also be used to increase or decrease the refractive index, depending on the metal dielectric function and the optical frequency. In another multi-implantation experiment, surface plasmon absorption from Ag nanocrystals in SiO_2 was found to become weaker with an increasing concentration of implanted Sb [21]; whereas, Ag + Cd-implanted SiO_2 exhibits absorption that is consistent with a superposition of the surface plasmon peaks from both Ag and Cd phases [22]. Similar effects have been observed in a variety of other metallic alloy nanocrystals. By implanting various concentrations of Ag + Cu into SiO_2, a transition can be induced from the nonlinear absorption to nonlinear saturation regime [23]. Other experiments have aimed at forming bimetallic "core-shell" nanoparticles, where changes in the electronic properties across the core-shell boundary add an additional degree of freedom for the experimental reduction of α and for increasing $\chi^{(3)}_{\text{eff}}$.

The nonlinear optical properties of ion-implanted nanoclusters composites are attractive for a variety of reasons, including high values of $\chi^{(3)}_{\text{eff}}$, short response times, relative ease and flexibility of synthesis, and compatibility with materials processing technology widely employed in industry. Research is focusing on multi-implantation means to enhance the figure of merit and on narrowing the size distributions (thereby providing the double benefit discussed above). Both aspects of metal-quantum-dot nanocomposite research will be critical with respect to future device applications. Surface-plasmon effects may find applications as optical filters or in future devices based on the "plasmon waveguide" concept [24]. Surface-plasmon oscillations of closely spaced nanoparticles are coupled in the near field, giving rise to a coherent energy propagation that can be guided along chains or arrays of nanoparticles. This could represent a new way to transmit and guide optical signals on dimensions smaller than the diffraction limit. Surface-plasmon resonance may also find applications in submicrometer "plasmon printing" lithography, where a broad optical stimulus excites the surface-plasmon resonance of strategically placed nanoparticles, which in turn emit light at a wavelength that develops a photoresist in their immediate vicinity [25]. Other potential areas of use that may not depend as strongly on the control of the size distribution are optical recording [26], sensitizers in erbium-doped fibers [27] and enhanced photoconductors [28]. However, for broad acceptance of ion implantation to applications in many of these areas, it will be necessary to control the size, size distribution and lateral spacing of the metal nanoparticles.

4.2 Light-Emitting Materials

The effects of particle size on optical properties are also pronounced in semiconductor nanoparticles. Many review articles [29, 30] and books [31, 32] deriving in detail the basic physics of nanoscale semiconductors have been published. In semiconductor nanocrystals, the energy bands become quantized at energies that are a function of the radius, r, of the crystal [33]:

$$E = \frac{\chi^2 h^2}{4r^2} \cdot \left[\frac{1}{m_{\mathrm{e}}^*} + \frac{1}{m_{\mathrm{h}}^*}\right] - \frac{1.8e^2}{\varepsilon_2 r} + \frac{e^2}{r} \cdot \sum_{n=1}^{\infty} a_n \left(\frac{S}{r}\right)^{2n}, \tag{3}$$

where χ is the root of the spherical Bessel function, m_{e}^* and m_{h}^* are the effective masses of electrons and holes, a_n is a term that depends on the dielectric properties of the particle and the host, and S is a position variable. The first term on the right-hand side of (3) is the electron-hole quantum localization term, the second is the Coulomb attraction term, and the third is due to polarization. The polarization term is small compared to the first two terms and can often be ignored.

The exciton radius, r_{e}, can be estimated with the following equation:

$$r_{\mathrm{e}} = r_{\mathrm{B}} \frac{\varepsilon \cdot m_{\mathrm{e}}}{\left[\frac{1}{m_{\mathrm{e}}^*} + \frac{1}{m_{\mathrm{h}}^*}\right]}, \tag{4}$$

where r_{B} is the Bohr radius (0.053 nm), ε is the dielectric constant of the nanocrystal, and m_{e} is the electron rest mass. Materials that have a large exciton radius (e.g., PbS: 20 nm; CdSe; 9 nm) should show pronounced quantum-confinement effects; whereas, if the exciton radius is small (e.g., Si: 4.3 nm;

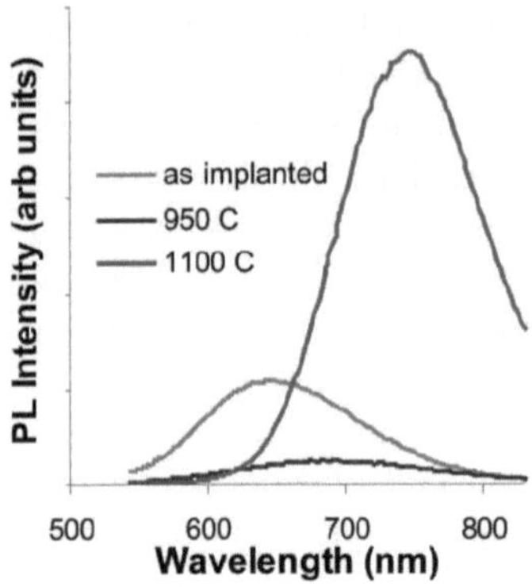

Fig. 5. Photoluminescence spectra from Si nanoclusters in SiO_2, after [59]

CdS: 3.5 nm; ZnS: 2.5 nm), the precipitates must be correspondingly smaller in order for strong confinement effects to be observed. Other effects controlling the optical properties of semiconductor nanocrystals involve the spatial isolation of nonradiative traps and the electronic states at the cluster/matrix interface.

Many types of single-element and compound semiconductor nanocrystals have been formed by ion implantation of either fused silica glass or sapphire wafers. Figure 2 shows that the microstructure of these composites can be complicated, and the size distributions are wide compared to semiconductor nanocrystals synthesized in solution, and compared to nanocrystals in films formed by physical or chemical vapor deposition. Nevertheless, these composites show clear evidence of quantum confinement and can demonstrate relatively intense light emission.

Silicon nanocrystals embedded in fused silica have been the most extensively investigated semiconductor nanocomposite formed by ion implantation. Although in a short chapter it is impossible to summarize the hundreds of papers published on this topic, some of the basic observations can be summarized. Shortly after the report of strong visible PL in porous silicon [34], Si nanocrystals formed by ion implantation of Si into SiO_2 glass were found to exhibit a strong and broad PL peak at wavelengths centered at ~750 nm [35]. Since the initial observation, many workers have reported and confirmed intense PL at 750–950 nm from Si nanocrystals embedded in silica glass (e.g., Fig. 5). The emission dynamics follow the stretched exponential function with characteristic decay times of several tens to hundreds of microseconds and an exponent, β, of ~0.7. Si nanocrystals produced by implantation of silicon into SiO_2 have potentially attractive properties with respect to possible silicon-based lasers [36, 37], light-emitting diodes [38, 39], and nonvolatile memory [40, 41]. Optical gain has been reported in ion-implanted silicon nanocrystal composites [42, 43]. The waveguiding properties of silicon nanocrystal layers have been extensively investigated [44] and the relevant optical cross sections are well known (e.g., [45]).

The ion doses and energies (and postimplantation annealing) required for forming highly luminescent silicon nanocrystals are interdependent parameters. The most intense PL appears to occur for film compositions close to $SiO_{1.5}$ [46]; at higher silicon concentrations the nanocrystals can become large and interconnected, while at lower silicon concentrations the number density of luminescent nanoclusters is low. For ion energies ranging from tens to hundreds of kV, implant fluences are typically in the range of 10^{16} to 10^{17} ions cm^{-2}. Generally, SRIM simulations are run in order to obtain the desired concentration gradient in the specimen. A major difficulty for the ion-implantation technique is the near-Gaussian concentration gradient of implanted silicon that causes a much wider size distribution than can be obtained, for example, by physical vapor deposition.

The microstructure of silicon nanocrystals produced by ion implantation has been extensively investigated by several groups [47–51]. There is evidence for the generation of amorphous silicon-rich regions when specimens are annealed at temperatures lower than 900°C, as confirmed by a detailed study of films grown by electron-beam evaporation [52]. At higher annealing temperatures, the silicon nanocrystals are well crystallized, demonstrating few lattice defects and good crystal faceting. For higher implanted-ion concentrations, larger nanocrystals have been reported to contain twin planes and stacking faults, with a critical diameter for the presence of these volume defects of ~5–6 nm [49]. These crystal shapes have also been investigated by Monte Carlo simulations [53]. Size distributions are narrower than for compound nanocrystals formed by ion implantation, and can be further narrowed (or involve the formation of platelet structures) by using low-energy implants, as has been done for silicon-nanocrystal-based nonvolatile memory devices [54, 55].

The sensitizing action of silicon nanocrystals for 4f-shell transitions in Er^{3+} is of particular technological interest (e.g., see [56, 57] and references therein). Ion implantation offers a versatile and flexible method to control the concentration of erbium within the layer of silicon nanocrystals, subject to the limitations of the concentration gradient discussed earlier. The erbium site has been determined to be outside the nanocrystals [58]. In order to form silicon-nanocrystal-sensitized erbium-luminescent films, one may coimplant silicon and erbium (the latter to a much lower implanted concentration in order to avoid undesirable erbium–erbium interactions) followed by thermal processing to precipitate the clusters and activate the erbium. Alternatively, one can first implant the silicon and treat at high temperature, followed by erbium implantation and a second postimplantation thermal processing to recover the crystallinity of the nanoclusters. The latter method has the advantage that the optimum temperature for activating the erbium ions can be selected independently of the initial high-temperature treatment required to nucleate and grow the nanocrystals.

The origin of the intense red and infrared luminescence characteristic of Si nanocrystals in fused silica has been a matter of considerable debate. Although not the main object of this short chapter, a very brief discussion of this contentious topic is given. Several effects probably play important roles, including quantum confinement, the effects of radiative centers and nonradiative traps at the nanocrystal/matrix interface, and transfer of carriers between closely spaced nanocrystals. Hydrogen annealing enhances the PL intensity from Si nanoparticles in SiO_2 [59, 60], probably due to effective passivation of nonradiative traps. The critical role of the interface was recently confirmed in a study using electron spin resonance to identify dangling Si bonds at the SiO_2/Si interface as the main nonradiative recombination site [61]. One model consistent with these observations is that light emission from SiO_2–Si composites is controlled by two types of surface "defects": the Si=O double bond, if present, may be the effective radiative recombination site, and the dangling Si bond, if present, is the preferred nonradiative site. A slightly more detailed model consistent with some of the existing observations has recently been proposed [62]. In this "reactive nanocluster" model, interactions between nearby particles separated by a thin oxide layer decreases the effective energy levels of the radiative interface defects. Others have shown the critical importance of migration of carriers from smaller clusters to larger ones [63]. The physical mechanisms responsible for the PL of silicon nanoclusters appear to be very complicated and continue to lead to much constructive debate in the literature.

Since germanium has a larger exciton radius than Si, quantum-confinement effects should, in theory, be even more pronounced for Ge nanocrystals. Like with Si nanocrystals, Ge clusters may also find nonvolatile memory applications [64]. Many experiments, of which only a small sampling can be provided here, have shown that Ge-implanted and annealed SiO_2 has a fairly pronounced and quite broad emission line peaking in the red or infrared, possibly with a defect-related emission in the blue. The red band has been ascribed to radiative recombination of quantum-confined excitons, but most studies have suggested oxide-related luminescent defects [65–67]. The interface defect hypothesis clearly illustrates the importance of detailed studies in order to determine the origin of the luminescence from Group IV nanocluster composites formed by ion implantation. The microstructure of Ge nanocrystals formed by ion implantation seems to be generally similar to that for silicon – i.e., narrower than for compound nanocrystals but still limited by the implanted-ion concentration profile. Recent work has demonstrated narrower distributions by using a low-energy Ge implantation protocol [68].

Direct-gap compound semiconductor nanocrystals formed by ion implantation can also be strongly luminescent. For example, GaAs nanocrystals formed by sequential implantation of Ga and As in fused silica luminesce in the red and near-IR portions of the spectrum [69]. The emission mechanism is not fully understood, although quantum confinement probably plays a role, since the emission energy is slightly higher than the bandgap of bulk GaAs. ZnS nanocrystals in fused silica are usually not luminescent (probably due to nonradiative defects and the small exciton radius); however, implantation of a manganese impurity into the ZnS nanocrystals creates a weak yellow luminescence band [70]. Excited carriers in the ZnS nanocrystals are probably trapped at levels associated with the incorporated Mn ions. The emission is at shorter wavelengths than for bulk Mn-doped ZnS and has been observed only for the smallest ZnS nanoclusters. Both observations suggest that quantum confinement is a likely origin for the yellow luminescence.

Al_2O_3–CdS composites seem to be more strongly luminescent and have been investigated using single-particle spectroscopy techniques. Composites annealed at 1000°C in a reducing atmosphere show a broad, relatively intense luminescence band centered at ~500 nm, near the bandgap of bulk CdS. This PL is nearly independent of particle size, suggesting the influence of defects. Using scanning near-field optical microscopy (SNOM), two-dimensional luminescence images of CdS nanoclusters in Al_2O_3 have been obtained [71]. These images show randomly distributed bright regions attributed to the presence of CdS nanocrystals. Individual spectra were obtained from each of these regions, which permitted the broad emission band from the entire composite to be resolved into separate sharper components from each bright region. The narrower lines in each single-particle spectrum were consistent with a variety of oxygen and aluminum defects and impurities within the CdS nanoparticles. Different particles contained different defects, so the broad luminescence of the entire composite is actually a sum of several sharper peaks. If this interpretation is correct, sharp emission bands will be difficult to obtain from ion-implanted samples even if the size distributions are narrowed, since impurities from the host material may always be present.

Recently, PbS nanocrystals formed by ion implantation have been found to have many interesting properties, including strong luminescence in the technologically relevant infrared wavelength range around 1.5 micrometers [72–74]. These investigations explained the complex microstructures that had been observed in previous reports (see Fig. 2), and greatly enhanced the luminescent properties of the PbS nanocrystals via better-controlled nucleation and growth techniques. This work even demonstrated the luminescence blinking properties and energy splitting of the ground-state exciton that had previously been observed almost exclusively in solution-based nanocrystals. Due to the high degree of control over the particle-size distribution (at least, for ion-implantation studies), this work appears to represent one of the most promising routes toward feasible luminescent devices based on nanocrystals formed by ion implantation.

Several other potential applications for ion-implantation-produced semiconductor nanocomposites are being investigated. For example, semiconductor nanocrystals embedded in a gate oxide have charge-storage capabilities that may be useful in future high-performance nanocrystal-based floating-gate memory devices [75]. Continuous layers of superconducting nanoparticles can be formed in thin films on silicon, suggesting possible integration of superconducting materials compatible with silicon technology [76]. Ion implantation can also be used to create lightweight engineered nanocomposites with remarkable durability, where the embedded nanocrystals harden the surface and reduce crack formation. Photovoltaic devices based on the ion-implantation concept are being explored. If size-graded particles can be embedded in an appropriate matrix, the effective bandgap could be tailored to extend across the solar spectrum.

4.3 Magnetic Materials

Ion implantation can be used to create ferromagnetic nanocomposites consisting of transition-metal nanocrystals embedded in an insulating host. By injecting varying concentrations of Fe, Co, or Ni into dielectric host materials such as SiO_2, soft magnetic composites are created, with low coercivities and a magnetic moment per atom similar to that of bulk magnetic material. Superparamagnetism is often observed due to particle sizes well below that needed to prevent random thermal reorientation of the particle magnetization. Blocking temperatures have been calculated from the precipitate sizes and anisotropy constants, and seem to agree reasonably well with experimentally observed blocking temperatures obtained from field-cooled and zero-field-cooled measurements of the magnetization. Nevertheless, the size distributions are relatively wide and the thermal blocking temperature is accordingly smeared over a range of temperatures.

One motivation for research on ferromagnetic nanocomposites produced by ion implantation has been the potential for creating materials with applications in magnetic recording (see chapter "Magnetic Properties and Ion Beams: Why and How" by Devolder and Bernas). Single-nanocrystal-per-bit data recording could represent an important step forward for the information storage capacity of such media [77]. Nevertheless, several stringent requirements must be satisfied in order to achieve device-quality performance. For single-particle bits to be written individually, the precipitates must be discrete, magnetically isolated, single-domain ferromagnetic nanoparticles that are larger than the superparamagnetic limit and whose coercivity, size, orientation, and position can be controlled. Ferromagnetic nanoclusters produced by ion implantation are single domains, they can be made larger than the superparamagnetic limit, the coercivity can be controlled by postimplantation treatment or by creating magnetically hard alloy particles, and the crystallographic orientation (and, therefore, magnetic anisotropy directions) can be selected. However, the question of magnetic isolation is debatable, particle-size distributions are wide, and there has been, until recently, no inplane spatial control over the distribution of nanoclusters.

In ferromagnetic nanocrystalline composites, experimental data is usually explained by assuming an ensemble of particles that do not interact magnetostatically and whose anisotropy axes are randomly oriented. Ion implantation offers an excellent opportunity to test these assumptions, because nanoparticles can be formed that are *not* randomly oriented, and that have symmetrical crystal shapes including cubes (e.g., Fig. 3) or hexagonal prisms. Since the nanoparticles are single domains that show coherent magnetization reversal, only three main energy contributions control the magnetic properties of the composite [8]: (i) the Zeeman energy that couples the particle magnetization to the applied field, (ii) the net anisotropy energy of the particles, and (iii) the interparticle interaction energy. The second term is often the most difficult to estimate, because there are several contributions to the effective anisotropy, including crystalline anisotropy, shape anisotropy, surface anisotropy, and magnetostriction.

Strong arguments have been put forth that support interparticle interactions in ion-implanted composites [8, 78]. For example, a [110]-oriented specimen of YSZ was implanted with Fe to a dose of 8×10^{16} ions cm^{-2} and annealed in Ar + 4%H_2 in a two-step process (1 h at 1000°C with a slow cool, followed by 2 h at 1100°C, after which the sample was quenched). This produced a layer of Fe nanocrystals that were crystallographically aligned in one of three possible orientations. A single [100] axis of the Fe was oriented with one of the three [100] axes of the YSZ, and the other two equivalent $[100]_{\mathrm{Fe}}$ axes were aligned with the [110] directions in the YSZ [8]. Computer simulations of the magnetic hysteresis of a layer of interacting, crystallographically oriented "nanocubes" of α-Fe closely matched the experimental results, whereas simulations for noninteracting particles did not [78]. The orientation dependence of the magnetic hysteresis was found to be consistent with the concept of frustration among groups of particles.

Further experimental evidence in favor of the interacting-particle model comes from the temperature dependence of the coercivity. For small particles, the relationship between temperature and coercive field, H_{c}, is:

$$\frac{H_{\mathrm{c}}(T)}{H_{\mathrm{c}}(0)} = 1 - \left(\frac{T}{T_{\mathrm{B}}}\right)^{n}. \tag{5}$$

Here, T_{B} is the blocking temperature and the exponential term n is close to 0.5 (it is exactly 0.5 for uniaxial anisotropy). T_{B} is proportional to the particle volume (V) and the anisotropy constant (K) and can be estimated for a specific experimental condition (for example, $T_{\mathrm{B}} \cdot KV/100k_{\mathrm{B}}$ for cubic anisotropy, a measurement interval of 100 s, and a flip attempt frequency of $10^9\,\mathrm{s}^{-1}$). Sorge et al. [8] found that the temperature dependence of oriented α-Fe nanoparticles deviated strongly from the expected $T^{0.5}$ relationship, further supporting the interacting nature of the composite.

Estimates were made of the ratio $4\pi M_{\mathrm{sat}}^2 f : 2K_{\mathrm{eff}}$, where M_{sat} is the saturation magnetization, f is the volume fraction of particles in the implanted layer, and K_{eff} is the effective anisotropy constant. The first term is an estimate of the nanoparticle interaction, and the second is the energy contribution from the magnetic anisotropy. Making reasonable estimates for the contributions to K_{eff} leads to a ratio of almost 4:1, further evidence in favor of strongly interacting particles [8]. Schulthess et al. [78] also point out that the changes in the hysteresis loops from magnetostatic interactions are qualitatively similar to those that result from random orientations of the magnetic anisotropy. Nevertheless, the story is probably not complete, nor by any means universally accepted.

With respect to magnetic recording media, ion implantation offers the possibility of fine control over the magnetic properties of the nanoclusters. The first clear example of high-coercivity ferromagnetic nanoparticles made by ion implantation was demonstrated for Fe nanoparticles formed by implantation of high-purity copper [79]. In that work, coercivities as high as 1150 G were reported, and were attributed to stress in the particles, interstitial impurities, and increased anisotropy due to the cubic shape of the precipitates. More recently, it has become possible to obtain experimental control over the coercivity of ion-implanted nanocomposites (Fig. 6). By implanting 140-keV Co^+ at ambient temperature to a fluence of 8×10^{16} ions cm^{-2} into [0001]-oriented sapphire and subsequently annealing for two hours at 1100°C, well-faceted and crystallographically oriented Co nanoclusters were produced [80]. X-ray diffraction showed that the Co was a nearly 50:50 mixture of the cubic and hexagonal phases. Nanoparticles with the hcp Co structure were aligned with their c-axis parallel to the c-axis of the host sapphire; whereas, the ccp nanocrystals were aligned with a [111] axis parallel to the sapphire c-axis. This composite demonstrated the low coercivities more typical of specimens produced by ion implantation.

The magnetic hysteresis of this Al_2O_3–Co composite can be modified by subsequent ion irradiation. By implanting Xe to various fluences through the Co nanoparticle layer, the coercivity of the composite can be increased from 50 to 460 G (Fig. 6) [80]. Implantation of Pt has an even more dramatic effect, with the coercivity increasing up to 1500 G. These results allow the nanocomposite to be tailored for a specific coercivity ranging from 50 to 1500 G. Further work has used coimplantation of Co and Pt to create CoPt alloy particles with the tetragonal $L1_0$ structure. This is an extremely hard magnetic material, and the specimens reach coercivities as high as 10,000 G at 5 K [81].

The properties of ferromagnetic nanocomposites produced by ion implantation are only just beginning to be understood from a fundamental perspective. A considerable amount of work will be required to obtain a more detailed understanding. The magnetic simulations recently performed by Schulthess et al. [78] are a step forward in this direction. A more detailed model that in-

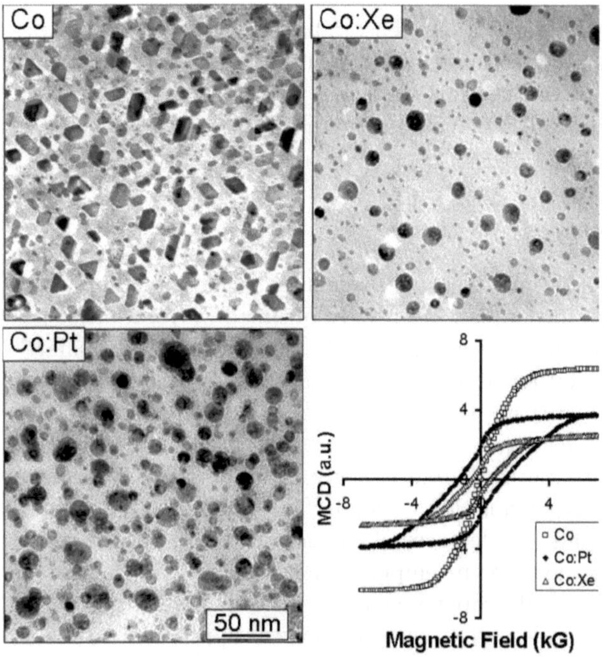

Fig. 6. Plan-view TEM images of Co nanocrystals in sapphire and corresponding magnetic hysteresis, modified after [1]. (Co): 140-keV Co ions implanted to a fluence of 8×10^{16} ions cm^{-2} at 30°C. The specimen was subsequently processed at 1100°C for 2 h under flowing Ar + 4%H_2, (Co:Xe): same as in (Co) but with a subsequent implantation of Xe (244 keV, 2×10^{16} ions cm^{-2}), and (Co:Pt): same as in (Co) but with an additional implantation of Pt (320 keV, 6.4×10^{15} ions cm^{-2}, 30°C). The different particle number densities in the TEM images are due to differing specimen thickness. After [80]

corporates a dynamical spin-rotation method based on the Landau–Lifshitz–Gilbert equation and that includes a larger number of particles interacting in three dimensions, while computationally challenging, would provide considerable further understanding of the magnetic properties of these materials. Some of this work has already been started, and some unexpected results have been found, including ultrafast magnetic relaxation in iron nanocrystals formed by ion implantation that was attributed to cluster–cluster interactions [82], and a very large Faraday rotation [83]. With respect to applications, the ability to control the orientation and crystal structure of the ferromagnetic precipitates and the durable nature of the composites themselves are important advantages of the ion-implantation technique. Control over the magnetic hardness of the composite has now been obtained, although the exact mechanisms responsible for the increased coercivity have not been unambiguously identified. Another critical challenge will be in the creation of uniformly sized particles patterned into a 2D array. Nanosensors based on the magnetostrictive effect have also been proposed (e.g., [84]). If ion implantation is to have applications in magnetic recording, this challenge will almost certainly have to be met. In Sect. 5, some of the future research directions toward these objectives are discussed.

4.4 Smart Nanocomposites

In the previous sections, some of the most promising applications for nanocomposites synthesized by ion implantation were outlined. Several other potential applications are being actively explored. One example of "smart" nanocomposite material developed by ion implantation is nanocrystalline VO_2 embedded in a SiO_2 matrix (Fig. 7) [85]. VO_2 is a transition-metal oxide that shows abrupt changes in electronic and optical properties at 67°C. On heating above this critical temperature the material undergoes a semiconductor-to-metal phase transition with a 10,000-fold increase in electrical conductivity and a significant decrease of the near-infrared optical transmission. It is precisely this near-room-temperature transition and the associated large optical and electrical property changes that have also made VO_2 a candidate material for a wide variety of technological applications such as thermochromic coatings, optical or holographic storage, laser scanners, and ultrafast optical switching. It is considered a particularly "smart" material because the switching/sensing functions are integrated in the intrinsic nature of the oxide without the need of external circuitry, control, or modular design.

VO_2 nanocomposites can be fabricated using the ion-implantation techniques described above. In particular, small VO_2 particles in the nanoscale regime have been produced in SiO_2 and Al_2O_3 substrates by coimplanting vanadium and oxygen ions at room temperature in a stoichiometric proportion. The implantation energies were 150 keV for vanadium and 55 keV for

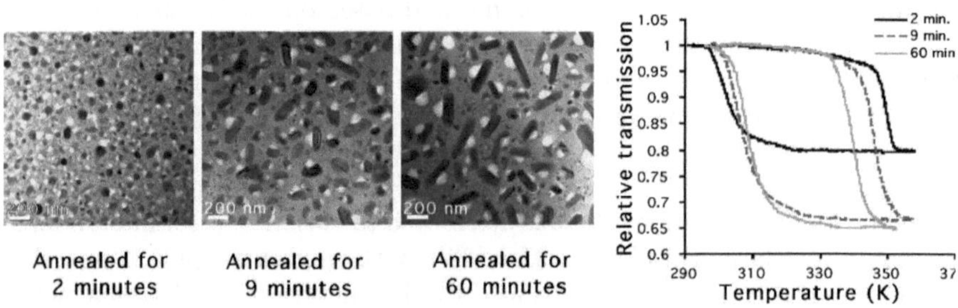

Fig. 7. Transmission electron microscopy images of VO_2 precipitates in SiO_2 and their optical transmission vs. temperature at a 1.5-μm wavelength for selected annealing times. Shorter annealing times produced smaller precipitates and produce wider hysteresis loops. The samples were prepared by implanting SiO_2 with 1.5×10^{17} V ions cm^{-2} at 150 keV and 3.0×10^{17} O ions cm^{-2} at 55 keV and then annealing in argon at 1000°C. After [85]

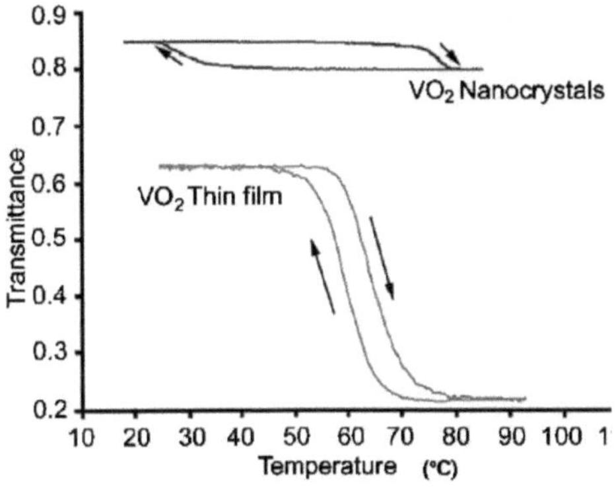

Fig. 8. Linear optical transmittance at $\lambda = 800\,\text{nm}$ of a nanocrystalline VO_2/SiO_2 composite and a standard VO_2 thin film cycled through their phase transition. After [89]

oxygen ions, to ensure overlapping concentration profiles within 200 nm from the specimen surface. Thermal processing at 1000°C in a flowing high-purity argon atmosphere induced the precipitation of the VO_2 nanocrystals whose size was controlled by varying the growth time in the furnace from 2 to 60 min.

Figure 8 illustrates the linear optical transmittance at $\lambda = 800\,\text{nm}$ for both VO_2 nanoparticles and a 210-nm thick standard VO_2 thin film undergoing the same phase transition. The VO_2 phase transition is manifested by a decrease in optical transmission when the metallic phase is formed in both thin-film and nanocrystalline samples; however, the magnitude of the change is larger in the film since it contains five times more VO_2 by mass than the nanocrystalline composite. When normalized to an equal mass, the VO_2 nanoparticles present an enhanced optical contrast compared with thin films [86] and exhibit a significantly wider hysteresis than the thin-film sample. Systematic size-controlled studies have shown that the width of the hysteresis loop is, in fact, a size-dependent property. The origin of this behavior is rooted in the heterogeneous nature of the phase transition. Smaller precipitates show larger hysteresis due to the statistically small probability of finding a phase-transition nucleation site within a smaller nanoparticle volume [87].

When the nanoparticles are in the metallic phase, a strong surface-plasmon resonance is present – a feature that is absent in bulk VO_2. According to (1), the large absorption coefficient at the surface-plasmon frequency can be calculated from the classical Mie formula for the polarizability of a spherical particle, for particle sizes much smaller than the wavelength of light. The surface-plasmon resonance in VO_2 nanoparticles can be observed only when the particles are in the metallic phase, in which the real part of the dielectric constant is negative. The relative change in transmission, $\Delta T/T$, as a function of wavelength in the case of large-aspect-ratio nanoparticles is well described by Mie scattering theory (solid curves of Fig. 9). According to Mie theory, the resonance condition for nonspherical particles depends on their orientation relative to the electric-field polarization. For ellipsoidal nanoparticles with their major axes parallel to the plane of incidence, Mie

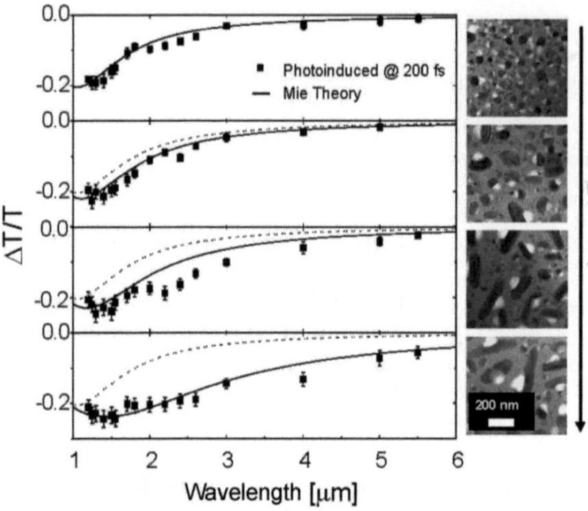

Fig. 9. Relative change of transmission as a function of wavelength for several aspect ratios (*top to bottom*: 1, 1.5, 2, 3.5) as obtained by varying the annealing time (3, 7, 10, and 45 min, respectively). *Squares*, measured at a delay of 200 fs after photoexcitation; *solid curves*, calculations based on Mie scattering theory. As a comparison, *dashed curves* show the prediction of Mie theory for spherical particles (aspect ratio, 1; the same as in the topmost figure). Representative TEM pictures of the corresponding samples are shown *at the right*. After [88]

theory predicts absorption features at longer wavelengths for polarization parallel to the major axis. Therefore, the continuous angular distribution of randomly oriented rods in the ion-implanted samples results in an overall broadening of the absorption spectrum toward longer wavelengths.

Additional interesting features of the switching behavior of these ion-implanted nanocomposites include room-temperature operation, compatibility with fiber-optic environment, and high efficiency at telecommunication wavelengths (1.3–1.5 μm). Furthermore, the phase transition can be induced by infrared light pulses of less than 200 fs duration [88]. Indeed, the phase transition can be optically triggered by electronic excitation on a timescale sorter than that of a thermal path. The threshold for the photoinduced phase transition is 300 $mJ\,cm^{-2}$, which is equivalent to a 150-pJ pulse for a typical 50-μm^2 size single-mode fiber, making possible the use of nanoscale VO_2 in all-optical switching schemes. In the particles with the highest aspect ratio, a switching efficiency of ~25% was observed at 1.55 μm over a propagation length of less than 100 nm and for a VO_2 filling factor of less than 10% [89]. One might achieve higher switching contrasts by increasing aerial filling factors or interaction length, although the nanosecond lifetime of the metallic state may limit the suitability of VO_2 for all-optical switching at high bit rates.

Ion implantation allows easy doping of the VO_2 nanoparticles with other ions by coimplanting them with the oxygen and vanadium prior to the heat thermal processing. The addition of other species is useful since they can help to tailor the critical temperature and hysteresis width of the phase transition. Tungsten in low atomic concentrations can decrease the phase-transition temperature below room temperature without producing a significant deterioration of the switching characteristics. In contrast, while titanium has little effect on the onset value of the transition temperature, it is found that it can significantly reduce the width of the hysteresis loop. Ion implantation may be one of the most attractive ways to produce nonequilibrium doping of smart nanocomposites in order to further control their properties.

5 Controlling Nanocrystal Size, Spacing, and Location

Despite the many potential applications, there are currently no commercial nanoparticle-based devices that make use of the ion-implantation concept. New ideas and advances are in various stages of development, and in some cases prototype devices have been built and tested. The major and most critical experimental difficulty yet to be overcome is control over the size distribution and spacing of the precipitates. Key research in a number of areas related to these problems is underway. This research is making use of newly developed focused ion beams, multiple-beam implanters, and submicrometer lithographic masking.

Focused ion beams (FIBs) are a readily available technology, with manufacturers claiming beam diameters of under 10 nm. Commercial FIBs have been developed primarily as thinning and cutting tools and are typically limited to Ga sources with a maximum beam energy of 35 kV, which are currently major limitations to their use as ion implanters. At this low energy, sputtering is the major difficulty since the implanted ions cannot be injected sufficiently deep into the host, and the beam acts more like an "ion drill". Accessories can now be obtained that permit a wider range of ions to be implanted. Some laboratories do have high-energy focused ion-beam machines, although these are often home-built and are not readily available commercially. Even if the voltage and source issues are overcome, FIB implantation is still a serial technique. Implantation of relatively large areas will, therefore, be a slow and costly procedure.

Lithographic masking is another technique that offers promise for controlling the distribution of nanoparticles within an implanted layer. The technique has been employed for decades in industry for precision doping of silicon wafers. The main difference in nanoparticle research is that the film material will have to withstand high heavy-ion-implantation fluences without physical degradation. A first example of patterned ferromagnetic nanoparticles was recently demonstrated [90], and subsequently the patterning of luminescent silicon nanocrystals via an ion-implantation mask was shown (Fig. 10) [91].

Fig. 10. Luminescent silicon nanocrystals array produced by high-fluence implantation through a metal mask. The letters in the "U of A" sign in the inset at the lower right are ∼68 micrometers high. The image color images were obtained with a fluorescence microscope using an Ar ion laser as the pump. From [91]

In this example, a thick molybdenum mask was patterned directly on a fused silica wafer, using conventional deep UV lithography. The specimen was subsequently implanted $>10^{17}$ ions cm^{-2} (silicon and iron were tested in the references cited above). The results showed that the masking material did not observably degrade due to ion implantation, and micrometer-scale spatial control was obtained. By using finer-scale lithographic techniques, it may be possible to create considerably smaller features, with the ultimate goal of creating one particle at each mask hole, although certain technical challenges will have to be overcome in order to accomplish this objective (e.g., lateral straggling of the implanted ions).

One of the current technological advancements pursued in several laboratories is the production of a multibeam implantation instrument. Parallel implantation using many focused ion beams on a single machine would significantly speed the pattern-generating capabilities of the current generation of FIB instruments. Ion-beam projection through a stencil mask is a further promising idea under investigation. A conventional ion beam first impinges on a durable patterned mask, so that only the patterned regions are "projected" down the beamline. The subsequent "beamlets" are focused and demagnified until they impinge on a specimen. Other multifocused-beam instruments will be maskless, instead using a stacked multiaperture electrode–insulator structure to accelerate each parallel beam with the same accelerator electrodes. As these technologies are developed, it will become possible to control the spacing, and even possibly the particle size, in a relatively inexpensive, reproducible fashion that avoids time-consuming lithography and makes practical high-volume manufacturing of ion-implanted nanostructured materials.

6 Conclusion

In the past decade, ion implantation has been demonstrated to be a versatile and flexible method for creating single-element or compound nanocrystals embedded in one of many possible host materials. To date, numerous nanocrystal–host combinations have been synthesized and their optical, electronic, magnetic, and microstructural properties investigated. A variety of possible applications has been identified that are being actively explored. The utility of the technique for creating an enormous variety of nanocrystalline composites is now well established. The research directions in the next decade will focus less on demonstrating the creation of new nanocrystal–host combinations, and more on solving the well-known difficulties associated with the technique and on developing new *classes* of materials with new properties (e.g., smart materials, micromechanical materials, or buried superconducting layers and wires). In the great majority of potential applications, it will be critical to narrow the size distributions. Advanced nucleation and growth techniques such as inverse Ostwald ripening or temporal separation of the nucleation and growth stages offer promise for improving size distributions. Lithography and parallel focused ion-beam instruments provide another means for potentially narrowing size distributions and patterning nanocrystal arrays on the submicrometer scale.

References

1. A. Meldrum, R.F. Haglund, L.A. Boatner, C.W. White, Adv. Mater. **13**, 1431 (2001) 256, 263, 275
2. E. Valentin, H. Bernas, C. Ricolleau, F. Creuzet, Phys. Rev. Lett. **86**, 99 (2001) 257
3. J.F. Ziegler, www.srim.org 258
4. J. Bosbach, D. Martin, F. Stretz, T. Wengel, F. Träger, Appl. Phys. Lett. **74**, 2605 (1999) 259
5. A. Halabica, R.H. Magruder III, R.F. Haglund Jr., *Photon Processing in Microelectronic and Photonics*, ed. by T. Okada, M. Meunier, A.S. Holmes, F. Träger, Proceedings of SPIE, vol. 6458 (2007) 259
6. V. Belostotsky, J. Non-Cryst. Solids **202**, 194 (1996) 260
7. A. Meldrum, E. Sonder, R.A. Zuhr, I.M. Anderson, J.D. Budai, C.W. White, L.A. Boatner D, O. Henderson, J. Mater. Res. **14**, 4489 (1999) 260, 261
8. K.D. Sorge, J.R. Thompson, T.C. Schulthess, F.A. Modine, T.E. Haynes, S. Honda, A. Meldrum, J.D. Budai, C.W. White, L.A. Boatner, IEEE Trans. Magn. **37**, 2197 (2001) 262, 273, 274
9. A. Meldrum, S. Honda, C.W. White, R.A. Zuhr, L.A. Boatner, J. Mater. Res. **16**, 2670 (2001) 262
10. B.G. Fernandez, M. Lopez, C. Garcia, A. Perez-Rodriguez, J.R. Morante, C. Bonazos, M. Carrada, A. Claverie, J. Appl. Phys. **91**, 798 (2002) 262
11. R.E. de Lamaestre, H. Bernas, Phys. Rev. B **73**, 125317 (2006) 263

12. R.F. Haglund Jr., in *Optics of Small Particles, Surfaces and Interfaces*, ed. by R.E. Hummel, P. Wißmuth (CRC Press, Boca Raton, 1997), pp. 151–186 263
13. P. Chakraborty, J. Mater. Sci. **33**, 2235 (1998) 263
14. U. Kreibig, M. Vollmer, *Optical Properties of Metal Nanoclusters.* Springer Series in Materials Science (Springer, Berlin, 1995) 264
15. C.F. Bohren, D.R. Huffman, *Absorption and Scattering of Light by Small Particles* (Wiley, New York, 1983) 264
16. R.F. Haglund Jr., L. Yang, R.H. Magruder III, C.W. White, R.A. Zuhr, L. Yang, R. Dorsinville, R.R. Alfano, Nucl. Instrum. Methods B **91**, 493 (1994) 265
17. L. Yang, D.H. Osborne, R.F. Haglund Jr., R.H. Magruder, C.W. White, R.A. Zuhr, H. Hosono, Appl. Phys. A **62**, 403 (1996) 265
18. A.E. Neeves, M.H. Birnboim, J. Opt. Soc. Am. B **6**, 787 (1989) 265
19. C. Flytzanis, F. Hache, M.C. Klein, D. Ricard, Ph. Roussignol, Prog. Opt. **29**, 321 (1991) 266
20. R.H. Magruder III, R.A. Zuhr, D.H. Osborne Jr., Nucl. Instrum. Methods B **99**, 590 (1995) 266
21. R.H. Magruder III, T.S. Anderson, R.A. Zuhr, D.K. Thomas, Nucl. Instrum. Methods B **108**, 305 (1996) 266
22. T.S. Anderson, R.H. Magruder, D.L. Kinser, R.A. Zuhr, D.K. Thomas, Nucl. Instrum. Methods B **124**, 40 (1997) 266
23. R.H. Magruder, D.H. Osborne Jr., R.A. Zuhr, J. Non-Cryst. Solids **176**, 299 (1994) 266
24. S.A. Maier, M.L. Brongersma, P.G. Kik, S. Meltzer, A.A.G. Requicha, H.A. Atwater, Adv. Mater. **13**, 1501 (2001) 267
25. P.G. Kik, S.A. Maier, H.A. Atwater, in *2001 MRS Fall Meeting Abstracts* (Materials Research Society, Warrendale, 2001) 267
26. M. Sugiyama, S. Inasawa, S. Koda, H. Hirose, T. Yonekawa, T. Omatsu, A. Takami, Appl. Phys. Lett. **79**, 1528 (2001) 267
27. T. Hayakawa, S. Tamail, Selvan, M. Nogami, Appl. Phys. Lett. **74**, 1513 (1998) 267
28. H.R. Stuart, D.G. Hall, Appl. Phys. Lett. **73**, 3815 (1998) 267
29. U. Banin, O. Millo, Annu. Rev. Phys. Chem. **54**, 465 (2003) 267
30. A.L. Efros, M. Rosen, Annu. Rev. Mater. Sci. **30**, 475 (2000) 267
31. S.V. Gaponenko, *Optical Properties of Semiconductor Nanocrystals* (Cambridge University Press, Cambridge, 1998) 267
32. L. Jacak, P. Hawrylak, A. Wojs, *Quantum Dots* (Springer, Berlin, 1998) 267
33. L.E. Brus, J. Chem. Phys. **80**, 4403 (1984) 267
34. L. Canham, Appl. Phys. Lett. **57**, 1046 (1990) 268
35. T. Shimizu-Iwayama, M. Oshima, T. Niimi, S. Nakao, K. Saitoh, T. Fujita, N. Itoh, J. Phys. Condens. Matter **5**, 375 (1993) 268
36. L. Pavesi, in *Optoelectronic Integration on Silicon*, Proc. SPIE, vol. 4997 (2003), p. 206 268

37. L. Pavesi, L. Dal Negro, C. Mazzoleni, G. Franzo, F. Priolo, Nature **408**, 440 (2000) 268
38. N. Lalic, J. Linnros, J. Lumin. **80**, 263 (1999) 268
39. M. Kulakci, U. Serincan, R. Turan, Semicond. Sci. Technol. **21**, 1527 (2006) 268
40. M. Porti, M. Avidano, M. Nafria, X. Aymerich, J. Carreras, O. Jambois, B. Garrido, J. Appl. Phys. **101**, 064509 (2007) 268
41. C. Bonafos, H. Coffin, S. Schamm, N. Cherkashin, G. Ben Assayag, P. Dimitrakis, P. Normand, M. Carrada, V. Paillard, A. Claverie, Solid-State Electron. **49**, 1734 (2005) 268
42. K. Luterova, M. Cazzanelli, J.P. Likforman, D. Navarro, J. Valenta, T. Ostatnicky, K. Dohnalova, S. Cheylan, P. Gilliot, B. Honerlage, L. Pavesi, I. Pelant, Opt. Mater. **27**, 750 (2005) 268
43. L. Dal Negro, L. Pavesi, G. Pucker, G. Franzo, F. Priolo, Opt. Mater. **17**, 41 (2001) 268
44. P. Janda, J. Valenta, T. Ostatnicky, E. Skopalova, I. Pelant, R.G. Elliman, R. Tomasiunas, J. Lumin. **121**, 267 (2006) 268
45. N. Daldosso, M. Melchiorri, L. Pavesi, G. Pucker, F. Gourbilleau, S. Chausserie, A. Belarouci, X. Portier, C. Dufour, J. Lumin. **121**, 344 (2006) 268
46. A. Meldrum, A. Hryciw, A.N. MacDonald, C. Blois, K. Marsh, J. Wang, Q. Li, J. Vac. Sci. Technol. A **24**, 713 (2006) 269
47. B.G. Fernandez, M. Lopez, C. Garcia, A. Perez-Rodriguez, J.R. Morante, C. Bonafos, M. Carrada, A. Claverie, J. Appl. Phys. **91**, 798 (2002) 269
48. F. Iacona, C. Bongiorno, C. Spinella, S. Boninelli, F. Priolo, J. Appl. Phys. **95**, 3723 (2004) 269
49. Y.Q. Wang, R. Smirani, G.G. Ross, Nanoletters **4**, 2041 (2004) 269
50. Y.Q. Wang, R. Smirani, G.G. Ross, Nanotechnology **15**, 1554 (2004) 269
51. Y.Q. Wang, R. Smirani, G.G. Ross, Appl. Phys. Lett. **86**, 221920 (2005) 269
52. J. Wang, X.F. Wang, Q. Li, A. Hryciw, A. Meldrum, Philos. Mag. **87**, 11 (2007) 269
53. G. Hadjisavvas, I.N. Remediakis, P.C. Kelires, Phys. Rev. B **74**, 165419 (2006) 269
54. E. Kapetanakis, P. Normand, D. Tsoukalas, K. Beltsios, J. Stoemenos, S. Zhang, J. van den Berg, Appl. Phys. Lett. **77**, 3450 (2000) 269
55. P. Normand, D. Tsoukalas, E. Kapetanakis, J.A. Van den Berg, D.G. Armour, J. Stoemenos, C. Vieu, Electrochem. Solid State Lett. **1**, 88 (1998) 269
56. L. Pavesi et al. (eds.), *Towards the First Silicon Laser*. NATO Science Series II, vol. 93 (Kluwer Academic, Dordrecht, 2003) 269
57. M. Fujii, M. Yoshida, Y. Kanzawa, S. Hayashi, K. Yamamoto, Appl. Phys. Lett. **71**, 1198 (1997) 269
58. C. Maurizio, F. Iacona, F. D'Acapito, G. Franzo, F. Priolo, Phys. Rev. B **74**, 205428 (2006) 269

59. S.P. Withrow, C.W. White, A. Meldrum, J.D. Budai, D.M. Hembree, J.C. Barbour, J. Appl. Phys. **68**, 396 (1999) 268, 270
60. M. López, B. Garrido, C. García, P. Pellegrino, A. Pérez-Rodríguez, J.R. Morante, C. Bonafos, M. Carrada, A. Claverie, Appl. Phys. Lett. **80**, 1637 (2002) 270
61. M. López, B. Garrido, C. Bonafos, A. Pérez-Rodríguez, J.R. Morante, A. Claverie, Nucl. Instrum. Methods B **178**, 89 (2001) 270
62. T. Shimizu-Iwayama, T. Hama, D.E. Hole, I.W. Boyd, Solid State Electron. **45**, 1487 (2001) 270
63. R. Lockwood, A. Hryciw, A. Meldrum, Appl. Phys. Lett. **89**, 263112 (2006) 270
64. C.J. Park, H.Y. Cho, S. Kim, S.H. Choi, R.G. Elliman, J.H. Han, C. Kim, H.N. Hwang, C.C. Hwang, J. Appl. Phys. **99**, 036101 (2006) 270
65. J.-Y. Zhang, Y.-H. Ye, X.-L. Tan, X.-M. Bao, Appl. Phys. Lett. **86**, 6139 (1999) 270
66. S.H. Cho, S.C. Han, S.T. Hwang, Thin Solid Films **413**, 177 (2002) 270
67. S.N.M. Mestanza, E. Rodriguez, N.C. Frateschi, Nanotechnology **17**, 4548 (2006) 270
68. S. Duguay, J.J. Grob, A. Slaoui, Y. Le Gall, M. Amann-Liess, J. Appl. Phys. **97**, 104330 (2005) 270
69. Y. Kanemitsu, H. Tanaka, T. Kushida, K.S. Min, H.A. Atwater, J. Appl. Phys. **86**, 1762 (1999) 270
70. C. Bonafos, B. Garrido, M. López, A. Romano-Rodríguez, O. González-Varona, A. Pérez-Rodriguez, J.R. Morante, R. Rodríguez, Nucl. Instrum. Methods B **147**, 373 (1999) 271
71. M. Ando, Y. Kanemitsu, T. Kushida, K. Matsuda, T. Saiki, C.W. White, Appl. Phys. Lett. **79**, 539 (2001) 271
72. R.E. de Lamaestre, H. Bernas, D. Pacifici, G. Franzo, F. Priolo, Appl. Phys. Lett. **88**, 181115 (2006) 271
73. R.E. de Lamaestre, H. Bernas, J. Appl. Phys. **98**, 104310 (2005) 271
74. R.E. de Lamaestre, J. Majimel, F. Jomard, H. Bernas, J. Phys. Chem. B **109**, 19148 (2005) 271
75. S. Tiwari, F. Rana, H. Hanafi, A. Hartstein, E.F. Crabbe, K. Chan, Appl. Phys. Lett. **68**, 1377 (1996) 272
76. H.Y. Zhai, H.M. Christen, C.W. White, J.D. Budai, A. Meldrum, D.H. Lowndes, Appl. Phys. Lett. **80**, 4786 (2002) 272
77. F.J. Himpsel, J.E. Ortega, G.J. Mankey, R.F. Williams, Adv. Phys. **47**, 511 (1998) 272
78. T.C. Schulthess, M. Benakli, P.B. Visscher, K.D. Sorge, J.R. Thompson, F.A. Modine, T.E. Haynes, L.A. Boatner, G.M. Stocks, H. Butler, J. Appl. Phys. **89**, 7594 (2001) 273, 274
79. G.L. Zhang, S. Yu, J. Phys., Condens. Matter **9**, 1851 (1997) 274
80. S. Honda, F.A. Modine, T.E. Haynes, A. Meldrum, J.D. Budai, K.J. Song, J.R. Thompson, L.A. Boatner, Mater. Res. Soc. Symp. Proc. **581**, 71 (2000) 274, 275

81. S. Withrow, C.W. White, J.D. Budai, L.A. Boatner, K.D. Sorge, J.R. Thompson, R. Kalyanaraman, A. Meldrum, in *2001 MRS Fall Meeting Abstracts* (Materials Research Society, Warrendale, 2001) 274
82. K.S. Buchanan, X.B. Zhu, A. Meldrum, M.R. Freeman, Nanoletters **5**, 383 (2005) 275
83. K.S. Buchanan, A. Krichevsky, M.R. Freeman, A. Meldrum, Phys. Rev. B **70**, 174436 (2004) 275
84. S. Honda, F.A. Modine, A. Meldrum, J.D. Budai, T.E. Haynes, L.A. Boatner, L.A. Gea, Mater. Res. Soc. Symp. Proc. **540**, 225 (1999) 276
85. R. Lopez, L.A. Boatner, T.E. Haynes, L.C. Feldman, R.F. Haglund Jr., J. Appl. Phys. **92**, 4031 (2002) 276
86. R. Lopez, T.E. Haynes, L.A. Boatner, L.C. Feldman, R.F. Haglund Jr., Opt. Lett. **27**, 1327 (2002) 277
87. R. Lopez, T.E. Haynes, L.A. Boatner, L.C. Feldman, R.F. Haglund Jr., Phys. Rev. B **65**, 224113 (2002) 277
88. M. Rini, A. Cavalleri, R. Lopez, L.A. Boatner, R.F. Haglund Jr., T.E. Haynes, L.C. Feldman, R.W. Schoenlein, Opt. Lett. **30**, 558 (2005) 278
89. R. Lopez, R.F. Haglund Jr., L.C. Feldman, T.E. Haynes, L.A. Boatner, Appl. Phys. Lett. **85**, 5191 (2004) 277, 278
90. K.S. Beaty, A. Meldrum, J.F. Franck, K. Sorge, J.R. Thompson, C.W. White, L.A. Boatner, S. Honda, Mater. Res. Soc. Symp. Proc. **703**, V9.38.1 (2002) 279
91. A. Meldrum, K.S. Buchanan, A. Hryciw, C.W. White, Adv. Mater. **16**, 31 (2004) 279, 280

Index

Metal Nanoclusters for Optical Properties

Giovanni Mattei[1], Paolo Mazzoldi[1] and Harry Bernas[2]

[1] Department of Physics, University of Padova, via Marzolo 8, 35131 Padova, Italy, e-mail: giovanni.mattei@unipd.it
[2] CSNSM-CNRS, Université Paris-Sud XI, Orsay, 91405, France

Abstract. This chapter is focused on the use of ion-beam processing for controlling the linear and nonlinear optical properties of different nanostructures based on metallic clusters embedded in silica. Three case studies will be presented: the first is the modification of the surface-plasmon resonance of metallic nanoclusters by direct implantation or irradiation. The second is the far-field and local-field modification by means of controlled ion irradiation of bimetallic nanoclusters. The last example deals with the synthesis of ordered plasmonic nanostructures using ion implantation and/or irradiation. These three examples are a sort of hierarchical approach to the control of the optical properties of nanostructures based on nanoparticles, starting from the synthesis of the functional *building block* of our approach, i.e., randomly positioned metal nanoparticles in silica, to end at the last level with the formation by ion-implantation-ordered arrays (chains or planar 2D assembly). In all these examples, ion implantation demonstrates to be not only a simple synthesis technique, but a very powerful processing tool for obtaining new functional properties. A brief description of the theoretical approach to the comprehension of the light-particle interaction from a semiclassical point of view will be given to aid the interpretation of the experimental results.

1 Introduction

Probably the best-known optical property of noble-metal nanoparticles (NP) embedded in glass is the intense absorption band in the visible spectrum arising from a collective excitation (i.e., oscillation) of the free electrons, which is called the surface-plasmon resonance (SPR). Its wavelength position in the spectrum is a sort of fingerprint for each metal and it is responsible for the color exhibited by the glass, as it has been exploited in artistic masterpieces based on multicolor glasses or decorated lusters.

The first glasses containing metal clusters were indeed fabricated by Roman glassmakers in the fourth century AD and the most celebrated example of their artistic use is the Lycurgus Cup, made of a soda-lime glass containing nanoclusters of a Au–Ag alloy. The puzzling point is that the cup changes its color from opaque green when illuminated in reflection to strong red when illuminated from inside. The formal explanation of this remarkable phenomenon was given in the fundamental work published in *Annalen der Physik* in 1908 [1] by Gustav Mie, who analytically solved the electrodynamics of the

H. Bernas (Ed.): Materials Science with Ion Beams,
Topics Appl. Physics **116**, 287–316 (2010)
DOI: 10.1007/978-3-540-88789-8_10,

interaction of a plane wave with an isolated metal particle embedded in a dielectric medium. The most relevant factors controlling the position and intensity of the SPR absorption band are: (i) the nanoparticle size, shape and composition, and (ii) the refractive index of the matrix. Also, interactions among nanoparticles can strongly affect the SPR, as the extensions of the Mie theory have showed, as we will see in the following.

Therefore, the aim of this chapter is to demonstrate the possibility of controlling the linear and nonlinear optical properties of different nanostructures based on metal nanoclusters embedded in silica by using ion-beam processing techniques, as a tool for growing, modifying and coupling metallic nanoparticles.

We will follow a sort of hierarchical approach presenting three case studies of increasing complexity: the first, basic level is the modification of the surface-plasmon resonance of metallic NPs by high-fluence direct implantation. The second is the far-field and local-field optical properties modification by means of controlled ion irradiation of bimetallic NPs. The last level deals with the synthesis of plasmonic nanostructures (chains or 2D arrays of NPs) by means of ion implantation and/or irradiation. The level of complexity in the interpretation of the optical properties increases at each level, starting from the noninteracting case and reaching the full coupling between plasmonic and photonic modes for the ordered plasmonic structures. Therefore, a brief description of the theoretical approach to the comprehension of the light-particle interaction from a semiclassical point of view will be given to aid the interpretation of the experimental results.

At each level of this hierarchy, ion implantation demonstrates to be not only a simple synthesis technique, but a very powerful processing tool for making new materials with new functional optical properties.

2 Optical Properties of Metal Nanoclusters

The electrodynamic problem of the interaction between light and a NP was posed just after the foundation of Maxwell equations in the middle of the nineteenth Century. Without entering the fascinating history of this problem (see for instance, [2]), we just recall here three fundamental milestones: (i) the first attempt to explain the nature of the color induced in the glasses by nanosized metal precipitates of Faraday in 1857 [3], (ii) the work of Maxwell-Garnett, who introduced in 1904 the effective medium description of a composite with metallic nanoinclusions [4], and (iii) the work of Mie, who solved in 1908 the Maxwell equations describing the scattering of a plane wave by a spherical isolated inclusion in a nonabsorbing medium [1].

The general expression of the extinction cross section (i.e., the sum of absorption and scattering) given by Mie [1] contains all the multipolar expansions of the fields needed for describing the dephasing effects of the field oscillation within the NPs when their size is comparable to or larger than the light wavelength λ. Such a cross section, σ, can be written in terms of the polarizability, α, of the NP as:

$$\sigma = 4\pi \frac{\omega}{c} \varepsilon_{\mathrm{m}}^{1/2} \mathrm{Im}[\alpha], \tag{1}$$

where ε_{m} is the dielectric constant of the nonabsorbing host medium (i.e., whose refractive index is $n = \sqrt{\varepsilon_{\mathrm{m}}}$). In the case of spherical noninteracting NPs of radius R, much smaller than λ (Rayleigh or quasistatic regime), the polarizability and therefore the extinction cross section can be simplified in the dipolar approximation as:

$$\alpha = 3V_0 \frac{\varepsilon - \varepsilon_{\mathrm{m}}}{\varepsilon + 2\varepsilon_{\mathrm{m}}}, \tag{2}$$

$$\sigma = 9 \frac{\omega}{c} \varepsilon_{\mathrm{m}}^{3/2} V_0 \frac{\varepsilon_2}{(\varepsilon_1 + 2\varepsilon_{\mathrm{m}})^2 + \varepsilon_2^2}, \tag{3}$$

where V_0 is the cluster volume, c is the speed of light in vacuum and $\varepsilon = \varepsilon(\omega, R) \equiv \varepsilon_1(\omega, R) + \mathrm{i}\varepsilon_2(\omega, R)$ is the size-dependent complex dielectric function of the cluster. When $R \sim \lambda$, high-order multipolar expansions should be taken into account in the general Mie formula [5], because retardation effects inside the cluster make electrons oscillate not all with the same phase of the local field.

The SPR absorption resonance holds when the denominator in (3) exhibits a minimum. In particular for noble-metal NPs the equation:

$$\varepsilon_1(\omega, R) + 2\varepsilon_{\mathrm{m}} = 0 \tag{4}$$

has a solution in the visible range at the SPR frequency (also called the Frölich frequency [6], proportional to the square root of the electronic density), because the real part of the dielectric function ε_1 is negative. Therefore, the

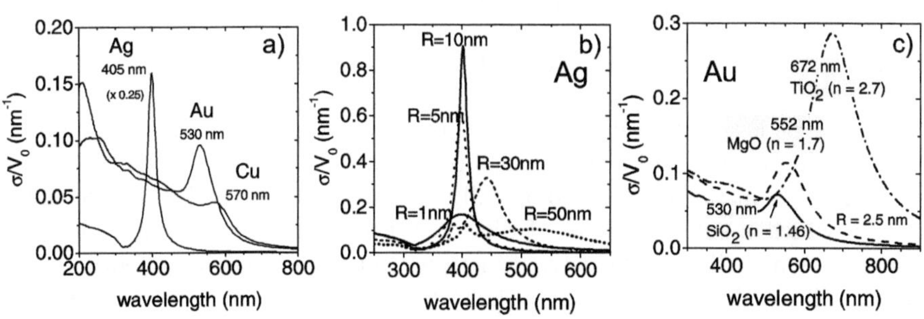

Fig. 1. Absorption spectra of spherical noninteracting NPs embedded in nonabsorbing matrices (Mie calculations): (**a**) effect of the NP composition for $R = 5\,\mathrm{nm}$ NPs (Ag, Au, Cu) in silica ($n = 1.46$); (**b**) effect of the size for Ag NPs in silica; (**c**) effect of the matrix for $R = 2.5\,\mathrm{nm}$ Au NPs (the refractive index $n = \sqrt{\varepsilon_{\mathrm{m}}}$ and the position of the surface plasma resonance are reported for each considered matrix)

position of the SPR absorption band is controlled by (4), whereas the imaginary part of the NP dielectric function affects the width of the resonance. One relevant point concerns the size dependence of the NP dielectric function $\varepsilon(\omega, R)$, which can be obtained as a quantum-mechanical correction to the original work of Mie, which of course disregarded the quantum confinement of the delocalized free electrons in the metals as the NP size becomes comparable with the electron mean free path (of the order of 10 nm, for typical metals). Following [5], the adopted strategy for achieving size correction is to modify the Drude-like part of the dielectric function related to delocalized free s-electrons by introducing a size-dependent damping frequency $\Gamma(R)$. It accounts for the increased electronic scattering at the NP surface as $\Gamma(R) = \Gamma_\infty + Av_F/R$, where Γ_∞ is the bulk damping frequency, v_F is the bulk Fermi velocity and A a numerical constant usually assumed to be 1.

To better illustrate the main factors influencing the position and the shape of the SPR absorption band, Fig. 1a shows the effect of the NP composition for $R = 5$ nm NPs of Ag, Au and Cu in silica ($n = 1.46$). The SPR position shifts from 405 nm (Ag), to 530 nm (Au), to 570 nm (Cu) with a decreased intensity due to the increasing overlap between SPR and interband transitions, which act as an effective damping system. In Fig. 1b the effect of the NP radius is shown for Ag NPs in silica: when the size is no longer negligible with respect to the wavelength there is a redshift of the SPR position with the appearance of multipolar peaks in the UV-blue region. The effect of the matrix for a fixed NP size is described for $R = 2.5$ nm Au NPs in Fig. 1c: the higher the matrix dielectric constant the more redshifted is the SPR, due to a better dielectric decoupling between SPR and interband transitions.

Another remarkable result of the Mie theory is the fact that the electric field $\mathbf{E}_{in}$ within the NP is different from zero and that near the SPR frequency it can be highly enhanced by the dielectric coupling with the matrix. In the dipolar approximation such field reads:

$$\mathbf{E}_{in} = \frac{3\varepsilon_m}{\varepsilon + 2\varepsilon_m}\mathbf{E}_0 = f_e\mathbf{E}_0 \quad \text{with } f_e \equiv \frac{3\varepsilon_m}{\varepsilon + 2\varepsilon_m}, \tag{5}$$

where the $\mathbf{E}_0$ is the external field and f_e is the local-field enhancement factor.

The Mie theory can be generalized to account for more complex NP morphologies. For instance, a technologically relevant class of NPs is the core-shell one, in which a spherical core of dielectric function ε and radius R is surrounded by a shell with dielectric function ε_s and thickness d embedded in a matrix with dielectric function ε_m [7]. The same approach can be used also for modeling the optical properties of partial core-shell NPs like the so-called *cookie-like* NPs made for one half of a metal and for the other of an oxide, e.g., Au–Cu_2O [8] or Au–NiO [9]. Further generalizations can take into account nonspherical shapes as in the theory developed by Gans in 1915 [10], or complete multipolar interactions among the NPs as in the generalized multiparticle Mie theory [11–13]. In particular, this complex interaction can be taken into account at least at the dipolar level in the effective medium theory

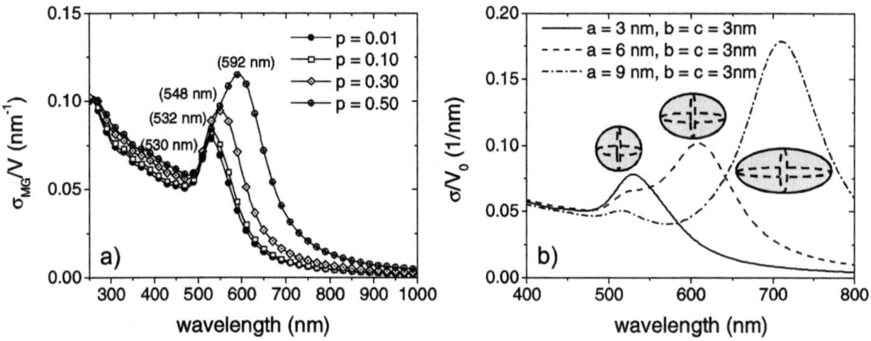

Fig. 2. (**a**) Extinction cross section of $R = 2.5\,\text{nm}$ Au NPs in silica, normalized to the NP volume, calculated with the Maxwell-Garnett effective medium theory as a function of the filling factor p; (**b**) effect of the NP aspect ratio for prolate ellipsoids within the Gans theory

of Maxwell-Garnett [4]. The approach is simple: instead of having a distribution of spherical monodispersed NPs with dielectric function $\varepsilon(\omega)$ embedded in a matrix with dielectric function $\varepsilon_{\mathrm{m}}(\omega)$, Maxwell-Garnett describes the whole system as a *homogeneous* material with an effective dielectric function $\varepsilon_{\mathrm{eff}}(\omega)$ obtained as:

$$\frac{\varepsilon_{\mathrm{eff}} - \varepsilon_{\mathrm{m}}}{\varepsilon_{\mathrm{eff}} + 2\varepsilon_{\mathrm{m}}} = p\frac{\varepsilon - \varepsilon_{\mathrm{m}}}{\varepsilon + 2\varepsilon_{\mathrm{m}}}, \tag{6}$$

with p the volumetric fraction (filling factor) of the NPs. The effect of the increased dipolar interparticle interaction (controlled by the filling factor p) on the extinction cross section is shown in Fig. 2a. Two main properties are worth noting: (i) the redshift of the SPR absorption with increasing filling factor and (ii) in the limit of small p the Maxwell-Garnett result reproduces that of the Mie theory, as it should. The redshift of the SPR band can be further enhanced by increasing the aspect ratio of the NPs as predicted by the Gans theory. Ellipsoidal metal particles randomly oriented with respect to the incident light remove the degeneracy of the single SPR band typical of spherical NPs, which becomes split in three subbands, at different wavelengths one for each of the three axes. Figure 2b shows for example the extinction cross section in the dipolar approximation for prolate ($a \neq b = c$) ellipsoids: as the aspect ratio increases, the extinction spectra transforms from spherical (single peak) to rod-like (two peaks), with a redshift related to the polarization parallel to the major axis.

As we have seen in this section, the control over the size, composition and morphology of the NPs in a system can produce dramatically different absorption features in the visible or near-infrared spectrum. The point is now: how can these structure be synthesized, and in particular within a matrix, in order to be used in a solid-state device? Nowadays, colloidal chemistry can produce a wealth of metal NP in solution with a variety of morphologies, from sphere to cube to complex core-shell [14], with a very good control of the size dispersion down to 5%. The main problem is that to produce solid-state devices such NPs have to be transferred from solution to a matrix, with problems of solubility, aggregation or coalescence. To overcome these problems advanced functionalization of the NP surface (e.g., thiols for Au) has to be introduced, whose effect can be, on the other hand, detrimental as it could promote a charge transfer from the NP to the capping agent with possible depression, for instance, of the SPR band [15]. Moreover, when patterning of the nanostructures is required, the use of colloidal chemistry is not always very practical. Therefore, this forces us to relax, to a certain extent, the size-dispersion issue, relying on more flexible techniques such as ion implantation. In the following section, we will give a short overview of the relevant physical and chemical aspects of NP nucleation and growth, specializing the discussion to the ion-implantation case.

3 Metal-Nanoparticle Synthesis by Ion Implantation

Ion implantation is a very efficient technique for both the synthesis and the processing of nanoparticle-based materials [16, 17], because it can introduce any desired amount of the guest phase in a host matrix, without thermodynamic limitations typical of other synthesis techniques, due to the fact that the ion implantation works under nonequilibrium conditions. The control of cluster size with dimensions in the nanometer range either during the synthesis process or after subsequent thermal annealing is one of the challenging issues of nanocluster technology, as the electronic and optical properties of the NPs depend strongly on their dimensions. In fact, ion implantation is usually coupled to postimplantation thermal treatments that have a twofold aim: (i) annealing of the implantation-induced defects and (ii) growth of the nucleated embryos by means of suitable combination of annealing atmosphere, temperature and time. In a silica matrix, for instance, annealing of the implantation damage requires temperatures near or above 600°C: at these temperatures the thermal diffusion of the implanted species (controlled also by atmosphere composition) can be quite effective in modifying the postimplantation dopant distribution, promoting either redistribution of implanted species or clustering around nucleated embryos. Therefore, a precise understanding of the microscopic mechanisms influencing the evolution of cluster size during thermal annealing is of paramount importance [18–20].

3.1 The Issue of Size Distribution

Considering that the electronic properties of the NPs are strongly size dependent one of the main issues in NP technology is to reduce (or at least to control) the size dispersion. Indeed, the main effect of size dispersion is to introduce a broadening (so-called *inhomogeneous*, different from the *homogeneous* one due to thermal effects) in the average properties. Typical size dispersions in ion-implanted systems are quite large (20–50%): this is due to the in-depth profile of the implanted ions that is roughly Gaussian-like, so the level of nucleation and growth is different when moving from the ion-projected range (where the maximum concentration is reached) toward the surface or the bulk of the sample. A simple strategy to produce a relatively box-like concentration profile is to combine implantations at different energies and fluences, whose convolution can approximate the constant profile required for producing a more homogeneous degree of supersaturation, i.e., a reduced NP size dispersion.

Some processes of NP ensemble synthesis lead to a well-defined shape of the size distribution. Examples are clustering by precipitation in liquid solutions, which often leads to very narrow Gaussian size distributions, or Ostwald ripening in specific cases (e.g., low solute concentrations, see, e.g., [18]) where the analytical Lifshitz–Slyozov–Wagner (LSW) [21, 22] shape accounts for the size distribution. If the synthesis mechanism is not well known, can we obtain information on its crucial features from the size distribution's shape? Consider a rather ubiquitous case: that of lognormal size distributions,

$$f_{\mathrm{LN}}(R=r) = \frac{1}{r\sqrt{2\pi}\ln\sigma}\exp\left(-\frac{(\ln(r/\mu))^2}{2(\ln\sigma)^2}\right), \tag{7}$$

where R is the NP radius, r the running variable, μ the geometrical average and σ the geometrical standard deviation of the size distribution. Such distributions are found in many different physical processes: coagulation, deposition, fractal aggregation, etc., as well as in NP synthesis, by ion implantation/irradiation or otherwise. A systematic study [23] showed that the parameters of the lognormal distribution remained unchanged under extremely different conditions. For example, they were identical for different implanted species (including sequential implantation to synthesize II-VI NPs), or when diffusion occurred outside of the implant profile, or when a solute reacted chemically with a glass host component such as oxygen.

The proposed explanation is as follows. In cases such as LSW-type Ostwald ripening, the system conserves the memory of its initial structure, and its evolution is uniquely determined. By contrast, the lognormal distribution is the result of processes that interfere which each other, thus progressively erasing any memory of the initial conditions. This was quantified by Rosen [24] who characterized the information contained in the size-distribution function by its entropy and applied the entropy maximization principle according to which two conservation equations determine our systems' evolution: matter conservation, and conservation of the size space population. The latter depends explicitly on the microscopic growth mechanism. When this mechanism is no longer defined, due either to the multiplication of different mechanisms or to interfering processes such as diffusion or chemical interactions, entropy maximization occurs under the sole – very general – constraint of matter conservation. This leads to an analytic asymptotic form that closely resembles the lognormal distribution function, and whose most probable geometrical standard deviation is generally in the range 1.4–1.5 as found in most experiments. Thus, a lognormal size distribution actually signals a loss of memory (hence a loss of control) of the nucleation and growth process.

3.2 Ion Implantation for Plasmonic Nanostructures

The synthesis of plasmonic structures based on metal NPs performed by using ion-beam processing can be exploited with three different methodologies:

- *Ion-beam direct synthesis*: mono- or bielemental NPs are produced by single or sequential implantation of the metal ions inside the host matrix [25–27], i.e., the ion beam is used to create a supersaturated solid solution that, either during implantation itself or after thermal treatments, nucleates and grows, producing well-defined NPs. Recently, the use of focused ion beams (FIB) has become more and more efficient [28–30], allowing, for instance, to directly pattern a substrate by *writing* metallic nanostructures in a way that is very similar to electron-beam lithography (EBL);
- *Ion-beam modification*: the energy (electronic or nuclear) released by the implanted ions is used to modify already-formed NPs by changing their size, shape, composition and topology. In particular, this mode can be conveniently applied to modify the electromagnetic environment around a NP by the formation of peculiar cluster-satellite topology or to promote a selective dealloying in bimetallic NPs [31];
- *Ion-beam indirect synthesis*: ion-beam irradiation in this case is used to tailor the energy deposited in the target (nuclear vs. electronic, in-depth distribution, etc.) so as to promote nucleation of NPs made of atoms already present in the host matrix [18], or introduced as dopants by auxiliary techniques, like for instance ion exchange [16].

The last two items can be globally considered as *ion irradiation*, meaning that the implanted ions are not intended themselves to form the nanostructure, but to act as *energy carriers* to the nucleating species. In the following we will give three case studies in which ion implantation demonstrates to be not only a simple synthesis technique, but a very powerful processing tool for making new functional materials.

3.3 Nucleation and Growth of Metal Nanoparticles

The most obvious way to synthesize NPs inside a matrix is to implant ions of an element M at a concentration that is above the solubility limit of M in the host to promote precipitation of a new phase from a supersaturated solid solution (ion-beam direct synthesis).

Ion implantation is not simply triggered by physical (ballistic) processes: chemistry also plays a relevant role due to the different reactivity of the implanted species with the elements in the matrix. In the case of silica or silicate glasses, for instance, this implies the formation of oxides or silicates instead of the metallic ones [32–34]. In particular, the effect of the redox potential of the matrix with respect to the implanted or irradiated species has been clarified in a series of experiments involving the Ag precipitation following irradiation in different glasses in analogy with photography [18, 35]. In insulators, chemistry always plays a major role in the evolution of such systems, via the complex interactions between both solute and host atoms and the charged defects created by the ion-beam impacts. The fact that such effects are often masked by solute–solute interactions in high-fluence implantation syntheses[1] should not lead us to neglect them, since in most cases they ultimately determine nucleation and influence growth, both of which are obviously crucial for NP properties.

Ionizing radiation interacting with insulators and semiconductors creates electrons and holes that become mobile upon annealing even at comparably low temperatures. The fact that the density and thermal evolution of such charges, and their interaction with defects, may drastically affect nanocrystal clustering and growth rates is long-known in the photographic process [36], where photon absorption by AgBr crystals creates electron–hole pairs whose components migrate and may be trapped by defect centers or silver ions in different charge states. Depending on the charge balance, the latter case can induce Ag clustering or, alternatively, Ag dissolution as Ag^+ in the host. The resulting "latent image" depends on the clustering efficiency, i.e., on the competition between electron scavenging by Ag^+ ions, hole scavenging by Ag^+ or electron recombination with holes, hence on redox interactions within the photographic emulsion, the developer then providing electrons to neutralize Ag^+ in the vicinity of a stable cluster [37]. The primary mechanism of metallic NP synthesis in glasses and silica under directed energy deposition is basically similar. The metal is dissolved in the glass as Ag^+, and the latter interacts with both electrons and holes (whose mobilities differ). Early on, [38] showed that Ag^+ in glass is neutralized when ion irradiated in the electronic stopping regime, but did not link the neutralized quantity to the final NP density. In [18] it has been demonstrated that the primary irradiation initiates nucleation with an efficiency depending in a simple way on the deposited energy density statistics, and that growth may occur by oligomer diffusion, leading to a size distribution whose long-term limit is close to the Lifshitz–Slyozov–Wagner (LSW) [21, 22] characteristics of Ostwald ripening. The latter effect was quantitatively understood, but the former remained obscure. Recent optical absorption (OA) and electron spin resonance (ESR)

[1] Complications are due, e.g., to high concentrations of moving metallic ions and defects during implantation that may lead to radiation-enhanced or -induced diffusion. Also, the collisional stopping power of ions produces light (notably oxygen) atom recoils that modify the glass composition and lead to strong chemical potential gradients in and around the implanted profile depth.

studies of the initial nanoclustering stages in glasses and silica under low (gamma-ray) and high (MeV ion) deposited energy density irradiation conditions show the complexity of the mechanism [39].

As in photography, the primary interaction leads to both neutral and charged metal species; e.g., Ag^0 diffuses via annealing and forms cluster nuclei; Ag^+ diffuses as well, but may either remain in solution or contribute to NP growth. The electrons and holes created by the irradiation (whose initial concentration depends on the deposited energy density) interact with other charged oligomers and with irradiation-induced defects [40, 41]. As a result, the final NP density depends on the irradiation fluence, but the precipitation mechanism and its efficiency differ considerably depending on the stability of cluster nuclei, on subsequent diffusion and recombination processes, and on the irradiation-induced defects that selectively trap either electrons or holes and release them upon annealing. Hence: (i) A high charge density in ion tracks enhances electron–hole recombinations. This should tend to reduce the clustering efficiency, but the same process speeds Ag-cluster nucleation and growth, forming stable NPs that are less susceptible to oxidation during annealing, so that clustering prevails; (ii) Differences in the energy deposition lead to differences in the surviving population of electron- and hole-trapping defects. This is a major source of differences in NP formation probability, since both the growth mechanism and the ultimate clustering efficiency are strongly affected by the interaction of the neutral and ionized Ag species' interactions with the charge populations released from the electron- and hole-containing defects. The formation of NPs in glass requires that the medium be sufficiently reducing to favor growth over dissolution: nanoclustering is therefore critically dependent on both the beam's deposited energy density and on the system's redox properties. This may be illustrated by a qualitative extension of the phase diagram for NP nucleation and growth in glasses (see Fig. 3). The usual phase diagram is drawn in terms of the temperature and a quantity (the "glass acidity") that qualifies the initial (equilibrium) redox state E_0 of the host. The latter depends on the nature and the number of electron donors it contains. Classical glass chemistry is portrayed in the (T, E_0) plane (e.g., effect of compositional changes to scavenge holes and thus enhance metallic NP formation). We add a third variable: the ionizing radiation's deposited energy density (DED). This ion-beam-related quantity only depends on the beam and target elements' mass and density, not on the host's thermodynamical or chemical state. It just parametrizes the energy deposited in the target's electron bath, and provides a measure of the initial charge (electron and hole) density injected into the glass by irradiation. As noted above, the DED influences nucleation and growth via both primary charge injection and defect creation, but defects can be treated simply as charge reservoirs, and the charge balance is modified by temperature-dependent defect kinetics, e.g., in the (DED, T) plane. The clustering probability also depends on the combined influence of the DED and average host redox po-

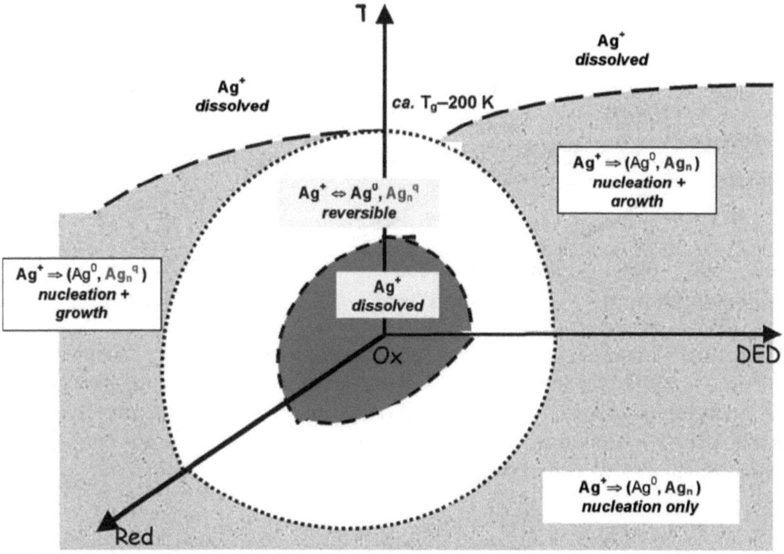

Fig. 3. Schematic phase diagram describing Ag oligomer and NP evolution in terms of temperature (T), ion-irradiation deposited energy (DED), and redox potential (Red or E_0) of the base host. From [39] with permission of the American Physical Society

tential, as shown, e.g., in the (DED, E_0) plane. The combination of all three parameters provides some rationalization of existing results, and may suggest new systematic approaches. In the lower part of the diagram, the metal (here, Ag) simply dissolves in the oxide glass as Ag^+. Raising the temperature and increasing the DED or/and the reducing efficiency leads to a balance between nucleation and redissolution. If the temperature is high enough for monomers to be mobile, and if the DED or the reducing efficiency are large, significant nucleation and growth set in. At higher temperatures, dissociation and dissolution reappear. The DED is very effective in biasing the system towards NP nucleation: combining control over the temperature, DED and glass acidity may allow NP tailoring by controlling the nucleation and growth speeds.

Another way to control the chemistry in the supersaturated solid solution is to modulate the reactivity of the external atmosphere (e.g., reducing, inert or oxidizing) upon postimplantation thermal treatments, which are normally required to anneal radiation-induced defects. This point is better evidenced in Fig. 4, which shows the optical absorption spectra of Au-implanted silica for different annealing temperatures in air, in comparison with those measured in samples annealed in Ar atmosphere. The use of gold is particularly suitable when attempting to decouple as much as possible the *chemical* from the *physical* phenomena triggering the precipitation of a metallic species during ion implantation. Indeed, Au has a reduced chemical interaction with the

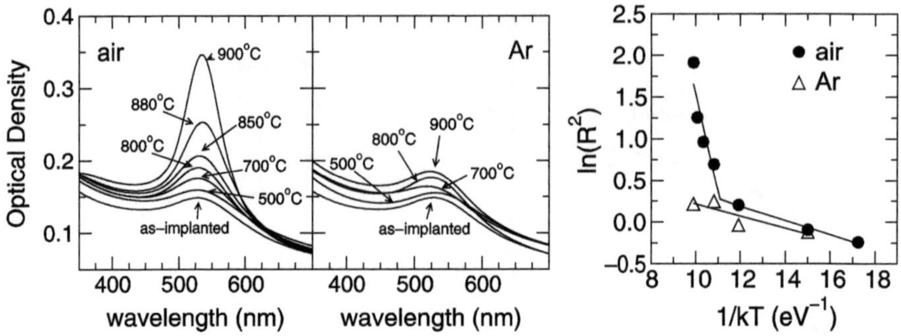

Fig. 4. Optical absorption spectra of Au-implanted silica samples at 3×10^{16} Au^{+} cm^{-2}, 190 keV, annealed in air (*left panel*) or Ar (*central panel*) for 1 h at different temperatures. (*Right panel*) Arrhenius plot of the squared average NP radius R^2 after 1 h annealing in air (*filled circles*) or argon (*empty triangles*). *Solid lines* are linear fits to the experimental data

elements constituting the matrix (Si and O) and a low diffusivity [42] in comparison with other noble metals like Ag for example: this minimizes the role of diffusion-controlled processes, which occur in post-implantation thermal annealings, allowing the NPs size and density distribution to be varied by means of the atmosphere in which post-implantation annealings take place. Figure 4 shows that in the case of air annealing a clear enhancement of the surface-plasmon resonance (SPR) band near 530 nm due to Au NPs in the matrix [43] is evidenced above 700–800°C, indicating the onset of faster NPs growth. On the contrary, a modest increase in the SPR intensity (i.e., NP growth) is exhibited by the samples annealed in Ar (or H_2–Ar), even at high-temperature, with a growth rate as small as the one observed in the low-temperature regime (<700°C) during air annealing. The Arrhenius plot of the squared average NP radius after annealing in air or Ar for 1 h is shown in the right panel of Fig. 4. The abrupt increase of NP size at about 800°C for air annealings suggests that two different growth regimes occur during thermal treatments in air up to 900°C. At temperatures above 700°C the oxygen permeation in silica is sufficient to overcome the purely thermal (or radiation-enhanced [44, 45]) Au diffusion, and, following the Kelvin–Onsager model [46], the thermodynamic interaction between oxygen and gold promotes a correlated diffusion, which is absent in the inert or reducing annealing. This active role played by external oxygen in promoting gold diffusivity is evidenced by the measured activation energy of the process (1.17 eV $atom^{-1}$). This is very different from the literature value of Au diffusion in silica (2.14 eV $atom^{-1}$ [42]) but compatible with the molecular oxygen diffusion mechanism through an interstitial mechanism (in the range from 1.1 to 1.3 eV $atom^{-1}$ [47, 48]).

As we are ultimately interested in the NP size control, it is worth following the kinetics of this process under high-temperature annealing as a function of time. From a thermodynamic point of view, the precipitation processes that occur during either implantation or annealing of ion-implanted materials may be schematically divided in three steps not necessarily strictly separated: (i) nucleation, (ii) noncompetitive or diffusion-limited growth, (iii) competitive growth (i.e., coarsening) or Ostwald-ripening regime.

Impurity implantation at fluences exceeding the solubility threshold in the matrix [43, 46] results, in the first stage of the precipitation process, in a system of new phase precipitates, i.e., gold NPs with radius exceeding the critical one, R^*. Although radiation-induced defects, both point and extended defects, may act as nucleating centers in implanted materials and *heterogeneous* nucleation may play a significant role [46], let us picture nucleation in terms of *homogeneous* nucleation from a supersaturated solid solution, the difference with respect to heterogeneous nucleation being just a decrease of the Gibbs free energy barrier $\Delta G^* \equiv \Delta G(R^*)$. During annealing, the NPs (already formed after implantation) with radius exceeding the critical one, grow directly by solute depletion of the surrounding matrix, without competing with the growth of any others. Then, the diffusion-limited aggregation regime sets in, where the radius of the spherical precipitates $R(t)$ evolves as $R^2(t) = R_0^2 + 2K_1Dt$ [21, 49, 50], where R_0 is the value of R at $t = 0$ (which accounts for the radius of the already-formed precipitates by implantation) and K_1 is a nondimensional constant related to the degree of supersaturation. In this process, diffusion is purely thermal: the radiation-enhanced diffusion (RED) and diffusion triggered by the correlation of dopant with the diffusing atoms coming from the annealing atmosphere [46] also occur. When the NP size is fairly large and the supersaturation becomes extremely small, coarsening (Ostwald ripening) takes place. In this process, the mass transfer from the matrix to the NP is controlled by the Gibbs–Thomson equation [22]: since smaller NPs have a higher solute concentration than larger ones, the diffusional balance promotes matter transfer from smaller to larger precipitates. Consequently, the average size R increases as $R^3(t) = R_0^3 + K_2Dt$ [21, 22].

Thus, we distinguish two different kinetic regimes of NP growth: (i) diffusion (occurring at the earlier stage of growth) characterized by a time dependency of NP radius scaling as $(Dt)^{1/2}$, where D is the diffusion coefficient and t the diffusion time; (ii) a coarsening regime (occurring at longer annealing times) with a radius scaling as $(Dt)^{1/3}$. The evolution of the NP size as a function of the air annealing time can be followed in the TEM cross-sectional images of Fig. 5. Spherical Au NPs of different size are formed. Due to low gold diffusivity [42], the centroid of gold concentration does not move appreciably during annealing and remains near R_P, where the largest NPs are formed. The R^2 evolution as a function of t, in Fig. 6a, clearly shows the occurrence of two regimes in the growth. A linear relation between R^2 and t is expected when the NP growth is only due to the precipitation process of a supersaturated solution [51]. The change of the slope may be explained by the onset of coarsening with the NP following a $t^{1/3}$ law of growth.

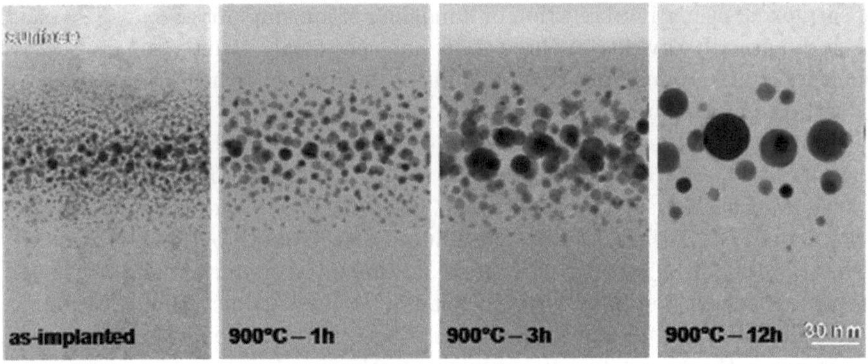

Fig. 5. TEM cross-sectional bright-field micrographs for the Au-implanted samples as a function of the annealing time from the as-implanted to samples annealed at 900°C for 1 h, 3 h, 12 h

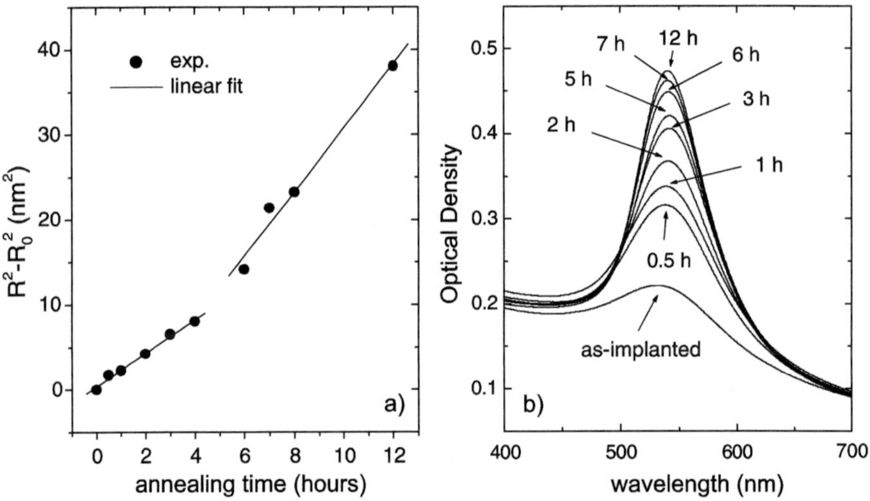

Fig. 6. (**a**) $R^2(t)$ evolution in Au-implanted silica samples annealed in air at 900°C for different time intervals. (**b**) corresponding evolution of the optical absorption spectra

These results provide a means of controlled change in the optical properties of the system. Figure 6b shows the optical absorption spectra of an Au-implanted sample for increasing annealing time intervals t in air atmosphere. The surface-plasmon resonance (SPR) near 530 nm, due to Au NPs in the matrix, is evident and its intensity increases from the as-implanted to the 12-h annealed sample, as expected when increasing the NP size [43].

3.4 Linear (LO) and Nonlinear Optical (NLO) Properties

Together with the size, one relevant parameter that can be used to control the linear optical properties of metal NPs is the composition. In particular, by growing bimetallic NPs the linear dependence of the SPR absorption band position on the electronic density of the metal, i.e., on the alloy composition, can be exploited. This can be done with ion implantation by tailoring ion energy and fluence to achieve concentration profile overlap under sequential implantation of two different ions. Using this approach, a *plasmon tuning* can be done, by synthesizing, for instance, alloy NPs of Au_xAg_{1-x} (SPR from 400 nm to 530 nm, for $x = 0$ to 1) and Au_xCu_{1-x} (SPR from 530 nm to 570 nm), in which the SPR position can be continuously shifted by changing the alloy composition, x. This can be seen in Fig. 7, which shows a comparison of absorption spectra either simulated with the Mie theory [1, 5] for 3-nm clusters of pure Au, Ag and $Au_{0.4}Ag_{0.6}$ alloy in silica (Fig. 7a) or measured for analogous systems [52] in ion-implanted silica (Fig. 7b).

At the nanoscale, the miscibility criterion valid for bulk systems as a constraint as alloy formation is less stringent, due to the incomplete onset of the bulk properties triggered by the large number of atoms at the surface, that makes a cluster more similar to a molecular than to a massive system [53]. This leads to new possible alloy phases, which may be thermodynamically unfavored in the bulk. In the case of noble-metal-based systems (Au–Cu, Au–Ag, Pd–Ag and Pd–Cu) perfect miscibility is expected from the bulk phase diagrams and in fact sequentially implanted samples exhibit direct alloying, without further annealing. For a more detailed description of the NPs obtained by noble-metal sequential ion implantation in silica, see [27]. On the contrary, systems like Co–Cu or Au–Fe that are not miscible in the bulk showed nanoalloy formation after sequentially implantation in silica [16].

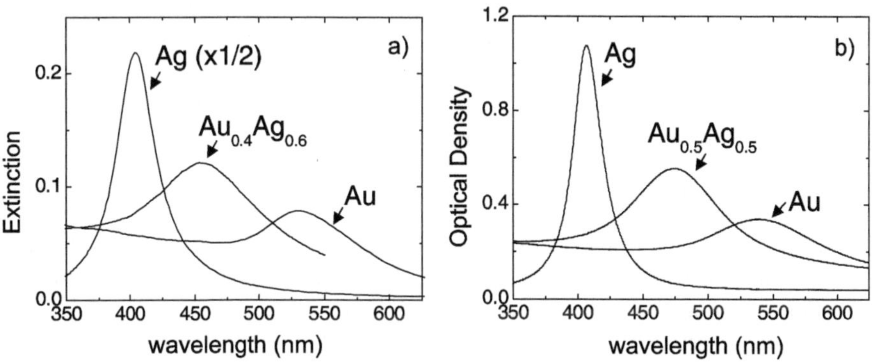

Fig. 7. Comparison between a simulation based on the Mie theory of the optical absorption in the UV-Vis range for 3-nm NPs of pure Au, Ag and $Au_{0.4}Ag_{0.6}$ alloy in silica (**a**), with the experimental optical density of the same systems in ion-implanted silica (**b**). From [16] with permission of the Italian Physical Society

A useful strategy for tuning the linear optical properties of monoelemental NPs near the infrared region is to increase their aspect ratio from spherical to ellipsoidal or rod-like. This can be done by colloidal chemistry [54, 55], but often contradicts thermodynamic stability [55]. Direct synthesis with high aspect ratios is feasible by using ion irradiation in the electronic regime (i.e., when the energy-loss of the implanted ions is dominated by the inelastic electronic contribution). An elongation of spherical NPs occurs along the ion track, provided that a certain energy loss threshold is attained (few keV/nm). For instance, 30-MeV Cu ions were used in [56] to irradiate Au–Silica core-shell NPs to produce Au nanorods. Similarly, 30-MeV Si ions at fluences in the range 10^{15} ions cm^{-2} were used in [57] to deform Ag–silica core-shell NPs. This approach has been used to modify the optical properties of 5-nm Ag NPs in silica by irradiation with 8-MeV Si ions [58]: elongated Ag NPs along the ion track produced a splitting of the linear optical absorption band measured under polarized light, in good agreement with the Gans theory. The mechanism for this elongation is still under debate [58–61].

The control of the linear optical properties is also of paramount importance when optical functionalities are required from metal NPs. In particular, glass matrices embedding metal NPs exhibit an enhanced optical Kerr susceptibility, $\chi^{(3)}$, whose real part is related to the n_2 coefficient of the intensity-dependent refractive index, usually defined as $n(I) = n_0 + n_2 I$, where n_0 and I are the linear refractive index and the intensity of the light, respectively. Correspondingly, a nonlinear absorption also takes place that can be described macroscopically by the intensity-dependent absorption coefficient $\alpha(I) = \alpha_0 + \beta I$, where α_0 and β are the linear and nonlinear absorption coefficients, respectively. The intraband and interband electronic transitions that contribute to the effective $\chi^{(3)}$ turn out to depend on the type of metal and the form and size of the clusters, as well as on the metal–dielectric bonds [62, 63]. Suitable methodologies are therefore needed for tailoring the formation of small metal clusters within the glass, with the aim of fabricating nonlinear glasses with prescribed optical performances. The third-order optical Kerr susceptibility of nanocomposites, $\chi^{(3)}_{\mathrm{eff}}$, formed by a non-absorbing matrix, with dielectric constant ε_{m}, containing metal NPs with a small volume fraction (i.e., filling factor) p is given by [64]:

$$\chi^{(3)}_{\mathrm{eff}} = p\chi^{(3)}_{\mathrm{NP}} |f_{\mathrm{e}}|^2 f_{\mathrm{e}}^2 , \tag{8}$$

where $\chi^{(3)}_{\mathrm{NP}}$ is the nonlinear contribution of the NPs and f_{e} the local-field enhancement factor defined in (5). Mutual electromagnetic interactions among NPs determine an increase of the modulus of f_{e}.

The inclusion of metal NPs in a silica matrix increases by several orders of magnitude the negligible third-order nonlinearity of the pure silica (of the order of $5 \times 10^{-16}\,\mathrm{cm^2\,W^{-1}}$) [62, 64–73]. For instance, AuAg and AuCu NPs produced by ion implantation in silica exhibit a NLO n_2 of about $10^{-10}\,\mathrm{cm^2\,W^{-1}}$ at a 527 nm wavelength [64], i.e., 6 orders of magnitude larger than that of pure silica. Therefore, composites made of metal NPs could be used in all-optical switching devices, which should exhibit commutation times (fs) faster than the fastest HEMT electronic devices, without the need for optical-to-electronic signal conversion. For device-oriented applications it is important to distinguish the true electronic (fast) nonlinearity from the mixed electronic-thermal (slow) one, which can be excited when using high repetition-rate and/or ns lasers [64, 68, 74]. The advantage of using ion implantation with respect to other synthesis techniques is that in principle one can make linear or nonlinear devices like passive or active (i.e., amplifying) waveguides by patterning, i.e., by *writing* devices through suitable masks, which are sufficiently thick to stop some of the energetic ions.

4 Core-Satellite for Nonlinear Optical Properties

In the previous section we have seen how the metal NP can be considered as a functional optical *building block* with its linear and nonlinear properties, and how ion implantation can be used, in combination with thermal treatments, to control both of them. The next level of our hierarchical approach is the one in which the interaction between NPs can be used to enhance the optical performances of the materials. Indeed, when solving the Mie problem of the scattering by a single isolated particles we have seen that the field within or at the surface of the NP is enhanced with respect to the external field by the f_e factor (5). When the interparticle distance is reduced to such an extent that NPs can no longer be considered as independent, each of them feels not only the external incoming field but also the sum of the fields scattered by all the others.

In this case, to properly model the optical response of strongly interacting spherical NPs, the generalized multiparticle Mie (GMM) theory has to be used [11–13]. The simplest case of interaction is a NP dimer, whose optical properties strongly depends on the interparticle distance and on the orientation with respect to the polarization of the incoming field.

Ion implantation can be used to obtain nanostructures that exhibit a dimer-like interaction. Such structures have been called *nanoplanets* and are composed by a large central NP (5–30 nm in diameter) whose surface is surrounded by a halo of smaller (1–5 nm) NPs (the satellites) extending up to 10–20 nm from the central NP. These are obtained for instance by irradiating Au_xCu_{1-x} or Au_xAg_{1-x} alloy NPs with 190-keV Ne^+ ions [31]. The main advantage of ion irradiation is that the nanoplanet topology can be tailored by tuning the irradiation parameters (fluence, energy, ions, flux), as discussed in [75, 76]. The irradiation effect is shown for Au_xAg_{1-x} alloy NPs ($x = 0.6$) in Fig. 8b in comparison with the unirradiated reference, Fig. 8a [8]. The most evident result is the new topology that is obtained: around each original NP a set of satellite NPs of about 1–2 nm are formed with an average distance of about 3–5 nm from the NP surface. Their size and density can be

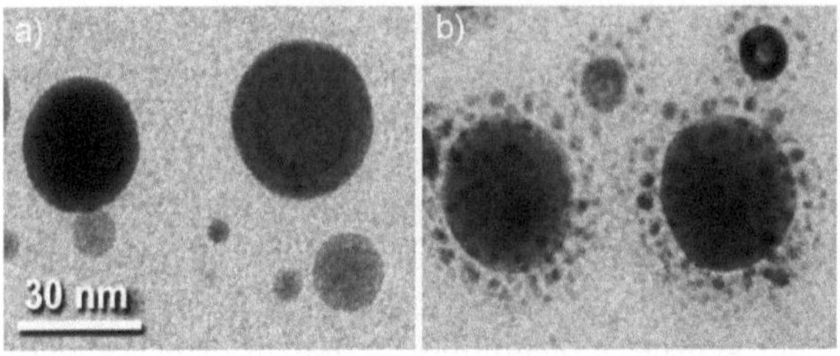

Fig. 8. Bright-field TEM cross-sectional micrographs of Au_xAg_{1-x} alloy NPs ($x = 0.6$, annealed in air at 800°C for 1 h) before (**a**) and after irradiation at room temperature with 190-keV Ar^+, at a current density 2.5×10^{16} ions cm^{-2} and fluence of $0.84\,\mu A\,cm^{-2}$ (**b**)

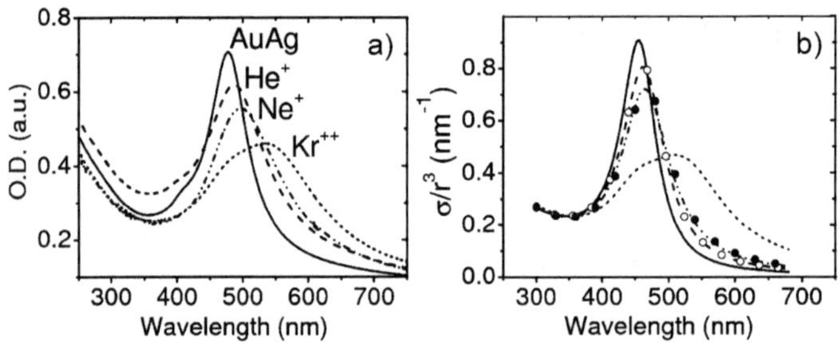

Fig. 9. Redshift of the optical absorption of Au_xAg_{1-x} alloy NPs after ion irradiation with different ions: (**a**) measured spectra; (**b**) simulated with the generalized multiparticle Mie theory. From [13] with permission of the Optical Society of America

increased by increasing the nuclear fraction (S_n) of the energy loss by using different ions, like He, Ne, Ar or Kr ions, at different fluence and energy (to deposit the same energy and power density on the sample) [77, 78]. Similar structures have been obtained in a set of ion-beam mixing experiments of Au films or islands embedded in silica and irradiated by 4.5-MeV Au ions [76, 79, 80]. In those studies, the NPs halo observed around the larger Au or Ag precipitates for the larger fluences was explained in terms of the ballistic process, which dissolves the films/islands into the SiO_2 host by ion mixing, and of the large increase in the solute concentration in the matrix with a subsequent precipitation.

The peculiar topology of the core-satellite NPs allows controlling the optical properties in the ion-irradiated samples by producing a redshift of the SPR absorption band, due to a strong coupling between the core and the satellite NPs, which strongly affects the local field near the core surface. This is shown in Fig. 9, in which the measured spectra are compared to the simulated ones within the generalized multiparticle Mie (GMM) theory, which confirmed the lateral coupling between neighboring nanoplanets (parallel to

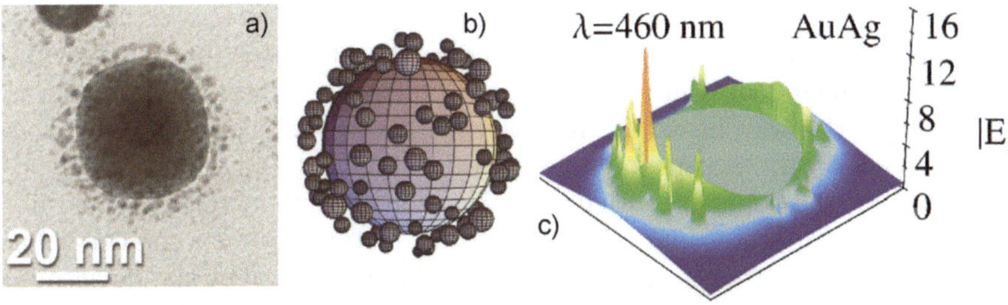

Fig. 10. (**a**) The experimental core-satellite nanostructure obtained upon He irradiation on a AuAg NP, (**b**) its corresponding model, (**c**) the computed modulus of the local field on a section in the equatorial plane of the system shown in (**b**) and calculated at the SPR position of the system (460 nm). From [17] with permission of Elsevier

incoming field) due to the overlapped satellites halos in a direction parallel to the sample surface [12, 13]. The local-field properties of the nanoplanet configuration are interesting also for the presence of hot spots. In Fig. 10 we show a GMM calculation of the field distribution around the central NP due to the coupling with its satellites. The experimental core-satellite nanostructure obtained after He irradiation on AuAg NP (see Fig. 10a) is modeled as in Fig. 10b, producing the modulus of the local field on a section in the equatorial plane of the system shown in Fig. 10c and calculated at the SPR position of the system (460 nm). A field enhancement of about a factor 15 is obtained, which could be exploited for controlling the nonlinear optical properties of these systems or to drive, acting as a nanoantenna, the external field to suitable emitters located in close proximity to the nanoplanet, similarly to resonant Raman spectroscopy [81], for which the cross section enhancement is proportional to $|f_e|^4$.

Further GMM simulations indicated additional strategies for obtaining even larger local-field enhancements: (i) increasing the size of the satellites (for instance, by increasing the irradiation fluence or by irradiating within a more nuclear regime) or (ii) changing the NP composition to pure Ag, which has a larger separation between the collective oscillation and interband de-excitation channels.

5 Plasmonic Nanostructures

At the last level of our proposed hierarchy for controlling the optical properties of NP-based materials we will consider ordered arrays of interacting metal NPs, either 1D (linear chains) or 2D (planar arrays). This is one of the possible approaches within a wider class of so-called *plasmonic* nanostructures, in which a regular metallic pattern is used to couple photonic with plasmonic modes in an optical device [82].

One of the challenging issues in contemporary nanophotonics is the lateral confinement of an electromagnetic wave below its diffraction limit. Achieving such confinement will overcome the limitation of conventional waveguides, by changing the way in which the energy of the wave is transported. The plasmon-polariton, i.e., the coupling between the incident field and the field generated by the coherently oscillating electrons at the metal/dielectric interface, can propagate nonradiatively within distances much smaller than the associated wavelength, allowing the development of photonic devices in the submicrometer range. This required the development of ad-hoc synthesis techniques able to pattern in a top-down approach a substrate with nanometer resolution. Among them, the best performing is electron-beam lithography (EBL) due to its high spatial resolution down to 10–20 nm [83], or X-ray lithography (XRL, with resolution of about 100 nm) [84]. EBL and XRL are currently replacing conventional photolithography when truly nanometric resolution has to be achieved. Other newly developed lithographies, like nanoimprint lithography [85], relies on the polymeric replica of suitable masks prepared by EBL or XRL and are much less expensive.

The confinement of the electromagnetic field, due to near-field plasmonic coupling between NPs, has been demonstrated by photon scanning tunneling microscopy in linear chains of $100 \times 100 \times 40\,\mathrm{nm}^3$ Au NPs made by EBL [86].

Recently, the possibility of guiding an electromagnetic field along chains of Ag metal NPs has been suggested [87] and demonstrated [88] giving a strong impulse to the research in this field. Of course, metals are very absorptive materials in the visible range, so a proper optimization of the NP size and distance in the chains has to be performed. The two main constraints to be fulfilled are: (i) the size of the NPs should be larger than the electronic mean-free path to reduce the scattering losses at the NP surface producing resistive heating and (ii) the interparticle separation should be not so large as to produce dephasing in the wave confinement and reduction of the group velocity. By proper optimization, with 50-nm Ag NPs displaced by 75 nm along a linear chain, a propagation group velocity of $0.1c$ with about $5\,\mathrm{dB}\,\mu\mathrm{m}^{-1}$ losses [87] can be obtained. Although such losses per length are quite high, they can be feasible for new photonic devices whose dimensions are less than or of the order of 1 μm, i.e., a size comparable to the electronic devices.

The fabrication of linear chains of metal NPs is in general achieved by means of lithographic techniques like EBL [86, 88]. Some attempts have been made to use organic templates to induce NP self-assembly (protein [89] or DNA [90, 91]), but with a modest degree of order. Recently, after the development of DNA origami [92], new ordered self-assembled organic templates are being developing, to which metal NPs can be attached to form ordered 2D arrays [93] with plasmonic properties.

Techniques based on moving a focused ion beam (FIB) [29, 94] along a prescribed path to obtain direct nanometric *writing* on the substrate hold great promise with the development of liquid-metal ion sources (LMIS) [94, 95].

In the meantime, to produce metal NP linear chains a simpler strategy is to use swift heavy-ion (SHI) irradiation of already formed NP to promote their alignment along the ion track due to the electronic energy loss. In [96, 97] Ag NPs prepared in a glass by ion exchange followed by Xe-ion irradiation (to promote their growth) were further ion irradiated at an angle of 60° with respect to the surface normal with 30-MeV Si ions at fluences up to 3×10^{15} ions cm^{-2}. The optical transmission properties of the composite changed as a function of the Si fluence according to the incident light polarization (longitudinal or transversal with respect to the arrays) indicating a correlated interaction along the chains of the Ag NPs.

To understand the redshift exhibited in the case of linear chains, in Fig. 11 the far-field extinction properties (calculated within the GMM theory) of linear chains of 10-nm Ag or Au NPs displaced by 21 nm are shown as a function of the chain length. Of course, in the case of ion irradiation there is no control on the correlated lateral interaction among the chains because there is no control on the ion-track position. This limits to 1D structures the applicability of this approach. The common technique for obtaining 2D ordered arrays is lithography: arrays of Au or Ag NPs [98–101], nanowires [102] or chains [87], can be synthesized and their optical properties can be used as highly sensitive sensors [103, 104].

The physics of the interaction between an electromagnetic wave and a 2D NP array is strongly dependent on the interparticle distance *d*. Two coupling regimes are active [100, 104, 105]: (i) the near-field one, which dominates for nearly touching particles due to the short range of the electromagnetic near fields of the order of some tens of nm; (ii) far-field (dipolar) coupling, with

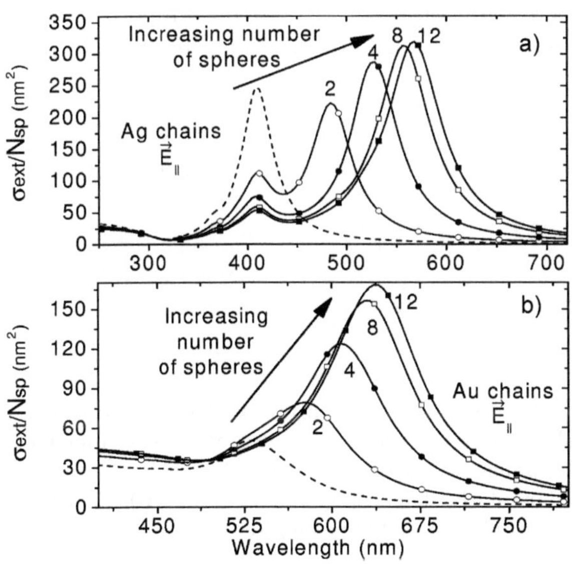

Fig. 11. Simulation within the GMM theory of the extinction cross sections of 10 nm Ag (**a**) or Au (**b**) NPs displaced by 21 nm along linear chains as a function of the chain length for a polarization of the incoming field along the chain axis. *Dashed lines* represent the extinction of a single noninteracting NP

its $1/d$ dependence, which is dominating when d becomes comparable with or larger than the incident wavelength. This indicates that a careful control of the interparticle spacing is needed.

Ion implantation can be a possible alternative to lithography, especially when extended plasmonic structures have to be synthesized, e.g., for sensing applications. To achieve nanopatterning of a substrate, ion implantation has to be performed through some mask that is able to stop incident ions in prescribed regions. A simple approach is the use of hybrid schemes in which self-assembled structures can be deposited on the surface of the substrate to be implanted and then (chemically) removed after the implantation, before additional thermal treatments suitable for promoting cluster growth, similar to nanosphere lithography [106, 107]. Figure 12 shows an example of a hexagonal close-packed 2D colloidal crystal formed by monodispersed silica nanospheres (NS, obtained as in [108]). Instead of evaporating the metal over the mask, low-energy (20–50 keV) ion implantation may be performed through the mask, so that most of the ions are stopped within the nanospheres and the remaining ones are buried just beneath the substrate surface, where they can grow under postimplantation treatments. Additional control on the exposed area can be obtained by suitable thermal treatments of the masks, promoting partial sintering and reduction of the mask "holes". This technique is generally restricted to the hcp configuration, therefore producing a honeycomb, graphitic-like arrangement sketched in Fig. 12.

Plasmonic nanostructures can be used for SERS [81], for optical-sensing applications [104] or as active devices for modifying the efficiency of emitters in close proximity with them. In particular, an increased radiative lifetime from different emitters (Si QD or Er ions, for instance) through the interaction with a plasmonic structure made of metal NPs has been reported [109–113]. The possibility of increasing spontaneous emissions lifetimes by the

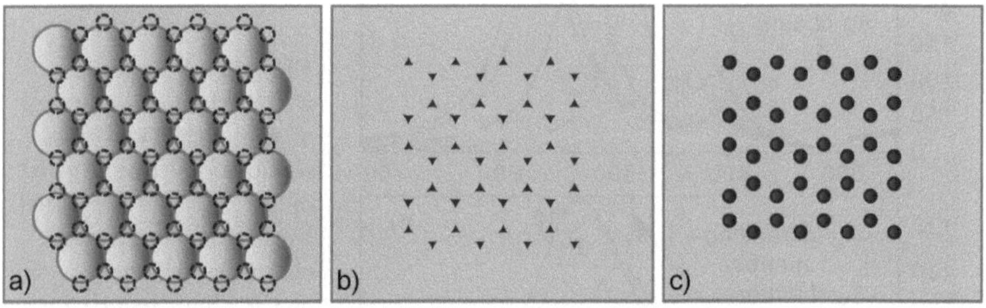

Fig. 12. Sketch of the mask-assisted ion-implantation process (top view): the 2D colloidal silica nanosphere mask is self-assembled on the substrate to be implanted. The *dashed circles* indicate the unmasked regions (**a**). After the ion implantation, the mask is removed leaving triangularly shaped implanted regions (**b**). Spherical metal NPs are then growth by suitable thermal annealing resulting in a graphitic-like 2D array (**c**)

interaction with external structures is of great practical and technological importance, since bad emitters characterized by intrinsic low quantum efficiencies could be converted into efficient ones just by the coupling with an external photonic/plasmonic device. In general, the spatial dependence of the interaction between emitter and metal NP indicates that the lifetime of the excited state of the emitters decreases when it is closer than the emitted wavelength to the metal surface, due to nonradiative energy transfer [114]. Moreover, a stronger enhancement of the nonradiative decay rate Γ_{NR} is obtained when the frequency of the emitter matches the surface-plasmon modes of the NP. Nevertheless, the local-field enhancement at the NP surface has been shown to produce an increased emission yield [109, 115], which is therefore a decreasing function of the emitter–NP distance. These two competing mechanisms (field enhancement and quenching due to nonradiative energy transfer to the metal) produce an optimal distance at which the quantum yield (i.e., the radiative decay rate Γ_{R} normalized to the total decay rate $\Gamma_{\mathrm{R}} + \Gamma_{\mathrm{NR}}$) is maximum [109].

This can be further extended when coupling the emitters not simply to a single metallic structure (a spherical particle [114, 116] or a surface [117]) but to a plasmonic array. In this last case, indeed, the photonic modes of the arrays become relevant. Preliminary calculations in the GMM theory [118], following the approach of [119], indicate that it is possible to achieve a tunable and wavelength-selective enhancement in the quantum efficiency by coupling the plasmonic-photonic modes of a linear chain of Ag dimers (size 100 nm, separation 20 nm) coupled to a dipolar emitter in between and displaced by 500–800 nm. Enhancements up to 50% even for poor emitter (with an intrinsic efficiency of 1%) can be achieved and the tunability of the resonant enhancements in the visible–near-infrared region can be controlled by the dimer spacing in the chain, which therefore can act as an efficient and tunable nanoantenna. Of course, the synthesis of such optimized nanostructures will require new strategies if ion-implantation techniques have to be used.

6 Conclusions

In this chapter we have presented a hierarchical approach to the synthesis of *plasmonic* nanostructures based on metal nanoparticles embedded in silica by using ion implantation as the reference technique. Starting from the simplest case (i.e., the *building block* of our scheme) of noninteracting nanoparticles obtained by direct implantation, we rised of one level through the formation by ion irradiation of metal nanoplanets (a big central nanocluster surrounded by a halo of close-lying smaller nanosatellites), to end our approach by synthesizing ordered 1D (chains) or 2D arrays of nanoparticles. At each level the degree of interaction between the structural building blocks is increased giving a further degree of freedom for controlling the optical properties of the materials. This requires a great experimental effort in tailoring the implantation/irradiation parameters in order to achieve the required level of reproducibility and control on the arrangement of the different components in new nanophotonic devices (plasmonic waveguides, nanoantennas, optical amplifiers, nonlinear systems,...) with size in the subwavelength, nanometric regime. Although competing techniques like colloidal chemistry or lithography are at present more widely diffused for their better control over size and spatial resolution, respectively, ion-implantation-based techniques (eventually coupled in some hybrid schemes to self-assembly strategies) can be fruitfully applied to obtain a large class of optically interesting materials and devices, embedded in transparent matrices, which can preserve their structure and functionality, allowing chemical, thermal and mechanical stability. Moreover, ion implantation, coupling its highly desirable intrinsic feasibility for large-area substrate patterning to its compatibility with contemporary silicon nanotechnology, will still be one of the reference techniques in the next operation of nanophotonics and nanoplasmonic devices.

References

1. G. Mie, Ann. Phys. (Leipzig) **25**, 377 (1908) 287, 288, 301
2. G. Mattei, P. Mazzoldi, in *Highlights on Spectroscopies of Semiconductors and Nanostructures*, ed. by G. Guizzetti, L. Andreani, F. Marabelli, M. Patrini (Societá Italiana di Fisica, Bologna, 2007), p. 349 288
3. M. Faraday, Philos. Trans. R. Soc. **147**, 145 (1857) 288
4. J.C. Maxwell-Garnett, Philos. Trans. R. Soc. A **203**, 385 (1904) 288, 291
5. U. Kreibig, M. Vollmer, *Optical Properties of Metal Clusters* (Springer, Berlin, 1995) 289, 290, 301
6. H. Frölich, Physica **4**, 406 (1937) 289
7. J. Sinzig, U. Radtke, M. Quinten, U. Kreibig, Z. Phys. D **26**, 242 (1993) 290
8. G. Mattei, G. Battaglin, V. Bello, G. De Marchi, C. Maurizio, P. Mazzoldi, M. Parolin, C. Sada, J. Non-Cryst. Solids **322**, 17 (2003) 290, 303
9. G. Mattei, P. Mazzoldi, M. Post, D. Buso, M. Guglielmi, A. Martucci, Adv. Mater. **19**, 561 (2007) 290
10. R. Gans, Ann. Phys. **47**, 270 (1915) 290
11. Y.L. Xu, Appl. Opt. **34**, 4573 (1995) 290, 303
12. G. Pellegrini, G. Mattei, V. Bello, P. Mazzoldi, Mater. Sci. Eng. C **27**, 1347 (2007) 290, 303, 305
13. G. Pellegrini, V. Bello, G. Mattei, P. Mazzoldi, Opt. Express **15**, 1097 (2007) 290, 303, 304, 305
14. Y. Xia, N.J. Halas, MRS Bull. **30**, 338 (2005) 292

15. M.A. Garcia, J. de la Venta, P. Crespo, J. LLopis, S. Penadés, A. Fernández, A. Hernando, Phys. Rev. B **72**(24), 241403 (2005) 292
16. P. Mazzoldi, G. Mattei, Riv. Nuovo Cimento **28**, 1 (2005) 292, 294, 301
17. P. Mazzoldi, G. Mattei, in *Synthesis of Metal Nanoclusters upon Using Ion Implantation* (Elsevier, Amsterdam, 2007), p. 281 292, 305
18. E. Valentin, H. Bernas, C. Ricolleau, F. Creuzet, Phys. Rev. Lett. **86**, 99 (2001) 292, 293, 294, 295
19. K.H. Heinig, B. Schmidt, A. Markwitz, R. Grötzschel, M. Strobel, S. Oswald, Nucl. Instrum. Methods B **148**, 969 (1999) 292
20. D. Ila, E.K. Williams, S. Sarkisov, C.C. Smith, D.B. Poker, D.K. Hensley, Nucl. Instrum. Methods B **141**, 289 (1998) 292
21. I. Lifshitz, V. Slezof, J. Phys. Chem. Solids **19**, 35 (1961) 293, 295, 299
22. C. Wagner, Z. Elektrochem. **65**, 581 (1961) 293, 295, 299
23. R.E. de Lamaestre, H. Bernas, Phys. Rev. B **73**(12), 125317 (2006) 293
24. J.M. Rosen, J. Colloid Interface Sci. **99**, 9 (1984) 293
25. A. Miotello, G. De Marchi, G. Mattei, P. Mazzoldi, Appl. Phys. A **67**, 527 (1998) 294
26. G. De Marchi, G. Mattei, P. Mazzoldi, C. Sada, A. Miotello, J. Appl. Phys. **92**, 4249 (2002) 294
27. G. Mattei, Nucl. Instrum. Methods B **191**, 323 (2002) 294, 301
28. L. Bischoff, J. Teichert, Appl. Surf. Sci. **184**, 336 (2001) 294
29. C. Akhmadaliev, B. Schmidt, L. Bischoff, Appl. Phys. Lett. **89**(22), 223129 (2006) 294, 306
30. A. Portavoce, R. Hull, M.C. Reuter, F.M. Ross, Phys. Rev. B **76**(23), 235301 (2007) 294
31. G. Mattei, G. De Marchi, P. Mazzoldi, C. Sada, V. Bello, G. Battaglin, Phys. Rev. Lett. **90**, 085502/1 (2003) 294, 303
32. H. Hosono, Jpn. J. Appl. Phys. **32**, 3892 (1993) 295
33. H. Hosono, H. Imagawa, Nucl. Instrum. Methods Phys. Res. B **91**, 510 (1994) 295
34. R. Bertoncello, A. Glisenti, G. Granozzi, G. Battaglin, F. Caccavale, E. Cattaruzza, P. Mazzoldi, J. Non-Cryst. Solids **162**, 205 (1993) 295
35. R. Espiau de Lamaestre, H. Béa, H. Bernas, J. Belloni, J.L. Marignier, Phys. Rev. B **76**, 205431 (2007) 295
36. R. Gurney, N. Mott, Proc. R. Soc. A **164**, 151 (1938) 295
37. J. Belloni, M. Treguer, H. Remita, R. De Keyzer, Nature **402**, 865 (1999) 295
38. G. Arnold, F. Vook, Radiat. Eff. **14**, 157 (1972) 295
39. R.E. de Lamaestre, H. Béa, H. Bernas, J. Belloni, J.L. Marignier, Phys. Rev. B **76**(20), 205431 (2007) 296, 297
40. E. Friebele, D. Griscom, *Radiation Effects in Glass* (Academic Press, New York, 1979) 296
41. H. Hosono, N. Matsunami, Nucl. Instrum. Methods B **141**, 566 (1998) 296

42. D.R. Collins, D.K. Schroder, C.T. Sah, Appl. Phys. Lett. **8**, 323 (1966) 298, 299
43. G.W. Arnold, J.A. Borders, J. Appl. Phys. **48**, 1488 (1977) 298, 299, 300
44. G. Battaglin, G. Della Mea, G. De Marchi, P. Mazzoldi, A. Miotello, Nucl. Instrum. Methods B **7/8**, 517 (1985) 298
45. G. Arnold, G. Battaglin, G. Della Mea, G. De Marchi, P. Mazzoldi, A. Miotello, Nucl. Instrum. Methods B **32**, 315 (1988) 298
46. A. Miotello, G. De Marchi, G. Mattei, P. Mazzoldi, C. Sada, Phys. Rev. B **63**, 075409 (2001) 298, 299
47. M.A. Lamkin, F.L. Riley, R.J. Fordham, J. Eur. Ceram. Soc. **10**, 347 (1992) 298
48. F. Norton, Nature **191**, 701 (1961) 298
49. H. Yukselici, P.D. Persans, T.M. Hayes, Phys. Rev. B **52**, 11763 (1995) 299
50. S.A. Gurevich, A.I. Ekimov, I.A. Kudrayavtsev, O.G. Lyublinskaya, A.V. Osinnskii, A.S. Usikov, N.N. Faleev, Semiconductors **28**, 486 (1994) 299
51. H.B. Aaron, D. Fainstein, G.R. Kotler, J. Appl. Phys. **41**, 4404 (1970) 299
52. G. Battaglin, E. Cattaruzza, F. Gonella, G. Mattei, P. Mazzoldi, C. Sada, X. Zhang, Nucl. Instrum. Methods B **166–167**, 857 (2000) 301
53. H. Yasuda, H. Mori, Z. Phys. D **31**, 131 (1994) 301
54. L.G.C.J.M.N.R. Jana, Adv. Mater. **13**, 1389 (2001) 302
55. J. Perez-Juste, I. Pastoriza-Santos, L.M. Liz-Marzan, P. Mulvaney, Coord. Chem. Rev. **249**, 1870 (2005) 302
56. S. Roorda, T. van Dillen, A. Polman, C. Graf, A. van Blaaderen, B. Kooi, Adv. Mater. **16**, 235 (2004) 302
57. J. Penninkhof, T. van Dillen, S. Roorda, C. Graf, A. van Blaaderen, A. Vredenberg, A. Polman, Nucl. Instrum. Methods B **242**, 523 (2006) 302
58. A. Oliver, J.A. Reyes-Esqueda, J.C. Cheang-Wong, C.E. Roman-Velazquez, A. Crespo-Sosa, L. Rodriguez-Fernandez, J.A. Seman, C. Noguez, Phys. Rev. B **74**(24), 245425 (2006) 302
59. A. Meftah, F. Brisard, M. Costantini, E. Dooryhee, M. Hage-Ali, M. Hervieu, J. Stoquert, F. Studer, M. Toulemonde, Phys. Rev. B **49**, 12457 (1994) 302
60. H. Trinkaus, A.I. Ryazanov, Phys. Rev. Lett. **74**, 5072 (1995) 302
61. C. D'Orléans, J. Stoquert, C. Estournés, C. Cerruti, J. Grob, J. Guille, F. Haas, D. Muller, M. Richard-Plouet, Phys. Rev. B **67**, 220101 (2003) 302
62. F. Hache, D. Ricard, C. Flytzanis, J. Opt. Soc. Am. B **3**, 1647 (1986) 302

63. R.F. Haglund Jr., in *Handbook of Optical Properties II: Optics of Small Particles, Interfaces, and Surfaces*, vol. 2, ed. by R.E. Hummel, P. Wissmann (CRC Press, New York, 1997), p. 191 302
64. E. Cattaruzza, G. Battaglin, F. Gonella, G. Mattei, P. Mazzoldi, R. Polloni, B. Scremin, Appl. Surf. Sci. **247**, 390 (2005) 302, 303
65. R.H. Magruder III, D.H. Osborne Jr., R.A. Zuhr, J. Non-Cryst. Solids **176**, 299 (1994) 302
66. N. Skelland, P. Townsend, Nucl. Instrum. Methods B **93**, 433 (1994) 302
67. K. Uchida, S. Kaneko, S. Omi, C. Hata, H. Tanji, Y. Asahara, A. Ikushima, T. Tokizaki, A. Nakamura, J. Opt. Soc. Am. B **11**, 1236 (1994) 302
68. M. Falconieri, G. Salvetti, E. Cattaruzza, F. Gonella, G. Mattei, P. Mazzoldi, M. Piovesan, G. Battaglin, R. Polloni, Appl. Phys. Lett. **73**, 288 (1998) 302, 303
69. Y. Takeda, V.T. Gritsyna, N. Umeda, C.G. Lee, N. Kishimoto, Nucl. Instrum. Methods B **148**, 1029 (1999) 302
70. G. Battaglin, P. Calvelli, E. Cattaruzza, F. Gonella, R. Polloni, G. Mattei, Mazzoldi. Appl. Phys. Lett. **78**, 3953 (2001) 302
71. E. Cattaruzza, G. Battaglin, F. Gonella, R. Polloni, G. Mattei, C. Maurizio, P. Mazzoldi, C. Sada, C. Tosello, M. Montagna, M. Ferrari, Philos. Mag. B **82**, 735 (2002) 302
72. Y. Takeda, O.A. Plaksin, N. Kishimoto, Opt. Express **15**(10), 6010 (2007) 302
73. E. Cattaruzza, G. Battaglin, F. Gonella, R. Polloni, B. Scremin, G. Mattei, P. Mazzoldi, C. Sada, Appl. Surf. Sci. **254**, 1017 (2007) 302
74. P. Mazzoldi, G.W. Arnold, G. Battaglin, F. Gonella, R. Haglund Jr., J. Nonlin. Opt. Phys. Mater. **5**, 285 (1996) 303
75. G. Mattei, C. Maurizio, P. Mazzoldi, F. D'Acapito, G. Battaglin, E. Cattaruzza, C. de Julian Fernandez, C. Sada, Phys. Rev. B **71**, 195418 (2005) 303
76. G. Rizza, M. Strobel, K.H. Heinig, H. Bernas, Nucl. Instrum. Methods B **178**, 78 (2001) 303, 304
77. V. Bello, G.D. Marchi, C. Maurizio, G. Mattei, P. Mazzoldi, M. Parolin, C. Sada, J. Non-Cryst. Solids **345–346**, 685 (2004) 304
78. G. Mattei, V. Bello, P. Mazzoldi, G. Pellegrini, C. Sada, C. Maurizio, G. Battaglin, Nucl. Instrum. Methods B **240**, 128 (2005) 304
79. J.C. Pivin, G. Rizza, Thin Solid Films **366**, 284 (2000) 304
80. J.C. Pivin, Mater. Sci. Eng. A **293**, 30 (2000) 304
81. C.E. Talley, J.B. Jackson, C. Oubre, N.K. Grady, C.W. Hollars, S.M. Lane, T.R. Huser, P. Nordlander, N.J. Halas, Nano Lett. **5**, 1569 (2005) 305, 308
82. S. Maier, M. Brongersma, P. Kik, S. Meltzer, A. Requicha, H. Atwater, Adv. Mater. **13**, 1501 (2001) 305

83. S.D. Golladay, H.C. Pfeiffer, J.D. Rockrohr, W. Stickel, J. Vac. Sci. Technol. B **18**(6), 3072 (2000) 306
84. H. Smith, D. Spears, Electron. Lett. **8**, 102 (1972) 306
85. S.Y. Chou, P.R. Krauss, P.J. Renstrom, Science **272**(5258), 85 (1996) 306
86. J.R. Krenn, A. Dereux, J.C. Weeber, E. Bourillot, Y. Lacroute, J.P. Goudonnet, G. Schider, W. Gotschy, A. Leitner, F.R. Aussenegg, C. Girard, Phys. Rev. Lett. **82**(12), 2590 (1999) 306
87. M.L. Brongersma, J.W. Hartman, H.A. Atwater, Phys. Rev. B **62**(24), R16356 (2000) 306, 307
88. S.A. Maier, P.G. Kik, H.A. Atwater, S. Meltzer, E. Harel, B.E. Koel, A.A.G. Requicha, Nat. Mater. **2**, 229 (2003) 306
89. R.A. McMillan, C.D. Paavola, J. Howard, S.L. Chan, N.J. Zaluzec, J.D. Trent, Nat. Mater. **1**, 247 (2002) 306
90. J. Richter, R. Seidel, R. Kirsch, M. Mertig, W. Pompe, J. Plaschke, H. Schackert, Adv. Mater. **12**, 507 (2000) 306
91. K. Keren, M. Krueger, R. Gilad, G. Ben-Yoseph, U. Sivan, E. Braun, Science **297**(5578), 72 (2002) 306
92. P.W.K. Rothemund, Nature **440**, 297 (2006) 306
93. J. Zheng, P. Constantinou, C. Micheel, A. Alivisatos, R. Kiehl, N. Seeman, Nano Lett. **6**(7), 1502 (2006) 306
94. L. Bischoff, T. Ganetsos, J. Teichert, G.L.R. Mair, Nucl. Instrum. Methods Phys. Res. Sect. B, Beam Interact. Mater. At. **164–165**, 999 (2000) 306
95. L. Bischoff, Ultramicroscopy **103**, 59 (2005) 306
96. J.J. Penninkhof, A. Polman, L.A. Sweatlock, S.A. Maier, H.A. Atwater, A.M. Vredenberg, B.J. Kooi, Appl. Phys. Lett. **83**(20), 4137 (2003) 307
97. L.A. Sweatlock, S.A. Maier, H.A. Atwater, J.J. Penninkhof, A. Polman, Phys. Rev. B **71**(23), 235408 (2005) 307
98. H.G. Craighead, G.A. Niklasson, Appl. Phys. Lett. **44**(12), 1134 (1984) 307
99. W. Gotschy, K. Vonmetz, A. Leitner, F.R. Aussenegg, Appl. Phys. B, Lasers Opt. **63**, 381 (1996) 307
100. B. Lamprecht, G. Schider, R.T. Lechner, H. Ditlbacher, J.R. Krenn, A. Leitner, F.R. Aussenegg, Phys. Rev. Lett. **84**(20), 4721 (2000) 307
101. W. Rechberger, A. Hohenau, A. Leitner, J.R. Krenn, B. Lamprecht, F.R. Aussenegg, Opt. Commun. **220**, 137 (2003) 307
102. G. Schider, J.R. Krenn, W. Gotschy, B. Lamprecht, H. Ditlbacher, A. Leitner, F.R. Aussenegg, J. Appl. Phys. **90**(8), 3825 (2001) 307
103. T. Okamoto, I. Yamaguchi, T. Kobayashi, Opt. Lett. **25**(6), 372 (2000) 307
104. S. Enoch, R. Quidant, G. Badenes, Opt. Express **12**(15), 3422 (2004) 307, 308
105. M. Meier, A. Wokaun, P.F. Liao, J. Opt. Soc. Am. B, Opt. Phys. **2**(6), 931 (1985) 307

106. J.C. Hulteen, R.P.V. Duyne, J. Vac. Sci. Technol. A, Vac. Surf. Films **13**(3), 1553 (1995) 308
107. M. Skupinski, R. Sanz, J. Jensen, Nucl. Instrum. Methods Phys. Res. Sect. B, Beam Interact. Mater. At. **257**, 777 (2007) 308
108. W. Stober, A. Fink, E. Bohn, J. Colloid Interface Sci. **26**, 62 (1968) 308
109. H. Mertens, A.F. Koenderink, A. Polman, Phys. Rev. B **76**(11), 115123 (2007) 308, 309
110. J.S. Biteen, N.S. Lewis, H.A. Atwater, H. Mertens, A. Polman, Appl. Phys. Lett. **88**(13), 131109 (2006) 308
111. H. Mertens, J. Biteen, H. Atwater, A. Polman, Nano Lett. **6**(11), 2622 (2006) 308
112. S. Kühn, U. Håkanson, L. Rogobete, V. Sandoghdar, Phys. Rev. Lett. **97**(1), 017402 (2006) 308
113. H. Mertens, A. Polman, Appl. Phys. Lett. **89**(21), 211107 (2006) 308
114. R. Ruppin, J. Chem. Phys. **76**(4), 1681 (1982) 309
115. J. Vielma, P.T. Leung, J. Chem. Phys. **126**(19), 194704 (2007) 309
116. Y.S. Kim, P.T. Leung, T.F. George, Surf. Sci. **195**, 1 (1988) 309
117. R. Ruppin, O.J.F. Martin, J. Chem. Phys. **121**(22), 11358 (2004) 309
118. G. Pellegrini, G. Mattei, P. Mazzoldi, Opt. Lett., submitted 309
119. P. Bharadwaj, L. Novotny, Opt. Express **15**(21), 14266 (2007) 309

Index

Ion Beams in the Geological Sciences

A. Meldrum[1] and D.J. Cherniak[2]

[1] Dept. of Physics, University of Alberta, Edmonton, Alberta, Canada, e-mail: ameldrum@ualberta.ca
[2] Dept. of Earth and Environmental Sciences, Rensselaer Polytechnic Institute, Troy, NY, USA

Abstract. This chapter discusses some of the uses of ion-beam technologies in the geological sciences. Most ion-beam methods used in the field are analytical, and we do not attempt to summarize all of these as many will be found in other chapters of this text. Instead, we focus on the applications of ion beams in several specific areas, including trace-element diffusion, alteration processes, and radiation effects in minerals. Of course, each of these specific topics is of importance in more general geological topics such as, for example, geochronology. Here, we focus specifically on some of the applications of ion beams as measurement or as "material-modification" tools.

1 Introduction

In recent years, ion-beam techniques have been used to simulate radiation damage, measure diffusion coefficients, and to investigate alteration mechanisms in natural minerals. These processes have implications in the modeling of major- and trace-element migration, evaluating mineral phases for geochronology, determining the chronology of the early solar system, investigating the solar abundance of noble gases in extraterrestrial environments, and in the immobilization of high-level nuclear waste in geological repositories.

In nature, mineral alteration and diffusion processes occur gradually over a long timescale, under conditions of variable temperature, pressure, and chemistry. The detailed history of geologic materials is generally difficult to determine or simulate by laboratory methods. For example, radiation damage and diffusion of radiogenic isotopes and their daughter products depends strongly on the pressure–temperature history of the specimen, as well on as the effects of impurities and various alteration processes. Since natural specimens have variable compositions and uncertain thermal histories, one is often faced with difficulties in obtaining generally applicable conclusions.

In contrast, ion-beam experiments are fast and the experiments are "clean" – that is, certain complications with natural specimens can be avoided. Ion implantation offers unparalleled flexibility to modify or control the composition and microstructure of geologic materials. In some cases, ion-beam techniques are the only efficient way to simulate certain processes

H. Bernas (Ed.): Materials Science with Ion Beams,
Topics Appl. Physics **116**, 317–343 (2010)
DOI: 10.1007/978-3-540-88789-8_11,

(e.g., the behavior of noble-gas bubbles and precipitates in crystals within meteorites). Ion-beam analysis techniques are nondestructive and have excellent depth resolution; they require no standards and provide direct multi-element analysis capabilities. Time–temperature conditions can be carefully controlled, so that the effects of interest can be isolated from other processes occurring in complex natural systems. Radiation-damage and dose-rate effects (i.e., flux) are two of the main difficulties associated with ion-beam analysis of geological materials.

We will focus on ion-beam modification and analysis of geological materials with emphasis on the type of information that can be gained from these experiments, on the use of ion-beam methods to simulate processes that occur in nature, and on the physics behind the processes that enable scientists to use ion-beam techniques to learn about the behavior of earth materials. Application of most ion-beam analysis techniques will be discussed, with the exception of particle-induced X-ray emission (PIXE); readers interested in this technique are referred to a review by Sie [1]. The use of ion beams to study materials for geologic disposal of nuclear waste will not be extensively discussed here; however, a concise review was provided by Matzke [2]. Rather, we will focus on the use of ion beams to simulate and analyze physical processes that occur in the geological environment.

2 Diffusion

2.1 Applications

Ion-beam techniques have been employed for several decades in the measurement of diffusion coefficients in minerals. Among the earliest studies was an investigation of oxygen diffusion in quartz using NRA with the nuclear reaction $^{18}O(p,\alpha)^{15}N$ [3]. Applications of ion-beam techniques to diffusion problems became more frequent by the 1980s. The study of Sneeringer et al. [4] in which Sr diffusion in diopside ($CaMgSi_2O_6$) was measured and results from RBS, SIMS, and radiotracer methods were compared, was also among the first significant applications of RBS to diffusion problems of geological interest. An advantage of ion-beam analysis methods, as illustrated in these early studies, is that nanometer-scale depth resolution permits the measurement of very small diffusivities (down to $\sim 10^{-23}\,m^2\,s^{-1}$) with reasonable annealing times. Therefore, diffusion coefficients can often be measured in the temperature range of geological interest, avoiding uncertainties inherent in large down-temperature extrapolations.

Measurement of diffusion coefficients has many applications in the geological sciences, especially in geochronology. Because many minerals incorporate trace amounts of radioactive impurities in their crystal lattices, radiogenic ages can be extracted by measuring the quantities of parent (e.g., ^{238}U) and daughter (e.g., ^{206}Pb, via a decay chain) nuclei in a given mineral grain if

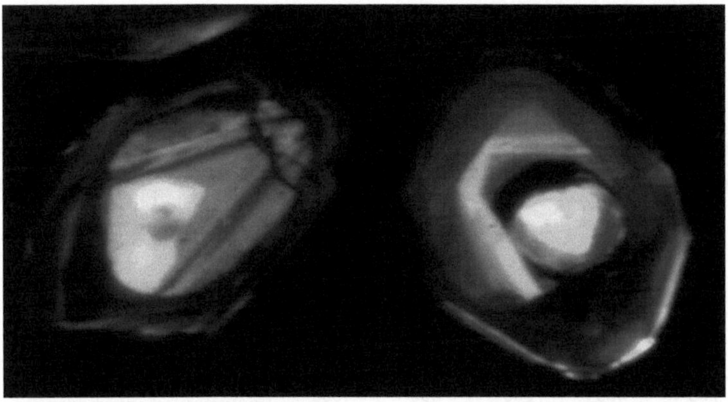

Fig. 1. Cathodoluminescence images of the fine-scale zoning in zircon ($ZrSiO_4$). These regions exhibit distinct differences in concentrations of trace and minor elements such as U, Th, Hf and the REE down to the micrometer scale, and can provide information about past geochemical environments. The preservation of zoning during thermal events subsequent to its formation is indicative of the very slow diffusion rates of these elements in zircon

the half-life is known. If the diffusion rates are rapid, however, erroneous ages can be obtained because the parent or daughter isotopes will be lost to the surroundings. A detailed understanding of the diffusion rates and their temperature dependence is, therefore, critical in the interpretation of radiogenic ages. Furthermore, many accessory minerals (minerals that are low in modal abundance but that tend to incorporate geochemical tracers such as lanthanides and actinides) are refractory and can survive episodes of crustal melting, thereby preserving a record of multiple stages of crystal growth over time. These stages are evident as isotopically and chemically distinct micrometer-scaled zones (Fig. 1). Understanding diffusion behavior becomes critical in such cases, in order to assess whether these distinct regions will preserve their identity through subsequent thermal events.

The utility of diffusion findings in interpreting isotopic ages in natural systems hinges on the concept of "closure temperature" ($T_{\rm close}$). If diffusion of either the parent or the daughter isotope is rapid at a given temperature, these isotopes may diffuse out of a mineral grain. Loss of the parent atoms will lead to an overestimate of the actual crystallization age and loss of the daughter product will result in anomalously young ages. At temperatures higher than $T_{\rm close}$, diffusion is sufficiently rapid that the daughter product will not accumulate. Once the mineral cools below this temperature, isotopic information will not be lost through volume diffusion unless there are subsequent thermal excursions above $T_{\rm close}$. For effective exploitation in geochronology, diffusivities must be sufficiently well known that diffusional loss of parent or daughter species can be correlated with thermal histories to provide informa-

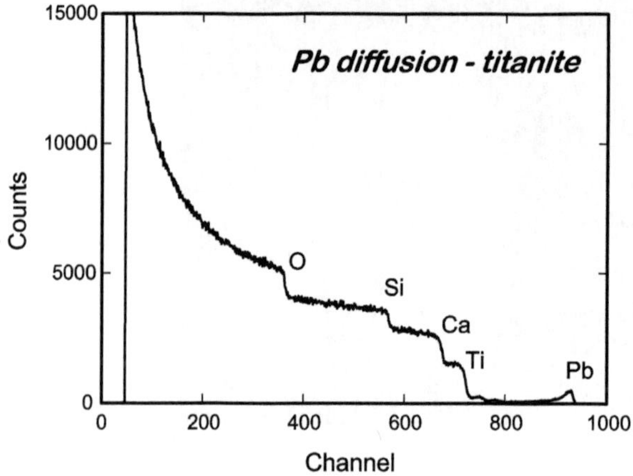

Fig. 2. RBS spectrum of titanite ($CaTiSiO_5$) following a Pb diffusion anneal. Major-element edges and Pb profile are indicated. The Pb profile is sufficiently removed from the signals due to the other elements constituting the titanite matrix so that it can be observed with minimal background interference. After Ref. [7]

tion on mineral crystallization and growth, and on the timing of geological events.

Dodson [5] developed expressions describing the mean closure temperature for a diffusing species in a mineral grain of effective diffusion radius (a) as a function of cooling rate ($\mathrm{d}T/\mathrm{d}t$):

$$T_{\mathrm{close}} = \frac{E_{\mathrm{a}}/R}{\ln\left(\frac{ART^2_{\mathrm{close}}D_0/a^2}{E_{\mathrm{a}}\mathrm{d}T/\mathrm{d}t}\right)}. \tag{1}$$

Here, R is the gas constant, and A is a geometric factor. This expression has since been modified by adjusting A to take into consideration cases with an arbitrarily small amount of diffusion. The Arrhenius parameters extracted from laboratory diffusion studies (i.e., the activation energy for diffusion (E_{a}) and the pre-exponential factor D_0), which describe the temperature dependence of the diffusion rate in the form $D = D_0 \exp(-E_{\mathrm{a}}/RT)$, permit calculation of the closure temperature.

Diffusion of elements of geochronologic interest can present ideal cases for RBS analysis, since these elements are generally of high mass, so their signals are separated in energy from those of the lighter constituents comprising the mineral matrix (e.g., Fig. 2). Pb diffusion has been measured in apatite [$Ca_5(PO_4)_3(OH,F)$] [6], titanite ($CaTiSiO_5$) [7], rutile (TiO_2) [8], zircon ($ZrSiO_4$) [9] and monazite [$(La,Ce)PO_4$] [10] by RBS. Pb closure temperatures calculated using the diffusion parameters derived in these studies follow, from highest to lowest, zircon $\sim$ monazite $>$ titanite $>$ apatite, a trend broadly consistent with closure-temperature determinations from field-based studies of natural systems. Pb diffusion has also been measured in more abundant phases such as the feldspars [11], which are often employed in corrections for "common Pb" in geochronology, and in pyroxenes [12] for which U–Pb systematics have considerable potential as a cosmochronometer.

Diffusion of U, Th, Hf, and the rare-earth elements has also been measured in zircon [13]. The rare-earths and hafnium are important indicators of geochemical processes and are involved in other radioactive-decay sequences employed in geochronology (^{147}Sm $\rightarrow$ ^{143}Nd; ^{176}Lu $\rightarrow$ ^{176}Hf). These studies have shown that the diffusion rate decreases with higher cation charge, and with larger ionic radius for cations of a given charge (among the lanthanides and tetravalent cations) – observations that are in concordance with trends predicted by simple elastic models for diffusion.

Measurement of diffusion coefficients in minerals using ion beams has a range of other applications. Workers have measured the diffusion coefficients for noble gases, notably Xe, in meteoritic minerals (e.g., olivine, feldspar and ilmenite) using ion implantation followed by RBS in order to determine their release rates and solar abundances. Xenon is a fission product of ^{238}U and is also obtained from beta decay of ^{129}I, itself another fission product with a 17-million-year half-life. Xenon accumulation in meteorites has been used to provide an early chronology of the solar system. More recently, the diffusion and solubility of Ar has been measured in quartz using RBS to investigate whether major rock-forming minerals have the potential to be reservoirs for noble gases within the earth.

RBS and NRA have both been used to measure self-diffusion in minerals, where an isotopically distinct tracer is used to characterize transport of a major constituent in a mineral. These measurements have application in geothermometry and geobarometry, and in understanding diffusion-controlled deformational processes, since solid-state creep rates may be limited by the slowest-diffusing species. NRA, because of its isotopic selectivity, is an effective analytical tool for many of these investigations, most notably for measuring oxygen diffusion. The nuclear reaction ^{18}O(p,α)^{15}N has been used to measure oxygen diffusion in accessory and other mineral phases [14], permitting clearer interpretation of the fractionation of oxygen isotopes among minerals, fluids, and vapor phases, which provides information about past earth temperatures and the presence of liquid water on the early earth. Silicon diffusion has been considered the possible rate-limiting species in diffusion-controlled processes such as exsolution and high-temperature solid-state creep in olivine and pyroxene. Si diffusion has been measured using a ^{30}Si tracer in quartz with the nuclear reaction ^{30}Si(p,γ)^{31}P [15], and in olivine and pyroxene using RBS. Cation partitioning in pyroxenes is often used for geobarometry, but diffusion rates of these species must be known in order to determine whether there has been complete chemical re-equilibration. Al, Ca, Fe, Mg and Mn diffusivities have been determined in clinopyroxene, using NRA to measure Al and RBS to measure the other elements.

In certain cases, combinations of methods can be employed in analysis of a single sample or set of samples to obtain added information about the diffusional process. For example, RBS measurements of Pb and Sr diffusion in feldspars were supplemented by NRA measurements of Al and Na in order to provide insight into substitutional mechanisms involved in Pb and Sr exchange in alkali feldspars and sodic plagioclase [11]. Similarly, measurements of phosphorus using the reaction $^{31}P(\alpha,p)^{34}S$ were made to accompany RBS measurements of REE diffusion profiles in zircon to investigate the role of the substitution $REE^{+3} + P^{+5} \rightarrow Zr^{+4} + Si^{+4}$ in rare-earth element diffusion in zircon [13]. Non-Rutherford scattering, in which scattering cross sections deviate from the classical formula and may be significantly enhanced for light elements, can also be used to advantage in this way. Cherniak [16] used the increased α scattering yields from ^{28}Si at 6.6 MeV to explore substitutional processes involved in REE diffusion in fluorapatite, and determined that REE diffusion rates differed depending on whether the process involved simple $REE^{+3} \rightarrow REE^{+3}$ exchange or a coupled exchange such as $REE^{+3} + Si^{+4} \rightarrow Ca^{+2} + P^{+5}$.

2.2 Experiments

Because ion-beam techniques have good depth resolution, and surface layers are not sputtered away as they are in SIMS, the means of introducing diffusants into minerals must be carefully chosen. The diffusant must be applied so that it is easily removed after annealing, or, if the deposited layer cannot be removed, it should be sufficiently thin that the ion beam can "see through" it to the underlying material. Four methods have thus far been widely employed in diffusion studies: (i) depositing a thin layer of polycrystalline material on the sample surface by RF sputtering; (ii) surrounding the samples by fine-grained polycrystalline sources, with sources selected such that samples can be removed after annealing with no residue remaining on the surface; (iii) using ion implantation to introduce the diffusant directly into the mineral specimen; (iv) placing the sample in a sealed system with the diffusant in the form of a gas.

Measuring diffusion profiles is generally done using RBS or NRA, applying standard techniques. For RBS, helium beams are usually employed, with energies typically ranging from 1 to a few MeV. Lower-energy beams can be used to slightly enhance sensitivity, whereas higher beam energies can improve mass resolution, which is often important in minerals containing high-Z elements (e.g., Fig. 3). NRA can be performed in two modes. In resonant mode, a spectrum is taken at a single incident beam energy, with depth scales determined by energy-loss rates of incident and outgoing product particles from the nuclear reaction. ^{18}O profiling, for example, is most often done in nonresonant mode at incident energies of around 800 keV, taking advantage of a region of fairly large and smoothly varying reaction cross section [17]. The resonant mode uses incident particle energies where there are regions of large increases in reaction cross sections spanning a narrow energy width. Resonant profiling in minerals has employed the $^{27}Al(p,\gamma)^{28}Si$ reaction using the 992-keV resonance, or $^{30}Si(p,\gamma)^{31}P$ using the 620-keV resonance.

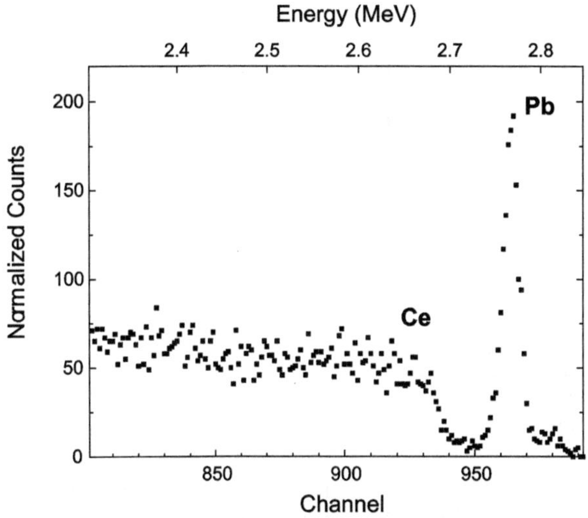

Fig. 3. Expanded region of an RBS spectrum of Pb-implanted natural fluorapatite, showing the Pb peak and a signal from rare-earth elements present in the apatite. RBS analysis is done in this case at 3.0 MeV to improve mass resolution and better separate the contributions to the spectra of these signals. Modified after Ref. [6]

Once the concentration of diffusant as a function of depth has been extracted from the ion-beam data, an appropriate solution to the diffusion equation must be obtained. In the case of sources considered semi-infinite where the sample surface will be maintained at constant concentration, a complementary error function will apply:

$$C(x,t) = C_0 + (C_s - C_0)\mathrm{erfc}\left(\frac{x}{\sqrt{4Dt}}\right), \tag{2}$$

where C_0 is the initial concentration of diffusant in the material, C_s is the surface concentration, and C is the concentration at distance x into the material (Fig. 4). In the case of thin sources of limited extent, the solution to the diffusion equation is:

$$C(x,t) = \frac{C_i - C_0}{2}\left[\mathrm{erf}\left(\frac{h-x}{\sqrt{4Dt}}\right) + \mathrm{erf}\left(\frac{h+x}{\sqrt{4Dt}}\right)\right] + C_0, \tag{3}$$

where C_i is the initial concentration of diffusant in the surface-deposited layer, h is the initial thickness of the layer, and C_0 is the initial concentration of diffusant in the material.

When ion implantation is used to introduce a diffusant, the model used assumes an initial Gaussian distribution of the implanted species:

$$C(x,0) = \frac{N_{\mathrm{imp}}}{\Delta r\sqrt{2\pi}}\exp\left[-\frac{(x-r)^2}{2\Delta r}\right], \tag{4}$$

where x is the distance from the implanted surface, $C(x,0)$ is the initial ion concentration at depth x, N_{imp} is the implanted ion fluence, r is the distance between the peak maximum and the sample surface, and Δr describes the range straggle. The boundary conditions are (i) the medium is infinite in the

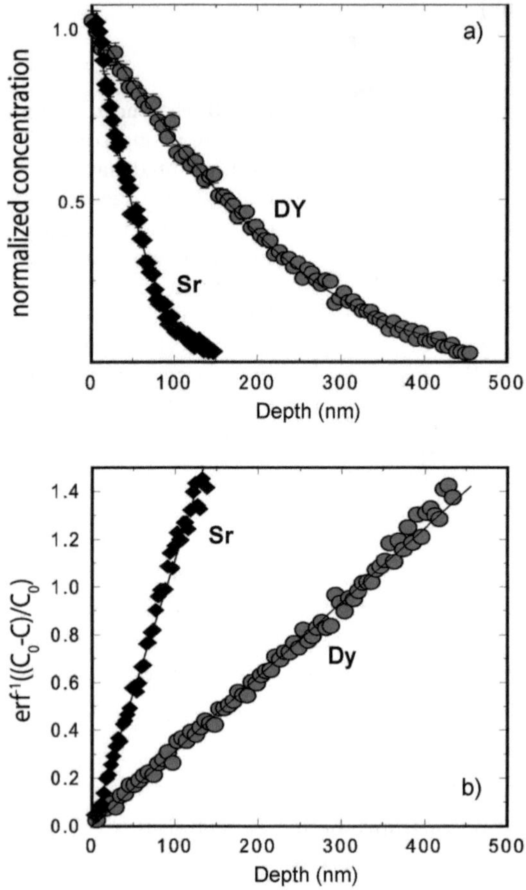

Fig. 4. (**a**) Typical diffusion profiles, with (**b**) linearization by inversion through the error function. The slopes of the linearized data are equal to $(4Dt)^{-1/2}$ when the diffusion profiles conform to the complementary error function solution to the diffusion equation (2). The profiles plotted are for Sr and Dy diffusion in fluorite, and illustrate typical ranges for lengths of depth profiles. Modified after D.J. Cherniak et al., Chem. Geol. **181**, 99 (2001)

positive x-direction; and (ii) that either there is no outdiffusion at the sample surface, i.e., $\partial C(x,t)/\partial x|_{x=0} = 0$, or that concentration goes to zero at the sample surface, i.e., $C(0,t) = 0$. Within (ii), the former constraint is valid when the implanted species has negligible vapor pressure at the corresponding thermal processing temperature, and the latter is appropriate when the vapor pressure is high. The general solution to the diffusion equation is then [18]:

$$C(x,t) = \frac{1}{\sqrt{4\pi Dt}} \int_0^\infty C(x',0) \left\{ \exp\left[-\frac{(x'-x)^2}{4Dt} \right] \pm \exp\left[-\frac{(x'+x)^2}{4Dt} \right] \right\} \mathrm{d}x', \tag{5}$$

where the minus sign in the expression applies when $C(0,t) = 0$, the plus sign when $\partial C(x,t)/\partial x|_{x=0} = 0$. Substituting in (4) for $C(x',0)$, we obtain the full expression:

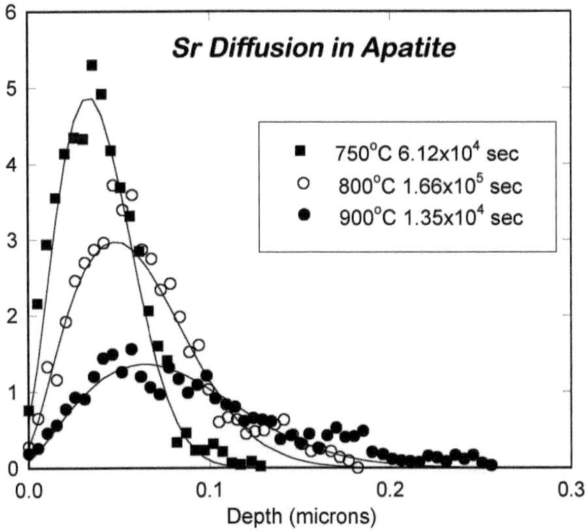

Fig. 5. Diffusion profiles for Sr in apatite for three different time–temperature conditions for annealing. Sr was introduced into the sample by ion implantation. Symbols represent data, lines are best fits to the data using the model described by (6), with boundary condition $C(0,t) = 0$. Modified after D.J. Cherniak and F.J. Ryerson, Geochim. Cosmochim. Acta **57**, 4653 (1993)

$$C(x,t) = \frac{\frac{N_{\mathrm{imp}}}{2\Delta r\sqrt{\pi}}}{\sqrt{1+\frac{2Dt}{\Delta r^2}}}\left\{\exp\left[-\frac{(x-r)^2}{4Dt+2\Delta r^2}\right]\left[1+\mathrm{erf}\left[\frac{\frac{r\sqrt{4Dt}}{\Delta r\sqrt{2}}+\frac{x\Delta r\sqrt{2}}{\sqrt{4Dt}}}{\sqrt{4Dt+2\Delta r^2}}\right]\right]\right.$$
$$\left.\pm\exp\left[+\frac{(x-r)^2}{4Dt+2\Delta r^2}\right]\left[1+\mathrm{erf}\left[\frac{\frac{r\sqrt{4Dt}}{\Delta r\sqrt{2}}-\frac{x\Delta r\sqrt{2}}{\sqrt{4Dt}}}{\sqrt{4Dt+2\Delta r^2}}\right]\right]\right\}. \qquad (6)$$

Again, the plus and minus signs are used depending on the surface boundary condition. Figure 5 illustrates fits of this model (with the boundary condition $C(0,t) = 0$) to Sr diffusion profiles with a range of time–temperature values.

Once diffusivities are extracted, the temperature dependence of D can be determined, which will permit the data to be used in addressing geological problems. Diffusivities can be described by an Arrhenius relation, where $\log D$ has an inverse dependence on absolute temperature, with the slope proportional to the activation energy for diffusion. An example Arrhenius plot is presented in Fig. 6, showing measurements of Pb diffusion in zircon obtained using RBS analysis [9]. Zircon is one of the most widely employed minerals in U–Pb geochronology, and considerable disagreement previously had existed regarding Pb transport rates in this material. Noteworthy in this figure is the agreement of diffusion measurements obtained by both RBS and electron-microprobe analysis, providing a remarkably self-consistent dataset spanning 500°C and some six orders of magnitude in D.

3 Alteration Processes

In recent years, ion-beam analysis techniques have been applied toward the understanding and modeling of geochemically important alteration processes.

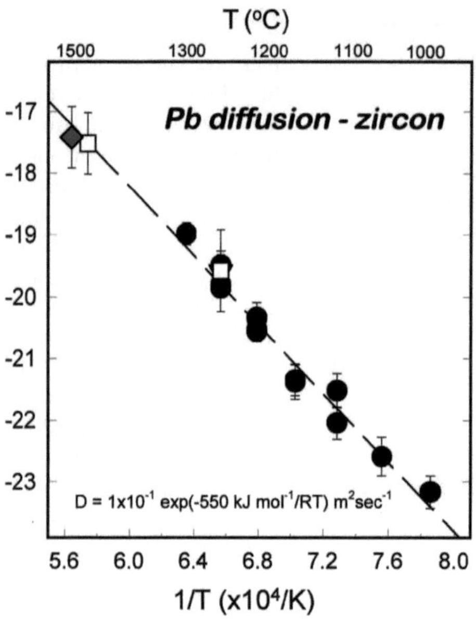

Fig. 6. Arrhenius plot of Pb diffusion in zircon from the study of Cherniak and Watson [9]. Plotted are diffusion experiments on synthetic and natural zircon with profiles measured using RBS (*circles*). Profiles in high-temperature experiments were measured with the electron microprobe. The line is a least-squares fit to the low-temperature data, yielding the Arrhenius parameters: $E_a = 550\,\mathrm{kJ\,mol^{-1}}$, and pre-exponential factor $1.1 \times 10^{-1}\,\mathrm{m^2\,s^{-1}}$. The high-temperature data fall along this line, indicating that a single Arrhenius relation can describe Pb diffusion in zircon over a temperature range of 500°C and diffusivities spanning 6 orders of magnitude

When minerals are exposed to fluids on the earth's surface (i.e., weathering) or in the subsurface during periods of hydrothermal activity, their chemical composition and crystal structure can be altered. Depending on the duration and intensity of the alteration process and on the chemical durability of a given mineral, a crystal may be entirely transformed to a new phase, or only the outer portion of the crystal may be altered while the core of the mineral remains pristine (e.g., see Fig. 7). Therefore, interface effects at the boundary between altered rims and pristine cores are critical in many types of mineral transformations.

Alteration processes can represent a key step in the geochemical cycle and mass redistribution of the major elements at the earth's surface. Chemical alteration of basaltic and rhyolitic glass has important implications for nuclear-waste disposal, since the composition of natural glass can be similar to that of nuclear-waste material. Alteration can have negative consequences in geochronology, since redistribution of parent and daughter elements will complicate the interpretation of isotopic results. Finally, the type and degree of alteration can be a sensitive measure of the intensity (temperature, dura-

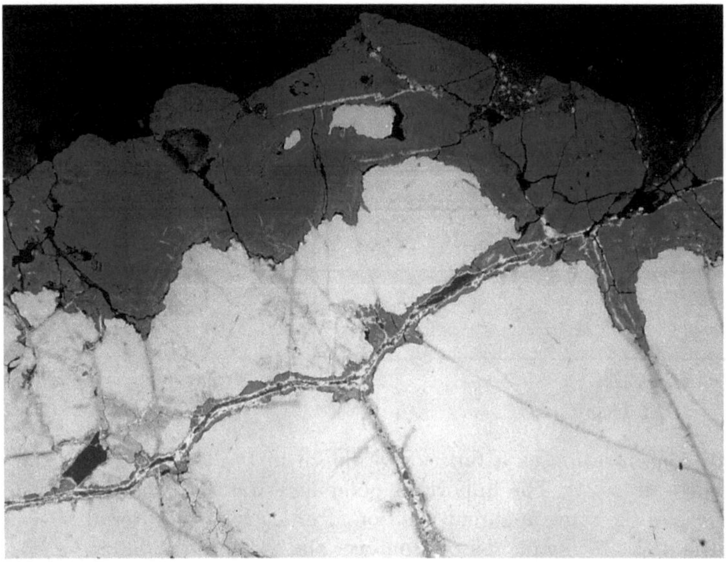

Fig. 7. Specimen of metamict brannerite [$(U,Ca,Y,Ce)(Ti,Fe)_2O_6$] from Cordoba, Spain. Alteration is apparent as *gray zones* along the crystal boundaries and within internal fractures. This alteration involved the loss of U and incorporation of Al, Si, P, and Fe. The width of the photo is 1 mm. Image contributed by Dr. G. Lumpkin (Australian Nuclear Science and Technology Organisation)

tion, and fluid composition) of the alteration event itself, thereby providing information on the local geology.

Since the alteration process initiates at the surface of a specimen and gradually progresses inwards, ion-beam analysis techniques are ideally suited for the study of chemical redistribution at the solution/mineral interface. Several ion-beam analysis methods have important applications for the study of alteration processes, including RBS, NRA, and elastic recoil detection analysis (ERDA) [19]. Ion-beam analysis has the important advantages that it is non-destructive, and the results are generally not difficult to quantify. Problems of specimen charging are usually insignificant, chemical and matrix effects are not observed, and the depth resolution is good [20]. On the other hand, ion-beam analysis generally requires large specimens ($>1\,mm^2$) (although RBS and some nuclear reactions can now be performed with microbeams), may suffer from significant peak overlap, can introduce unwanted radiation damage into the structure, and does not provide direct microstructural data. Despite the importance of alteration and the utility of ion-beam techniques in characterizing it, the first systematic studies were conducted as late as the 1990s by J.C. Petit and coworkers at CSNSM-Orsay, from whose work some of the following discussion is derived.

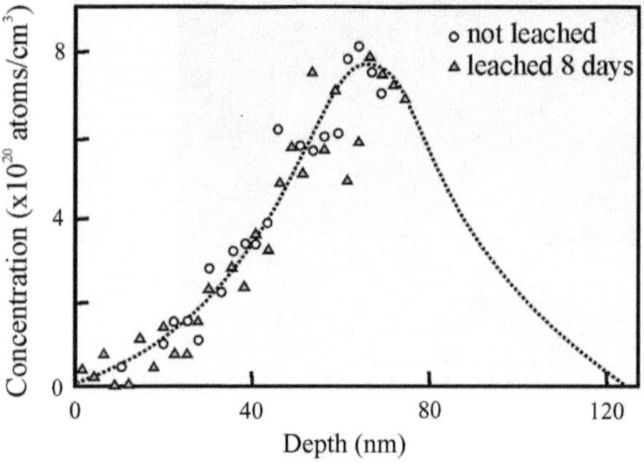

Fig. 8. Bismuth concentration as a function of depth in $LaPO_4$ extracted from RBS data, modified after [21]. The important point here was that the implanted Bi marker layer was in the same apparent position after leaching in water at 90°C, indicating that there was negligible dissolution over the 8-day duration of the experiment

Ion-beam techniques were used in an early investigation of dissolution and alteration effects in $LaPO_4$ (i.e., monazite – a common accessory mineral in many rock types). By implanting 4×10^{15} ions cm^{-2} of Bi, a well-defined marker layer was created at a depth of 65 nm, as determined by RBS analysis [21]. The surface of the monazite would almost certainly have been rendered amorphous by the implantation of the marker layer. The specimens were subsequently exposed to distilled water at 90°C for 8 days. If matrix dissolution occurs, the location of the marker layer would appear closer to the surface after leaching. However, the marker layer remained at the same depth after exposure (Fig. 8), indicating that this mineral, even if amorphized, is extremely insoluble in pure water. This represented an early but classic example of the use of ion implantation combined with ion-beam analysis to determine the stability of geological material in an aqueous environment.

Ion-beam analysis techniques have since made many contributions toward the understanding of water–rock interactions. For example, five separate processes that occur during dissolution of natural glasses were resolved, including ion diffusion and exchange, permeation of neutral species through the glass, network hydrolysis, adsorption of dissolved species, and condensation of Si–OH groups [22]. Glass dissolution was found to depend on all of these processes in varying degrees that depended on glass composition, solution pH, solution composition, and temperature. Separate corrosion stages were identified, and the glass composition was found to be an important parameter in determining the alteration rate and dominant dissolution mechanism. In basaltic glass (low silica and Fe-rich), alteration occurred almost stoichiometrically, and was mediated by the precipitation of Fe-hydroxide on the glass surface. On the other hand, in silica-rich rhyolitic glass, ion exchange of hydrogen species for alkali elements was a dominant process, resulting in significant nonstoichiometry of the altered layer. This was one of the first studies to resolve many of the separate mechanisms of hydration in geologically important materials.

Ion-beam analyses suggest that alteration processes can be quite different in crystalline and glassy materials. Petit et al. [23] showed that a hydrated layer enriched in heavy elements forms on the surface of basaltic glasses exposed to deionized water at 100°C; whereas the minerals olivine [$(Mg,Fe)_2SiO_4$] and zircon ($ZrSiO_4$) dissolve congruently. Differences between mineral phases are also observed; for example, an Fe-rich layer re-precipitates on the surface of olivine, whereas there are almost no differences in the RBS spectra of zircon before and after leaching. In contrast, K-feldspar ($KAlSi_3O_8$) shows incongruent dissolution, with a strong preferential depletion of potassium. These studies used Pb implantation to create a buried depth marker, possibly causing increased hydration and matrix destruction effects. Fortunately, the results also suggested that below an implanted fluence of 10^{13} ions cm^{-2}, the dissolution rate increases linearly with the defect concentration, and therefore it should be possible to the extrapolate the data back to zero dose.

In general, the ion-beam results demonstrate that network silicates show significant hydration, since a stable silica network still remains after cation release and exchange. In contrast, nesosilicates (structures with isolated SiO_4 tetrahedra) with exchangeable cations do not seem susceptible to hydration. This may be due to structural constraints in the nesosilicates, in which there is no stable hydrated state [20]. Ion irradiation enhances the alteration process by transforming the structure of nesosilicates into that of a glass, which is then susceptible to hydration without structural dependence. These investigations are important because they enable the prediction of which structures should be particularly susceptible to hydration, and shed light on the processes responsible for the more rapid hydrothermal alteration often observed in metamict minerals (the term *metamict* refers to U- and Th-bearing minerals that are amorphous due to the effects of accumulated α-decay damage).

As ion-beam techniques became more widely applied in the 1990s, individual alteration processes were identified for a variety of mineral and glass compositions. Nevertheless, a considerable amount of work remains to be done. The effects of temperature on the alteration rate have not been systematically studied. In general, cross-sectional transmission electron microscopy is an excellent complement to ion-beam analysis, since it provides microstructural and chemical information on the altered layers. Future technological advances in ion-beam technology will permit new types of nuclear analyses on geologically important elements (i.e., the transition metals), and focused ion beams will provide excellent spatial resolution. These techniques will have applications in many other areas where fine-scale compositional gradients are important, including mineral–mineral and mineral–fluid interactions during metamorphism or other high-temperature events.

4 Radiation Effects in Minerals

In 1893, W.C. Broegger coined the term "metamikt" to describe a class of minerals that maintained good external crystal morphology, but that demonstrated many of the characteristics of a glass. This term has remained in the geological literature, and is used today to describe minerals that have become amorphous due to natural α-decay processes (ion-irradiated specimens may, therefore, be amorphous but are not metamict). Many mineral phases are known to become partially or fully metamict as a result of α-decay damage, and the study of radiation effects in minerals is an active area of research.

The effects of radiation damage are especially important for U–Pb geochronology. With increasing damage levels, diffusion rates increase and elements in the radioactive-decay sequence can be lost to the surroundings. The daughter products of α-decay processes are generally located within a damaged or amorphous region where they are more susceptible to leaching or redistribution during low-temperature geologic events. Despite current techniques to minimize difficulties associated with the loss of daughter products, radiation-damage remains one of the largest problems for radiometric age dating. A knowledge of the mechanism of radiation-damage accumulation, its temperature dependence, its chemical and structural effects, and its ion mass and energy dependence is necessary to model the transition to the metamict state and to evaluate the reliability of various minerals for retaining their radiogenic isotope systematics in the natural environment.

All types of radioactive decay have the potential to produce structural damage. Each α-decay releases 4–6 MeV of energy, while β-decays and γ-decays generally release a few hundred keV. The energy released by radioactive decay is converted into kinetic energy of the decay products. Gamma-decays produce little, if any, structural damage unless the material has an efficient mechanism by which high-energy photons create atomic displacements (there is no mineral known to the authors that can be significantly damaged by γ-rays). Beta-decays also do not create significant radiation damage because the mass of the electron and the energy of the recoil nucleus are too low to produce atomic displacements by ballistic processes. Alpha-decays, on the other hand, do produce structural damage, and α-decay processes are responsible for the transformation to the metamict state. Spontaneous nuclear fission also produces considerable structural damage; however, fission events occur so infrequently that they do not contribute to the metamictization of minerals.

Each α-decay produces an α-particle with an energy of several MeV and a recoil nucleus with an energy of $\sim$100 keV. Because of their differing mass and energy, the damage mechanisms for these two particles are fundamentally different. Alpha-particles travel a distance of several micrometers, losing energy mostly by electronic excitations, and even for the highest decay energies they contribute less than 200 atomic displacements, mostly by nuclear collisions near the end-of-range. In some materials such as quartz, radiolysis (i.e., electronic-damage processes) can be significant; however, these processes do not seem to be important in most U-bearing minerals. In contrast, the recoil nuclei, with high mass and low energy, travel less than 100 nanometers and produce significant localized damage in the form of atomic displacement cascades. These are manifested as regions with high defect concentrations or as discrete amorphous zones. With time, the damaged regions accumulate and overlap and the mineral may become fully metamict.

Many investigations have focused on the structure and properties of natural minerals with variable U and Th content. Suites of minerals with variable damage levels provide information on the evolution of physical and microstructural properties as a function of increasing α-decay dose. Nevertheless, studies of α-decay-damaged minerals, while useful and informative, suffer from several problems. The thermal history of natural samples is poorly known by laboratory standards, and the effects of incorporated impurities, alteration, and pressure produce additional complications. While natural samples have provided useful generalizations (e.g., decreasing density, decreasing refractive index, increased fracture toughness, and enhanced leaching and alteration rates that occur with increasing α-decay dose), the actual mechanisms of radiation damage and recovery are difficult to determine. Alternatively, doping synthetic crystals with short-half-life actinides (e.g., ^{238}Pu; $t_{1/2} = 87.7\,\mathrm{yr}$) provides excellent experimental control and closely reproduces real α-decay processes. However, the experiments require years to be performed and advanced radiation-handling facilities are needed. A third technique uses ion beams to simulate radiation-damage effects. Due to the uncertainties associated with natural samples and the regulatory and experimental difficulties involved with radioactive specimens, ion beams are now widely used to simulate the effects of α-decay processes in the laboratory, with all the associated experimental advantages discussed earlier. The main drawback is that the ion-beam experiments employ a dose rate that is higher by a factor of 14–15 orders of magnitude compared to that experienced by natural minerals, so flux effects must be considered when extrapolating to geological specimens.

In order to compare materials with different chemical formulas and crystal structures, most experimenters report the amount of damage as an ion dose, in units of displacements-per-atom (dpa). A key advantage is that the number of α-decay events in minerals can also be converted to a dose in dpa, allowing for a direct comparison between ion-irradiated and α-decay damaged materials. To convert a measured ion fluence to units of dose, a simple formula is used:

$$D_{\mathrm{dpa}} = JF/n, \tag{7}$$

where D_{dpa} is dose in dpa, J is the number of atomic displacements caused by each ion per unit depth, F is the fluence, and n is the atomic density. J depends on a variety of parameters, including incident ion mass and energy, and the displacement energy (E_{d}) of the target atoms. Monte Carlo computer simulations (e.g., SRIM-2000) [24] are used to estimate J. J varies inversely with E_{d} (i.e., the higher the displacement energy, the fewer displacements are caused by each incident ion).

Atomic displacement energies are known for only a relatively few ceramic materials, most of which are simple oxides such as Al_2O_3, MgO, ZnO, CaO, and UO_2. Virtually all measurements of E_{d} use electron irradiation to produce isolated point defects, followed by optical spectroscopy or transmission electron microscopy to measure defect production and evolution. By conservation of momentum and energy, the displacement energy is simply the minimum ballistic energy transfer to the target atoms required to produce defects. Zinkle and Kinoshita [25] recently summarized the results of virtually all displacement-energy measurements and gave recommended values based on the various experimental results. These generally range from 10 to 70 eV. For example, in Al_2O_3 (corundum or sapphire), the recommended displacement energy values are 20 eV for Al and 50 eV for oxygen. More recently, molecular-dynamics simulations have extended the range of materials for which E_{d} can be estimated to complex silicates such as $ZrSiO_4$: 89 eV (Zr), 48 eV (Si), and 28 eV (O) [26]. Nevertheless, E_{d} is not known for most minerals, therefore it is generally estimated using a weighted average of the known values for various oxides.

In order to make quantitative comparisons of the amount of radiation damage in natural and ion-beam-irradiated specimens, the measured concentration of U + Th is also converted to a displacement dose. This can be done in two steps:

$$\begin{aligned} D_\alpha = {} & 8N_{238}\left[\exp(t/t_{1/2(238)}) - 1\right] + 7N_{235}\left[\exp(t/t_{1/2(235)}) - 1\right] \\ & + 6N_{232}\left[\exp(t/t_{1/2(232)}) - 1\right], \end{aligned} \tag{8}$$

where D_α = α-decay dose in units of α-decays/g, N_{238}, N_{235}, and N_{232} are the measured number of atoms/g of ^{238}U, ^{235}U, and ^{232}Th, $t_{1/2(238)}$, $t_{1/2(235)}$, and $t_{1/2(232)}$ are their respective half-lives, and t is the geologic age. Studies of radiation damage in minerals usually report D_α. To convert to a dose in dpa:

$$D_{\mathrm{dpa}} = \frac{D_\alpha \cdot W \cdot n}{a \cdot A}, \tag{9}$$

where, A is Avogadro's number, W is the molecular weight of the mineral, n is the average number of displacements per α-decay, and a is the number of atoms per formula unit. The n term is analogous to J in (7) and can be obtained using an identical computer simulation. For most minerals, n is between 800 and 1300 atomic displacements per α-decay event. A side benefit of the dpa unit is that the uncertainty in the displacement energy affects both calculations in exactly the same way, so that quantitative comparisons between ion-irradiated and α-decay-damaged materials are possible, in terms of the total number of displacements per atom. The observed amount of disorder may also depend on the flux.

There are two main classes of ion-beam experiments. Figure 9 gives an example of an ex-situ experiment on synthetic cadmium niobate pyrochlore (pyrochlores represent a large and diverse class of minerals). Single crystals of $Cd_2Nb_2O_5$ were implanted with 70-keV Ne^+ at ambient temperature. The implantation was periodically stopped and the specimen was transferred for RBS channeling analysis. The specimen was then returned to the implanter for an additional dose. This process was continued until the yield from the near-surface region of the specimen was equal to the random yield, at which point a 200-nm thick surface layer had become amorphous.

The rate at which the near-surface yield approaches the random yield can shed light on the amorphization mechanism. The lines in Fig. 9b were plotted with the simple Gibbons model:

$$\frac{A_{\mathrm{A}}}{A_0} = \left[1 - \left(\sum_{0}^{n} \frac{(A_{\mathrm{i}}D)^{n-1}}{(n-1)!} \exp(-A_i D)\right)\right], \tag{10}$$

where A_{A} is the total amorphous area, A_0 is the area of the specimen, A_{i} is the area of a single collision cascade, D is dose, and n is the number of cascade overlaps necessary to produce amorphization (e.g., for direct impact amorphization, $n = 1$, for single overlaps $n = 2$, and so on), and A_{A}/A_0 is the amorphous fraction. The data in Fig. 9 fit with (10) suggests that

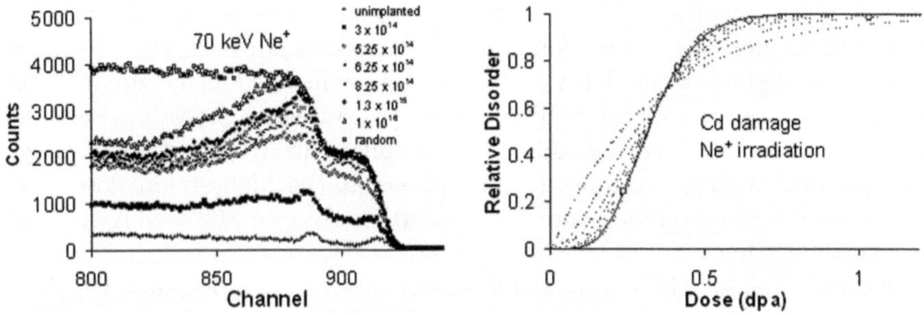

Fig. 9. (**a**) RBS-channeling spectra illustrating damage accumulation in $Cd_2Nb_2O_7$ pyrochlore irradiated with 70-keV Ne^+. The RBS spectra were deconvoluted to obtain the relative disorder as a function of Ne dose (**b**). The *curves* are the solution to the Gibbons' model with varying numbers of cascade overlaps. The best curve (*solid line*) corresponds to 7 cascade overlaps. Modified after A. Meldrum et al., Phys. Rev. B **64**, 103109 (2001)

amorphization occurs by a defect accumulation, multicascade overlap process for Ne-ion-irradiated $Cd_2Nb_2O_7$ pyrochlore.

The question of whether amorphization occurs directly within a collision cascade (i.e., direct impact amorphization) vs. a cascade overlap, defect accumulation process has been much debated in the literature – especially for the case of zircon. In zircon, a modified Gibbons model suggests that amorphization occurs via a double-cascade-overlap process. However, more recent models for direct impact amorphization can also accurately fit data such as that in Fig. 9. The question of the fundamental amorphization mechanism in minerals is still an open question, and will probably require detailed techniques such as ultrafine-probe EELS (electron energy-loss spectroscopy) on individual displacement cascades and molecular-dynamics cascade simulations.

The second main class of experiment uses ion irradiation with in-situ specimen analysis. For example, electron diffraction is often used to monitor the specimen as the ion-irradiation experiment proceeds. Figure 10 shows a set of electron-diffraction patterns for monazite ($CePO_4$) and $Cd_2Nb_2O_7$ pyrochlore irradiated to various fluences at room temperature. With increasing ion fluence, the initially sharp diffraction maxima fade away and an amorphous halo increases in intensity. The fluence at which the diffraction maxima completely disappear defines the critical amorphization dose (D_c). A major difference between the two phases is the disappearance of the pyrochlore (200) superlattice spots prior to the appearance of the amorphous halo. Monazite transforms directly to the amorphous state, but pyrochlore chemically disorders prior to or concurrently with amorphization, depending on the mass of the incident ions.

Figure 11 shows a set of high-resolution images for monazite, corresponding to the four electron-diffraction patterns in Fig. 10. At low dose, the specimen is mostly crystalline but isolated amorphous domains are also visible. As the dose increases, these domains grow and overlap until the specimen consists of slightly rotated crystalline islands embedded in an amorphous matrix. This rotation is probably due to stress caused by the volume expansion associated with the crystalline-to-amorphous (c–a) transition. Finally, the specimen appears completely amorphous at the highest ion dose. For monazite, the room-temperature amorphization dose for 800-keV Kr ions is just below 0.3 dpa.

Figure 12 shows the temperature dependence of the amorphization dose for monazite and zircon irradiated with 800-keV Kr ions. In both cases, the amorphization increases with temperature, although at vastly different rates. In the case of monazite, D_c begins to increase rapidly at room temperature, and above ~175°C, this mineral can no longer be fully amorphized. This is referred to as the critical temperature. The critical temperature is generally defined as the temperature at which the damage and recovery rates are equal and amorphization cannot occur [28]. In contrast, zircon appears to have a

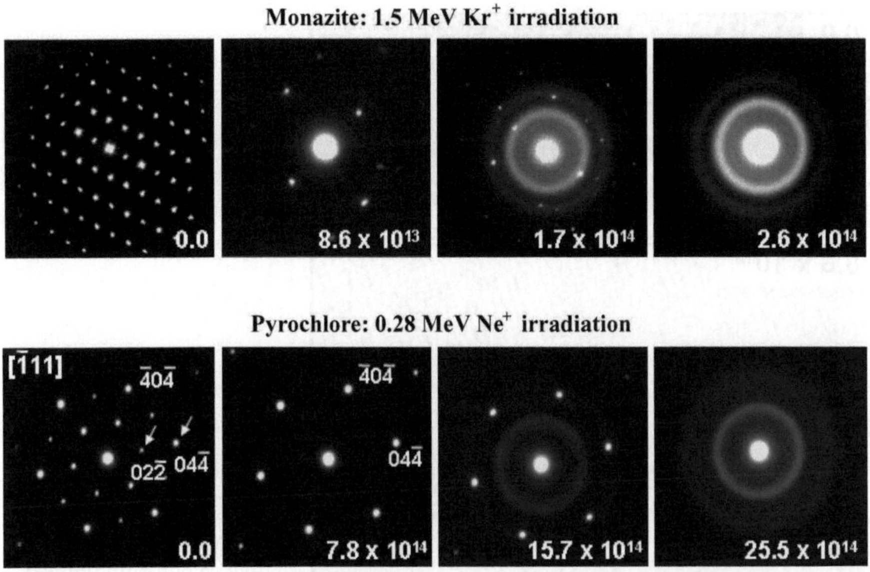

Fig. 10. Electron-diffraction patterns showing the effects of ion irradiation on monazite and pyrochlore. The ion fluence, in units of ions cm^{-2}, is given at the bottom of each diffraction pattern. Monazite transforms directly to the amorphous state, but pyrochlore chemically disorders prior to amorphization. This disordering is observed as a halving of the unit cell, with the corresponding loss of the 022 family of diffraction maxima. Modified after A. Meldrum et al., Phys. Rev. B **64**, 103109 (2001) and A. Meldrum et al., Nucl. Instrum. Methods Phys. Res. **116**, 220 (1996)

two-stage amorphization curve and has a much higher critical temperature. These curves show that both minerals can readily be amorphized, but that thermal recrystallization processes are considerably more rapid in monazite compared to zircon, and that more than one recovery stage may be resolvable in the temperature–dose curve for zircon. In nature, there are many examples of partially or fully metamict zircon; whereas, monazite, despite incorporating high concentrations of U + Th, is rarely found to be metamict.

Several models have been proposed that can fit temperature–dose data (e.g., see [27]). One frequently employed model describing the rate of change of the amorphous fraction during irradiation was derived by Weber et al. [28]:

$$\frac{\mathrm{d}f_a}{\mathrm{d}t} = \varphi\sigma_{\mathrm{a}} - \tau\exp\left(\frac{-E_{\mathrm{a}}}{kT}\right), \tag{11}$$

where f_{a} is the amorphous fraction, t is time, ϕ is the ion fluence, σ_{a} is the amorphization cross section, τ is a time constant, E_{a} is an activation energy for recrystallization, k is Boltzmann's constant and T is temperature. The solution to this equation under the condition that $f_{\mathrm{a}} = 0$ when $t = 0$ is:

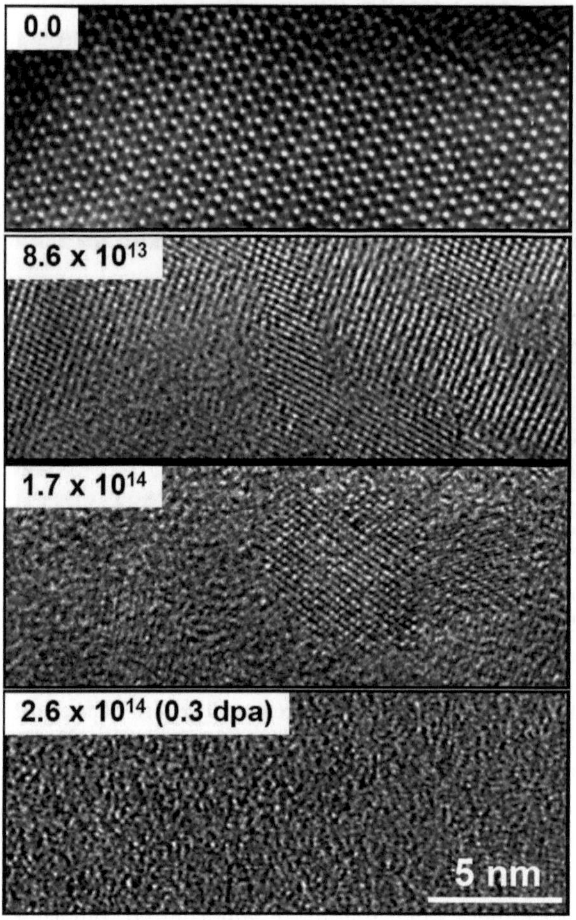

Fig. 11. High-resolution images showing damage buildup in monazite irradiated with 1.5 MeV Kr^+. The fluence (in ions cm^{-2}) is given in the top right of each micrograph. The four images correspond to the four diffraction patterns in Fig. 10. Modified after A. Meldrum et al., Nucl. Instrum. Methods Phys. Res. **116**, 220 (1996)

$$\ln\left(1 - \frac{D_0}{D_c}\right) = \ln\left(\frac{1}{\phi\sigma_a\tau}\right) - \frac{E_a}{kT}. \tag{12}$$

Here, D_0 is the amorphization dose at 0 K (determined by least squares fitting or by direct extrapolation in Fig. 12), D_c is the amorphization dose above 0 K, and t is time. $\mathrm{Ln}(1/\phi\sigma\tau)$, E_a, and D_0 can be varied to provide the best fit to the experimental data (Fig. 12). T_c is found by allowing D_c to go to infinity:

$$T_c = \frac{E_a}{k\ln(1/\varphi\sigma_a\tau)}. \tag{13}$$

Equation (11) makes certain assumptions about the mechanism of the amorphization process. Other models with different assumptions can give similarly good data fits. Regardless of the number of parameters or assumptions in the model, for comparing ion-beam-irradiated specimens to α-decay

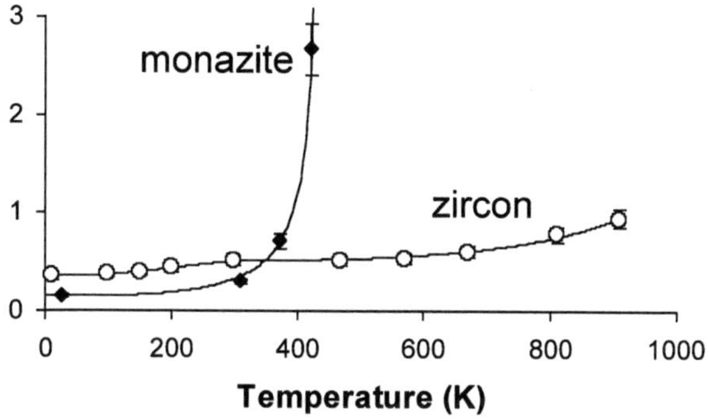

Fig. 12. Plot of the amorphization dose as a function of temperature for monazite and zircon irradiated with 1.5-MeV Kr^+. The curves represent a least-squares best fit solution for (12). Despite the structural similarity of monazite and zircon, these curves show several striking differences. Modified after Ref. [29]

damaged minerals the critical point is that the model fits the data closely. This allows for a reliable estimate of the amorphization dose at all intermediate temperatures irrespective of the amorphization mechanism.

The microstructural evolution with increasing α-decay dose is similar to that observed for the ion-beam experiments. For example, natural monazite from the 1.39 billion-year-old Petaca pegmatite deposit in New Mexico suffered highly uneven amounts of damage over its history. This monazite was inhomogeneous, and the local thorium content in particular was variable. For low Th concentrations, electron diffraction shows that the monazite is crystalline. In crystals with higher thorium concentrations, most of the monazite is metamict and only isolated crystalline islands remain. In the Petaca monazite, the amorphous state is reached at a Th concentration of corresponding to a displacement dose of $\sim$9 dpa.

The amorphization dose for the crystalline-to-metamict transition in the Petaca monazite is approximately 30 times higher than the room-temperature amorphization dose for 800-keV Kr ions. This large apparent discrepancy is the result of long-term thermal recrystallization that has occurred over the 1.39-billion-year history of the specimen. Over such long time spans, a significant amount of diffusion-driven recrystallization can occur at relatively low temperatures. This gradual thermal recrystallization offsets the damage buildup and is responsible for the high dose required for monazite to become metamict in nature. Other minerals such as zircon and pyrochlore may show a similar but less-pronounced effect. Therefore, by modeling the thermal recrystallization process, the ion-beam results could, in principle, be used to predict if and when a given composition would become metamict in nature, depending on the temperature and actinide concentration.

At least two separate effects must be modeled in order to extrapolate the ion-beam data to natural minerals. The first effect is the rate of damage

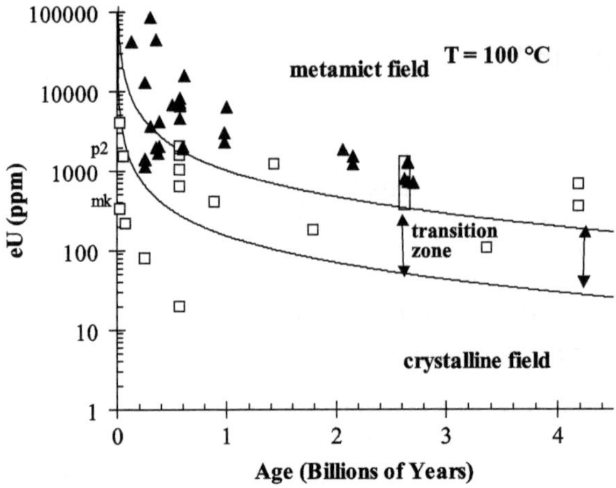

Fig. 13. Age vs. equivalent uranium relationship for metamictization of natural zircon. The upper solid line is the solution to (16) for $T = 100°C$. The lower line is the same solution multiplied by a factor of 0.15 (see text). Zircon plotting above the upper line should be metamict (at this temperature), and below the lower line it should be crystalline. The *region between the solid lines* represents the zone of partial metamictization. The *symbols* are zircon data obtained from the literature: *triangles* = metamict zircon; *squares* = crystalline zircon. Modified after Ref. [29]

accumulation, which itself depends on cascade size and on the amount of in-cascade crystal recovery. The second effect is long-term thermal recovery that can occur over geologic time. An equation that incorporates all these processes was recently derived [29]:

$$N_c(\text{ppm}) = \left[\frac{X}{[1 - e^{Y\cdot(1-T_c/T)}]\cdot(e^{\lambda\cdot t} - 1)\cdot n\cdot x}\right], \tag{14}$$

where $X = [6 \times 10^6 \times D_0]$, Y results from the model fit to the ion-beam data ($Y = E_a/kT_c$), n is the average number of atomic displacements per α-decay, T_c is the critical temperature, and x is the number of α-decays in the decay chain. N_c is the present-day equivalent U-concentration required for the mineral to be metamict (the unit "equivalent uranium" converts all incorporated actinides to an "equivalent" concentration of ^{238}U required to produce the same activity or dose). The damage accumulation rate is built into the Y term, and long-term thermal recovery (i.e., the effect of flux) is modeled by its consequence on T_c. For older specimens, T_c will be lower because of the extra time available for damage recovery. This effect on T_c was modeled according to the equation:

$$T_c = \frac{-E_a}{\ln\left(\frac{B\cdot A}{\nu}\right)\cdot k}. \tag{15}$$

Here, B is the damage volume per collision cascade, A is the activity of the specimen, ν is the atomic jump frequency, and E_a is the activation energy from (11).

For the specific case of zircon irradiated with 800-keV Kr ions:

$$N_c(\text{ppm}) = \frac{170}{[1 - e^{14.5\cdot(1-523/T)}]\cdot(e^{1.55\cdot10^{-10}\cdot t} - 1)}. \qquad (16)$$

In earlier work, X was 166, Y was 1.1, and T_c was 633 K. The differences in X and Y were due mainly to differences in the SRIM simulations and the data-fitting procedure.

Figure 13 shows a plot of N_c as a function of age for a temperature of 100°C using (16). The equation calculates the total equivalent uranium concentration required for the metamictization process to be completed. In fact, there is a range of N_c values at which *partial* metamictization should be observed. This can be accounted for by assuming that TEM specimens appear completely amorphous when the total amorphous volume fraction of the specimen (f_a) is equal to 0.95, and the onset of observable damage occurs for $f_a = 0.35$ [30]. If a direct impact amorphization model for zircon is assumed, as opposed to a multiple overlap of collision cascades, then the following equation relates the amorphous fraction, f_a, to the amorphization dose [31]:

$$f_a = 1 - \exp(-BD_c). \qquad (17)$$

All the terms have been defined previously. According to (17), if f_a is decreased from 95% to 35%, then D_c decreases by a factor of 0.85. A value of $0.15D_c$ can, therefore, be used to estimate the minimum fraction of radiation damage that is detectable by TEM. Since N_c is linearly related to D_0 (14), the minimum detectable amorphous fraction occurs at a value of $0.15N_c$. The upper line in Fig. 13 represents the transition to the completely metamict state (according to the electron-diffraction criterion). The lower line denotes the U concentration required for the observable onset of metamictization. Specimens that plot in the upper field should be completely metamict, in the middle field partially metamict, and in the lower field, completely crystalline (for $T = 100$°C). Data from several natural samples are also plotted in Fig. 13. The data agree relatively well with the model curve, despite the uncertainties in composition and thermal history.

Newer, more general models are at various stages of development (e.g., see [27, 32]). Nevertheless, the existing model shows a method by which the ion-beam irradiation data can be used to predict the crystalline-to-metamict transition that occurs in nature. There are several implications. On the most fundamental level, the ion-beam data combined with careful modeling can determine which minerals or mineral classes are susceptible to metamictization. The models show that, using the ion-beam data, metamictization can be employed as a qualitative geothermometer. Minerals that should be metamict, based on (14), but are found to be crystalline *must* have experienced higher temperatures. A minimum temperature can be calculated for which a given mineral is expected to remain crystalline for a given equivalent uranium concentration. The models can also be inverted to calculate the *future* crystalline-to-amorphous transition (and associated decrease in durability) for minerals such as monazite and zircon, which have been proposed for immobilization of various types of radioactive-waste products.

The above analysis also has inherent limitations. Different types of radiation damage are not individually modeled, and recent results suggest that different processes can result in different damage accumulation rates. For example, in the low α-decay dose regime, recovery in natural zircon is practically zero, even over geological time [33]. Raman results for zircon suggested that epitaxial recrystallization of amorphous zones is, however, more rapid than isolated defect recombination. The effects of flux are modeled as a change in T_c, however, this ignores the possibility of physical effects such as specimen heating and temporally overlapping damage cascades that can occur in the ion-beam experiments. Zircon appears to show two stages in its D_c vs. T curve, which considerably complicates the model results [34]. The potentially important effects of impurities are not modeled, and a relatively large experimental error exists in the in-situ irradiation experiments. Despite these various limitations, the model agrees with the available data for both crystalline and metamict samples.

As a result of these many applications, the list of minerals studied by ion-beam techniques has become long. A complete review was recently published by Ewing and coworkers [35]. At this point, more detailed experiments will be required to determine the atomic-scale mechanisms responsible for amorphization. Even for zircon, by far the most widely investigated mineral, the actual damage mechanism (e.g., direct-impact amorphization vs. defect accumulation and cascade overlap) is still debated. This will require the application of other experimental techniques, especially high-resolution XTEM and ultrafine-probe EELS analysis.

5 Conclusion

Ion-beam techniques are now used for a variety of applications in the earth sciences. Alteration, diffusion, and radiation damage are interacting processes that occur in nature, and have important implications in geochronology and mineralogy. Ion beams offer a number of important advantages over other more "conventional" experimental techniques. Ion-beam analysis techniques are nondestructive and have excellent depth resolution; they require no standards and provide direct multielement analysis capabilities. Techniques such as NRA and ERDA have found important applications in alteration and leaching processes, and for major-element diffusion. RBS is widely used to detect minor or trace elements and element migration due to diffusion or alteration. In the channeling mode, RBS is an excellent means to quantify radiation-induced disorder, particularly in combination with electron microscopy. Ion-beam modification due to radiation effects is, however, a common problem in studies of alteration and diffusion, unless the specific object is to study damaged or amorphous materials. Nevertheless, ion implantation and irradiation are excellent techniques for simulating radiation effects in actinide-bearing minerals. Ion-beam experiments have the important advantage that the many uncertainties associated with the study of α-decay-damaged minerals are avoided. Recent modeling has shown that, in principle, the results of ion-beam studies can be extrapolated to the behavior of actinide-bearing minerals, and can be used to predict the crystalline-to-metamict transition in nature.

The use and application of ion beams in the earth sciences has been a classic example of interdisciplinary research, with a strong overlap between solid-state physics and geology/mineralogy. New techniques are being developed and improved, including focused ion beams (FIB), combined FIB and SEM instruments, and dual- and triple-ion-beam facilities for in-situ implantation and analysis. New applications will no doubt be found as these techniques are refined and improved, and will provide the motivation for many new crossdisciplinary investigations.

References

1. S.H. Sie, Progress of quantitative micro-PIXE applications in geology and mineralogy. Nucl. Instrum. Methods B **75**, 403 (1993) 318
2. Hj. Matzke, Ion beam analysis of ceramics and glasses in nuclear energy. Surf. Interface Anal. **22**, 472 (1994) 318
3. A. Choudhury, D.W. Palmer, G. Amsel, H. Curien, P. Baruch, Study of oxygen diffusion in quartz by using the nuclear reaction 18O(p,alpha)15N. Solid State Commun. **3**, 119 (1965) 318
4. M. Sneeringer, S.R. Hart, N. Shimizu, Strontium and samarium diffusion in diopside. Geochim. Cosmochim. Acta **48**, 1589 (1984) 318
5. M.H. Dodson, Closure temperature in cooling geochronological and petrological systems. Contrib. Miner. Petrol. **40**, 259 (1973) 320
6. D.J. Cherniak, W.A. Lanford, F.J. Ryerson, Lead diffusion in apatite and zircon using ion implantation and Rutherford backscattering techniques. Geochim. Cosmochim. Acta **55**, 1663 (1991) 320, 323
7. D.J. Cherniak, Lead diffusion in titanite and preliminary results on the effects of radiation damage on Pb transport. Chem. Geol. **110**, 177 (1993) 320
8. D.J. Cherniak, Pb diffusion in rutile. Contrib. Miner. Petrol. **139**, 198 (2000) 320
9. D.J. Cherniak, E.B. Watson, Pb diffusion in zircon. Chem. Geol. **172**, 5 (2001) 320, 325, 326

10. D.J. Cherniak, E.B. Watson, T.M. Harrison, M. Grove, Pb diffusion in monazite, A progress report on a combined RBS/SIMS study. Eos Trans. AGU **81/17** (2000) 320
11. D.J. Cherniak, Diffusion of Pb in plagioclase and K-feldspar measured by Rutherford backscattering spectroscopy and resonant nuclear reaction analysis. Contrib. Miner. Petrol. **120**, 358 (1995) 320, 322
12. D.J. Cherniak, Pb diffusion in Cr diopside, augite, and enstatite, and consideration of the dependence of cation diffusion in pyroxene on oxygen fugacity. Chem. Geol. **177**, 381 (2001) 320
13. D.J. Cherniak, J.M. Hanchar, E.B. Watson, Diffusion of tetravalent cations in zircon. Contrib. Miner. Petrol. **127**, 383 (1997) 321, 322
14. E.B. Watson, D.J. Cherniak, Oxygen diffusion in zircon. Earth Planet. Sci. Lett. **148**, 527 (1997) 321
15. O. Jaoul, F. Béjina, F. Élie, F. Abel, Silicon self-diffusion in quartz. Phys. Rev. Lett. **74**, 2038 (1995) 321
16. D.J. Cherniak, Rare earth element diffusion in apatite. Geochim. Cosmochim. Acta **64**, 3871 (2000) 322
17. G. Amsel, D. Samuel, Microanalysis of the stable isotopes of oxygen by means of nuclear reactions. Anal. Chem. **39**, 1689 (1967) 322
18. J. Crank, *The Mathematics of Diffusion*, 2nd edn. (Oxford University Press, Oxford, 1975) 324
19. L.C. Feldman, J.W. Mayer, *Fundamentals of Surface and Thin Film Analysis* (Prentice-Hall, New York, 1998) 327
20. J.C. Petit, J.C. Dran, G. Della Mea, Energetic ion beam analysis in the earth sciences. Nature **344**, 621 (1990) 327, 329
21. B.C. Sales, C.W. White, L.A. Boatner, A comparison of the corrosion characteristics of synthetic monazite and borosilicate glass containing simulated nuclear defense waste. Nucl. Chem. Waste Manag. **4**, 281 (1983) 328
22. M.C. Magonthier et al., in *Proc. 2nd Int. Conf. Natural Glasses*, ed. by J. Konta (Charles University, Praha, 1987), pp. 57–64 328
23. J.C. Petit, J.C. Dran, G. Della Mea, A. Paccagnella, Dissolution mechanisms of silicate minerals yielded by intercomparison with glasses and rafiation damage studies. Chem. Geol. **78**, 219 (1989) 329
24. J.F. Ziegler, in *SRIM-2000* (IBM Research, Yorktown, 1999) 332
25. S.J. Zinkle, C. Kinoshita, Defect production in ceramics. J. Nucl. Mater. **251**, 200 (1997) 332
26. B. Park, W.J. Weber, L.R. Corrales, Molecular-dynamics simulation study of threshold displacements and defect formation in zircon. Phys. Rev. B **64**, 174108 332
27. W.J. Weber, Models and mechanisms of irradiation-induced amorphization in ceramics. Nucl. Instrum. Methods B **166**, 98 (2001) 335, 339
28. W.J. Weber, R.C. Ewing, L.M. Wang, The radiation-induced crystalline-to-amorphous transition in zircon. J. Mater. Res. **9**, 688 (1994) 334, 335

29. A. Meldrum, L.A. Boatner, W.J. Weber, R.C. Ewing, Radiation damage in zircon and monazite. Geochim. Cosmochim. Acta **62**, 2509 (1998) 337, 338
30. M.L. Miller, R.C. Ewing, Image simulation of partially amorphous materials. Ultramicroscopy **48**, 203 (1992) 339
31. J.F. Gibbons, Ion implantation in semiconductors – part II, Damage production and annealing. Proc. IEEE **60**, 1062 (1972) 339
32. A. Meldrum, Irradiation-induced amorphization of titanite. Mater. Res. Soc. Symp. Proc. **650** (2008, in press) 339
33. L. Nasdala, M. Wenzel, G. Vavra, G. Irmer, T. Wenzel, B. Kober, Metamictisation of natural zircon: Accumulation versus thermal annealing of radioactivity-induced damage. Contrib. Miner. Petrol. **141**, 125 (2001) 340
34. A. Meldrum, S.J. Zinkle, L.A. Boatner, S.X. Wang, L.M. Wang, R.C. Ewing, Effects of dose rate and temperature on the crystalline-to-metamict transformation in the ABO_4 orthosilicates. Can. Miner. **37**, 207 (1999) 340
35. R.C. Ewing, A. Meldrum, S.X. Wang, L.M. Wang, Radiation-induced amorphization, in *Transformation Processes in Minerals*, ed. by S.A.T. Redfern, M.A. Carpenter (Mineralogical Society of America, Washington, 2000), pp. 319–361 340

Index

Ion-Beam Modification of Polymer Surfaces for Biological Applications

G. Marletta

Dipartimento di Scienze Chimiche, Università degli Studi di Catania,
Viale A. Doria 6, 95125 Catania, Italy, e-mail: gmarletta@dipchi.unict.it

Abstract. We show how the controlled modification of material surfaces by ion beams may lead to biological applications ranging from biocompatible materials to biosensors and biological devices. Specifically, the present chapter addresses the use of low- and medium-energy ion beams to modify polymer surfaces. We provide a short introduction to those surface properties that determine the biological response and summarize basic features of ion–surface interaction processes in polymers, emphasizing the relation between ion irradiation and surface-property modification: the adsorption/organization processes of amino acids, peptide sequences, proteins and cells are all influenced by ion-irradiation treatments. The possibility of obtaining controlled interactions of biological systems (e.g., amino acids, peptide sequences, proteins and cells) with beam-modified polymers is described in terms of a few well-defined surface properties. These include the surface free energy (SFE), the surface morphology and topography, the surface polarity, surface termination and the mechanical properties of the outermost surface layers. In this context, the concept of "biocompatibility" is briefly explained as the ability of a material to provide specialized addressing of biological functions, including a message that a biological system can perceive as a proper "signaling mode" to prompt its appropriate response. Finally, the relevance of ion beams to induce spatially resolved adsorption/adhesion processes in biological systems is demonstrated: the scales range from tens of nanometers to tens of micrometers. The emphasis is on achieving nano(or micro)sized patterns of biological molecules to produce bioelectronics devices. Some examples are given of potential ion-beam applications in manipulating the organization of biological systems on surfaces.

1 Introduction

Exciting new directions in materials science involve not only the design of physical structures adapted to biological analysis techniques ("lab-on-a-chip"), but also attempts to use biological systems to build highly innovative devices. Fields as different as biocompatible surfaces, bioelectronics devices, biosensors, etc., are drawing wide interest to produce hybrid systems based on the controlled interaction of biological systems, including amino acids, peptide sequences, proteins and cells, as well as polymers and inorganic surfaces [1]. Accordingly, the concept of "biocompatibility" must be viewed in a broad sense as the ability of a material to maintain the "natural" biological functionality of the interacting biological system (biomolecules or cells).

H. Bernas (Ed.): Materials Science with Ion Beams,
Topics Appl. Physics **116**, 345–369 (2010)
DOI: 10.1007/978-3-540-88789-8_12,

This is indeed the key to fully exploit the high selectivity of bio-recognition processes, i.e., the capability of specialized function addressing. Although in general the interaction of two materials is critically determined by the physical and chemical structure of their surfaces, the critical problem when interacting with biological systems consists in what the latter can perceive as a proper "signaling mode" to prompt its appropriate response [2]. At the present state of knowledge, it appears that a limited number of basic properties may play a role in the primary signaling process:

(1) Surface free energy (SFE), or, less informative, the related "wettability", i.e., the balance between the hydrophilic and hydrophobic character of surfaces [3].
(2) Surface morphology and topography, including the average roughness, at the micro- and nanometer scale [4].
(3) Surface polarity, including both the overall charge borne by the biological medium and the existence of electrical domains or "patches" [5].
(4) Surface termination, involving the presence of specific chemical groups or functionalities at the surfaces, allowing for processes that may be grouped under the broad heading "molecular recognition" [6].
(5) Mechanical properties of the outermost surface layers, basically in terms of the Young's modulus and shear modulus [7].

Also, relevant applications in these fields are critically based on spatially resolved adsorption/adhesion processes of the biological systems of interest, the relevant dimensions ranging between tens of nanometers to tens of micrometers. Various types of cells, including neuronal ones, have shown the ability to create connected networks guided by topographical structures in the micrometer range in polymers, with a view to producing a kind of cell circuitry [8, 9]. In such cases a nontrivial dependence on the lateral dimensions is observed, different types of cells behaving differently according to spatial constraints [10]. Also in the micrometer range, the development of highly integrated multipurpose microsystems, including biosensor arrays, DNA-chips and related technologies, as well as lab-on-chip devices, critically depends on the ability to functionalize specific micrometer-scale areas in very complex polymer-based three-dimensional structures [11]. On the other hand, at the nanometer scale at least two different aspects of controlled biomolecule adsorption/organization must be considered, in order to achieve nanosized patterns of biological molecules for bioelectronics devices [12], or the fabrication of nanostructured surfaces able to promote protein (or enzyme) clustering. The latter opens new vistas on the way cells can "sense" nanometric surface features [13].

Ion beams have been one of the most effective methods to modify, in a simple and direct way, some or all of the above-mentioned surface properties for polymers, inducing drastic processes of chemical and physical modification of the outermost layers of the irradiated materials [14, 15]. Ion irradiation of polymers changes the chemical structure and composition [14], the SFE [16], the electrical properties [17], the roughness and morphology at the micro- and nanoscale [18]. In turn, all these properties play a role in determining the interactions of the modified polymer surfaces with biological systems, including the absorption and organization processes of oligopeptide sequences [19] and proteins [20, 21], the adhesion process, the spreading and proliferation of such different cells as fibroblasts, endothelial cells, astrocytomas, etc., onto very diverse polymeric surfaces such as polystyrene, segmented polyurethane, polyethylene, polysiloxane, etc., involving many different ion species (for instance, O_2^+, Na^+, Kr^+, N_2^+, N^+, Ar^+ or Ag^-) [22–28].

In the present chapter we address processes based on the use of low- and medium-energy ion beams to modify polymer surfaces. The reason for this is clearly related to the adequate understanding of beam-induced chemical modifications in polymers [14, 15, 29] and the outstanding capability of low-energy focused ion beams to achieve direct patterning of chemically different phases on such highly reactive materials [30]. Ion beams can be highly focused spatially, by properly tuning the energy and mass of the employed ions. Focused ion beam (FIB) facilities, with lateral resolution ranging from a few nanometers to several micrometers, are available to produce in a single step smart spatially resolved structures for cell adhesion at polymer surfaces [30, 31]. We will mainly discuss the approach based on controlled modification of the entire surface's chemical structure and the related properties. The main advantage of this approach is in the simultaneous modification of the various properties related to a given chemical structure, which act synergetically. After a short introduction to the main surface properties and to basic features of ion–surface interaction processes, we focus on the relationship between ion irradiation and surface-property modification, showing how the adsorption/organization processes of proteins and cells is influenced by ion-irradiation treatments. A few specific examples will be given to illustrate the potentialities of ion beams in manipulating the organization of biological systems on surfaces.

2 Surface Properties Drive Biological System Interactions

Most research on surface biocompatibility in the last 20 years was dedicated to modifications of surface chemistry, morphology, electrical charge and mechanical properties by "coating" a material in order to create a specific surface termination. Both deposition of a specific film or a modified layer are valid alternatives, since both methods imply surface "functionalization" based on grafting more or less complex molecules on the surface, from simple hydroxyl, amine, carboxylic groups to peptide sequences, oligomers, to polymers and proteins [32]. The leading rationale to modify a surface involves a search for the selective modification of one of the above-mentioned properties, the SFE modulation being the most popular, followed by the morphology and surface

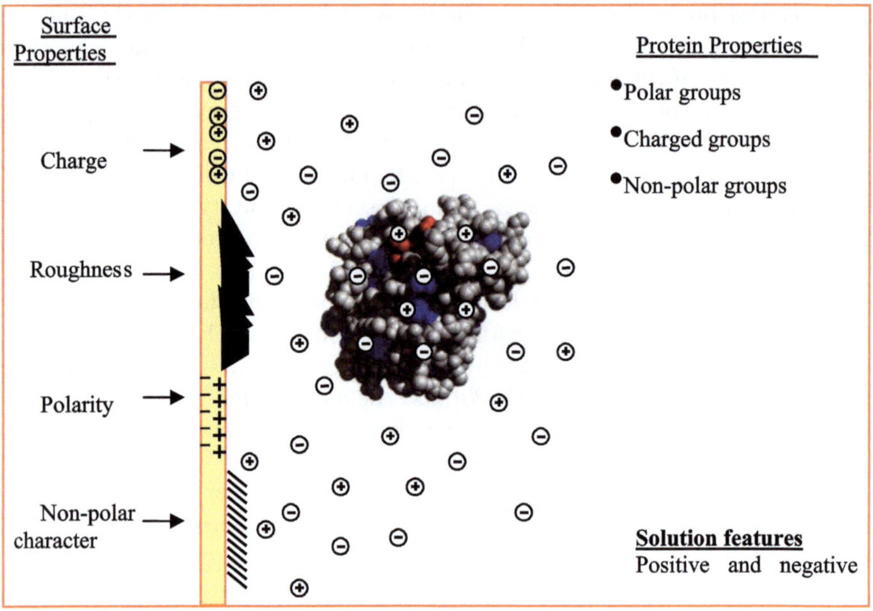

Fig. 1. Scheme of protein–surface interaction

polarity (Fig. 1). Mechanical properties are almost ignored, probably due to the experimental difficulties in producing measurable modifications by keeping all other properties unchanged. We briefly review the connection among the various factors and biological response.

2.1 Role of Surface Free Energy (SFE)

The SFE is an average property, essentially describing the work necessary to create an interface between two phases of unit area [33]. The SFE of a solid/liquid interface, labeled γ_{SL}, is measured by the relationship connecting the contact angle θ, formed by a sessile drop of a test liquid (of known γ_{LV} value, where LV indicates the liquid/vapor interface), usually water, on a solid substrate (Fig. 2), to the related liquid/vapor γ_{LV} and solid/vapor γ_{SV} tensions:

$$\gamma_{\mathrm{SL}} = \gamma_{\mathrm{SV}} - \gamma_{\mathrm{LV}} \cos\theta .$$

The general relationship that correlates wetting and interaction behavior at the interface of a ternary solid–liquid–vapor system (at thermodynamic equilibrium) is:

$$W_{\mathrm{a}} = \gamma_{\mathrm{SV}} + \gamma_{\mathrm{LV}} - \gamma_{\mathrm{SL}} = \gamma_{\mathrm{LV}}(1 + \cos\theta),$$

where W_{a} is the work of adhesion. The quantity γ, the surface tension or surface free energy in general, may be analyzed as the sum of dispersive (γ^{d})

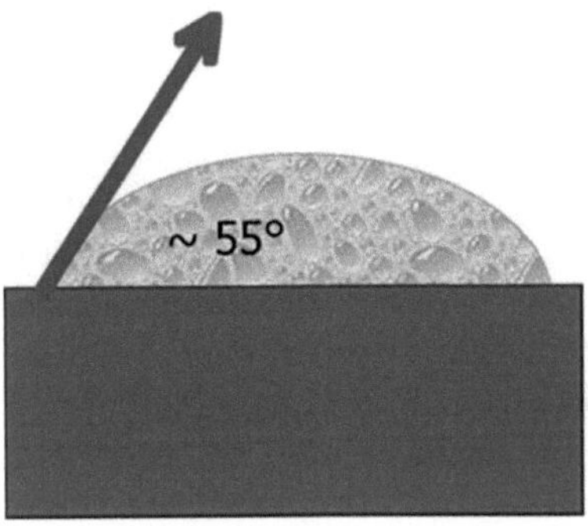

Fig. 2. Contact angle of a liquid droplet onto a solid surface

and polar (γ^{P}) contributions, resulting from a variety of intermolecular forces [31, 33]:

$$\gamma = \gamma^{\mathrm{d}} + \gamma^{\mathrm{P}}.$$

A general equation, providing the relationship among the polar and dispersive terms, can be derived from general arguments:

$$\cos\theta = -1 + \left[2\left(\gamma_{\mathrm{SV}}^{\mathrm{d}}\gamma_{\mathrm{LV}}^{\mathrm{d}}\right)^{1/2}/\gamma_{\mathrm{LV}}\right] + \left[2\left(\gamma_{\mathrm{SV}}^{\mathrm{P}}\gamma_{\mathrm{LV}}^{\mathrm{P}}\right)^{1/2}/\gamma_{\mathrm{LV}}\right].$$

The value of the two components can be obtained by using two liquids of known dispersive and polar components. The polar component is made up of the SFE terms γ^{+} and γ^{-}, respectively, due to electron-acceptor (Lewis acid) and electron-donor (Lewis base) functionalities [33]. The SFE controls the concentration of solute species i at a surface, determining its specific surface concentration, Γ_i (surface excess), as a function of the γ value at the interface. The relationship among Γ_i, a_i and γ is given by the well-known Gibbs adsorption isotherm:

$$\Gamma_i = -(1/RT)(\gamma/\ln a_i),$$

where Γ_i is in units of $\mathrm{mol\,m^{-2}}$, γ is the surface tension of the solution and a_i is the activity of the solute i. This expression indicates that a molecular species decreasing the surface tension will concentrate at the surface. The SFE is thus a phenomenological parameter of choice to express the interaction among biological systems and surfaces. It affects the way in which a protein (or in general, a biomolecule) is organized at interfaces in the following ways [34]:

(1) Selective migration and adsorption from the medium to the interface of specific proteins, i.e., building up of an extracellular matrix (ECM) with a peculiar composition at the interface.
(2) Conformational or functional changes of ECM proteins due to physical or chemical forces originating from the surface, affecting the occurrence and the extent of the protein-denaturation process, hence maintaining or not the biological functionality.

(3) Specific ways of aggregation of proteins at interfaces, i.e., formation of continuous layers or aggregates.
(4) Formation of charge double layers.

An extensive research effort has been dedicated to test the response of proteins and cells with respect to high (hydrophilic) and low (hydrophobic) SFE surfaces, to determine which conditions may preserve the protein native state and, more generally, ensure optimal interactions. For example, there is general consensus on the fact that hydrophilic surfaces prompt cell adhesion, spreading and proliferation, while hydrophobic surfaces tend to resist cell interaction. Based on experimental findings, it has been proposed that there must be a critical surface tension, roughly above 40 $\mathrm{mJ\,m^{-2}}$ [34, 35], in order to obtain good cell adhesion to a given material surface.

2.2 Surface Termination

Surface termination can be obtained by using a very wide class of strategies, all of which exploit specific chemical interactions such as acid–base or key–lock interactions. In particular, the covalent immobilization of bioactive compounds onto functionalized polymer surfaces has seen rapid growth in the past decade. The overall concept is illustrated in Fig. 3 [32]. In the process flow, the first step is critical, involving surface functionalization that must be adjusted in order to introduce the desired type and quantity of reactive functional groups. In this step mono- or polyfunctional compounds may be grafted onto the surface, modulating the number of available reactive functional groups per unit surface area. Tethering a bioactive compound to a solid substrate via a spacer molecule, as shown in Step 2, can substantially improve bioactivity by reducing steric constraints and shielding the biomolecule from surface-induced denaturation. The third and final step is then used to covalently attach a bioactive molecule, consisting in a peptide sequence or a protein or an enzyme, to the functionalized polymer surface, via an intermediary as needed. In the first level of this scheme, functionalization can involve the production of simple –OH or $-NH_2$ groups on the surface, inducing direct acid–base reactions or selectively reacting with a carboxylic group on the biomolecule, thus promoting effective grafting of peptides, proteins,

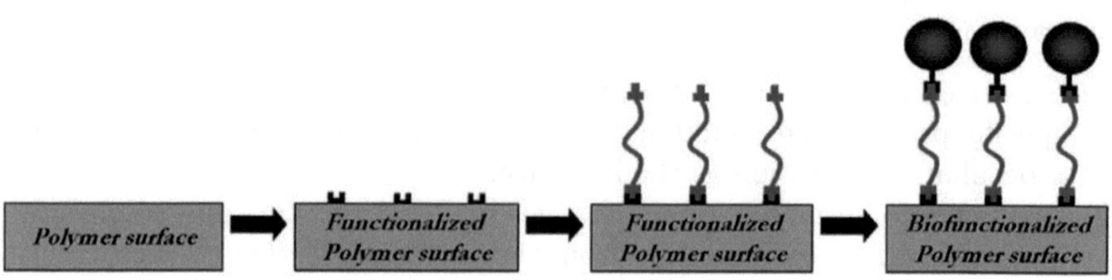

Fig. 3. Scheme for the functionalization sequence of polymer surfaces (from [32])

oligonucleotides for specific interactions with cell membranes [36]. As will be shown in the following, ion beams are one of the choice techniques to produce functionalized polymer surfaces by performing Step 2 in Fig. 3.

2.3 Electronic Structure and Electrical Properties of Surfaces

The electrical state of the surface can be primarily described in terms of the mean surface polarity, i.e., for instance, the capability to charge up the biological medium, forming electrically charged polar domains. Another possibility involves the formation of domains having different electronic structure, i.e., insulating and conducting domains. The surface electronic structure and the related electrical properties drastically affect the behavior of biomolecules near the surface itself. The surface charge may control the adsorption of biomacromolecules from the biological environment as well as their conformational state, via the electrostatic interaction of biomolecules with specific surface-charged sites. Also, the presence of electrically conducting domains may promote redox reactions across the interface. The first action, due to the surface-charge, occurs through the formation of a space-charge layer on both sides of the interface, forming an electrical double layer. The double-layer structure plays a "structuring" role for the extracellular fluid, which includes ions as well as proteins, prompting or preventing the adsorption of charged species as well as determining their equilibrium amount [37]. The structure and working mode of the double layer is well understood in terms of the ions' finite size and the surface adsorption properties, as is the space-charge region of the material. The charge within the space-charge region of the material can be modified, e.g., by a change of pH value. In this case, the orientation of the water dipole can also be made to change with the surface charge. The arrangement of the water dipoles, and their energy with respect to the charged surface, significantly affect the conformational structure of biomolecules [38]. The second effect involves the possibility of electron transfer from surface domains to biomolecules and vice versa. Such processes are in fact redox reactions modifying specific sites of the interacting biomolecules and changing their structure, their interaction and, eventually, their adsorption onto the electrical domains [39]. Charge exchange or charge transfer between a biomaterial surface and biomolecules is severely affected by double-layer-induced ion preadsorption from the medium and by the potential barrier for charge-carrier tunneling. The latter is linked to the detailed electronic structure of the polymer surfaces, including the presence of electronic states in the bandgap [39].

3 Ion Beams and Surface Properties

The use of ion beams to modify the above materials properties is based on unique features. In polymers, these are the following:

(1) The energy-deposition processes induce a complex "cascade" of chemical events, following thermalization of the primary physical events, basically consisting in bond breaking due to the decay of ionized and/or excited states. Primary events produce a large population of nonequilibrium precursors that recombine, modifying the initial chemical structure as well as the related physical properties of the irradiated layer [14];
(2) The basic scheme of ion-induced chemical reactions may be roughly tuned depending upon the predominant energy-deposition mechanism [14, 40, 41];
(3) It is possible to control, in a fairly simple and reproducible way, the type and amount of chemical and physical modification as well as the thickness of the modified layer, by setting a few characteristic beam parameters, such as ion fluence and total deposited energy;
(4) Efficient ion-beam focusing, together with the intrinsic in-track confinement of ion-induced material reorganization, allows direct writing of chemical modification and the related properties [30].

3.1 Ion-Dose-Dependent Chemistry

The chemical events induced by the energy deposition can be classified in terms of two relatively well-distinct regimes of energy deposition. The first involves ions in the keV range, depositing an average energy between 1–100 eV nm^{-1} in the target material by both electronic and collisional mechanisms, with ion tracks whose estimated average diameter is less than 1 nm [42]. The second involves ions in the MeV–GeV energy range ("Swift heavy ions"), losing energy only through electronic stopping (and by charge-exchange processes). Track sizes are then in the 5–50 nm range and typical energy depositions in the range 1–20 keV nm^{-1} [42]. As we only discuss surface-modification processes (thicknesses ranging from 1 to 100 nm) we will not discuss the latter energy regime, which mostly affects bulk properties. Chemical events in the keV regime can be simply rationalized in terms of fluence and total deposited energy (E_{d}).

Three distinct ion-fluence ranges may be considered. In the first, up to 5×10^{13} ions cm^{-2}, single tracks, shown in Fig. 4a, only modify the polymer primary structure in isolated impact regions, and relatively simple radiation-induced chemical processes, such as localized bond-breaking processes, produce distant primary reactive species. Their products, each in a different primary track area, generally do not interact. The creation of radical sites on a monomeric unit then implies a "local" reaction with a neighboring chain (or site on the same chain), or the rupture of the backbone ("chain scission") [43]. The modification of properties related to the molecular weight distribution (MWD) of the polymer, such as solubility (and the related resist properties) and crystallinity, occurs essentially in this first fluence regime [43, 44]. However, in this regime, the primary structure of the polymer and probably most of the secondary structure are still almost intact. The second

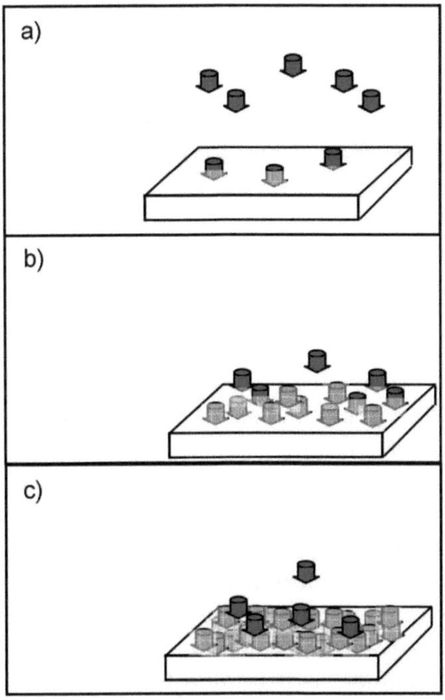

Fig. 4. Fluence regimes: (**a**) single track; (**b**) saturated track regime; (**c**) overlapping tracks or "reimplantation regime"

range roughly includes fluences between 5×10^{13} and 5×10^{14} ions cm^{-2}. In this range, an increasing fraction of the surface is covered by impact regions, until complete geometrical coverage by ion tracks (Fig. 4b). In this regime local modifications of the monomer chemical structure progressively overlap until the surface is geometrically saturated, the process depending linearly on the fluence. This has been termed the "mild chemistry regime" [15, 45]: a new material with completely different chemical and electronic structures is progressively produced, whose properties, including optical or electrical behavior and surface density of chemical groups, are determined by the ion-beam-induced changes [29]. A particularly important point is that the interactions of irradiated surfaces with biological objects such as cells, bacteria, proteins and amino acids are also strongly modified in this regime [19–28, 46, 47]. Note that this stage corresponds to comparatively "primary" effects of ion irradiation, before degradation by subsequent interactions. The third fluence regime includes processes occurring from 5×10^{14} to 1×10^{17} ions cm^{-2}. In this range, the areas initially modified by single-impact events undergo further drastic modification, due to reimplantation events occurring at increasing fluence (Fig. 4c). At 10^{15} ions cm^{-2}, each unit area on the surface has been struck several times, leading to a drastic reorganization and producing the well-known massive carbonization of carbon-based polymers [48, 49]. The properties of the materials obtained in this fluence range are essentially those of hydrogenated amorphous carbons, although the structure of these carbons may depend on the energy-deposition mechanisms [29, 45]. Materials obtained in this regime generally exhibit good biocompatibility, in agreement with what has been shown for pyrolytic carbons [50, 51].

We can now define the best irradiation conditions to drive biomolecule/cell adsorption/organization conditions. We consider mainly low-energy ions with energy ranging from 1–50 keV, leading to modifications within a few nanometers of the polymer surfaces. Furthermore, we only discuss, effects of irradiation with inert gases, in order to focus on energy-deposition effects or, more precisely, on intrinsic reorganization effects induced by ion energy deposition. Finally, we emphasize the intermediate fluence regime (5×10^{13}–1×10^{15} ions cm^{-2}), in which "primary" polymer modifications are operative. In this regime the desired chemical modifications can be quite simply obtained by tuning the fluence and the energy-deposition mechanism. We already know that the surface phases produced by ion irradiation consist in highly "reticulated" amorphous phases, which may exhibit signs of short-range chemical reorganization, producing clustering or local densification [29, 45]. Furthermore, in the selected ion-fluence range, the irradiated regions adhere strongly to the underlying unmodified region because they are intrinsically compositionally graded phases, due to the progressively decreasing chemical modifications. This in itself is one of the important advantages of ion-beam irradiation relative to other surface-modification techniques, as it implies high mechanical stability of the irradiated surfaces. Also, the surface recovery processes typical of plasma treatments linked to inward/outward diffusion of polymer segments and low molecular weight fragments, are hindered in the present case by the formation of the relatively thick and highly reticulated layer. Another important feature is that low-energy ion beams are easily adapted to patterning processes, either by using suitable masks, or by using focused ion beams. It is clear that in the saturation fluence regime of interest for our experiments (5×10^{13}–5×10^{14} ions cm^{-2}) patterning relies on spatially resolved chemical surface modifications, allowing specific studies of biomolecule adsorption/organization processes when interacting with irradiated surfaces.

3.2 Beam-Induced Modification of Surface Properties Relevant to Biological Interactions

Ion-induced chemical modification of polymer properties relevant to cell–surface interactions leads to the formation of specific chemical groups or chemical domains, modifying the wettability, interfacial energy and electrical behavior of surfaces. We briefly discuss how ion irradiation may affect the SFE (and related wettability) and the electrical behavior of surfaces. Ion irradiation may be used to modify the surface chemical structure following three basic strategies, involving: (1) the reaction of the ion-beam-produced carbon radicals at surfaces with oxygen and water in the residual vacuum or, after ion irradiation, in air ("extrinsic" mechanism); (2) the recombination processes of carbon, oxygen, nitrogen (if present) and hydrogen radicals and excited species, produced inside collision cascades in the irradiated layer ("intrinsic mechanism"); (3) "surface grafting" of polyatomic ions, such as OH^+, CF_x^+, etc., at very low energies (3–20 eV).

The "Extrinsic Mechanism"

The first mechanism is just an activation-and-reaction process, involving dissociation reactions of physi- or chemisorbing gaseous molecules such as O_2 and H_2O with radicalic or excited sites produced by the primary ion impact on the polymer chains [52]. Actually, most of the surface chemical modifications, and related wettability changes, observed for keV-irradiated polymers not containing oxygen, such as polyethylene, polypropylene, etc., are due to the postirradiation reaction of air-exposed surfaces. The simplicity and efficiency of the latter mechanism has prompted the development of a methodology aimed at achieving high surface oxidation levels, not obtainable by collision cascade reorganization. Figure 5 shows the drastic decrease of water contact angle obtained by using 0.3–1.0-keV Ar^+ with O_2 flowing in the irradiation chamber with respect to the "intrinsic" irradiation effect [52]. Surfaces modified by means of this activation-and-reaction mechanism are rather unstable, however: normal aging in air leads to partial recovery after a few hours, and complete recovery occurs after 2 weeks. This surface instability is due to polymer chain dynamics as well as to the Gibbsian surface segregation of low molecular weight oligomers formed during the irradiation [53].

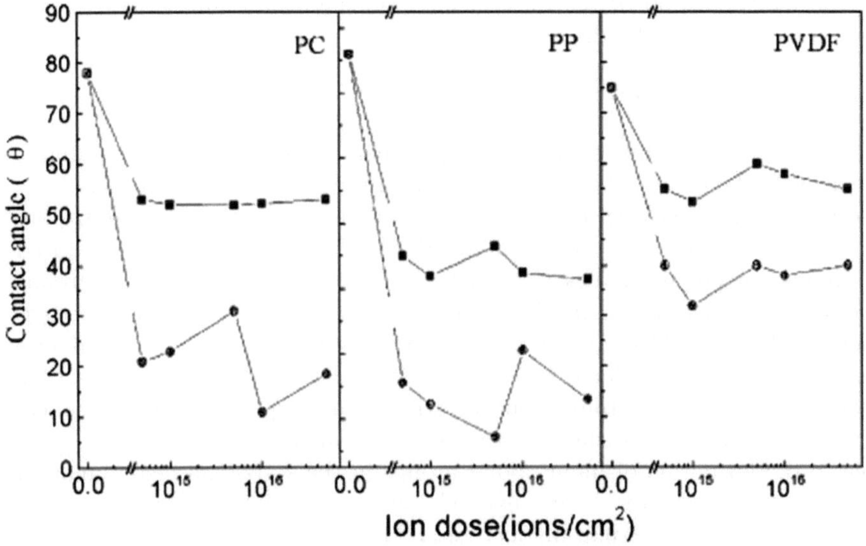

Fig. 5. Change in the static water contact angle without flowing gas (*full squares*) and with flowing gas during irradiation (*empty dots*) (from [52])

The occurrence of such striking so-called "hydrophobic recovery" suggests that (1) the modified layers are very thin, due to limited O_2 diffusion in the modified layers, (2) the oxidation process mainly involves chains that maintain their conformational degrees of freedom, suggesting that oxidation successfully competes with the crosslinking events.

The "Intrinsic Mechanism"

The "intrinsic" reorganization of ion-irradiated polymers is triggered by the radicals and excited species produced within the primary ion track's collision sequence. These active species diffuse freely around the track. The modified layer thickness is comparable to the energy-deposition range and the surfaces remain stable under aging if protected from environmental hydrocarbon contamination [24, 28]. The modified layer consists of very complex phases, whose composition depends on the ion energy and polymer structure. Two main types of effects can be observed within the ion-modified layers: the formation of new chemical structures and functionalities at the outer surfaces, or the formation of heterogeneous phases, possibly containing nanometric aggregates. The former is easily recognized by surface-sensitive techniques, including XPS and ToF-SIMS as well as the measurement of water contact angle. The latter is more difficult to identify, due to the very low thickness of the altered layer and to the difficulty of measuring chemical heterogeneity on the nanometric scale. Scanning probe microscopy (SPM), electrical and dynamic contact-angle measurements may provide information, in addition to the interpretation of XPS and ToF-SIMS data. Of course, these effects may coexist. We briefly discuss them in turn.

Surface Chemical Modification

The chemical structure of irradiated polymers undergoes drastic modifications. XPS shows that the chemical groups present in the main chain may be reduced or oxidized with high efficiency. For example, sulphonyl groups are reduced to sulfide in polyethersulfone (PES) [15], amide groups are reduced to amines in polyurethanes (PU) [24] and carboxylic or carbonate groups are reduced to carbonyls and hydroxyls in polycarbonate (PC), polymethylmethacrylate (PMMA) and polyethyleneterephtalate (PET) [44, 54]. Quite generally, ion irradiation induces a loss of heteroatom-containing groups, thus increasing the carbon atomic concentration. A consequence of new chemical structures forming at surfaces is that the SFE increases for most polymers, due to the formation of highly polar groups in the statistical recombination process of excited radicals and excited species. In polymers with a carbon-based backbone, the formation of ether and hydroxyl groups is privileged, followed by carbonyl formation and lower concentrations of carboxylic groups. All these processes can be understood in terms of simple reactions involving carbon, oxygen and hydrogen radical attachment to (or radical removal from) carbon sites on the main or lateral chains, these reactions competing with on-chain recombination and gaseous molecule formation and escape [15, 29]. Typically, an average change of the water contact angle (WCA) between 20–30 degrees is found for oxygen-, nitrogen- and sulfur-containing polymers [24, 55]. The WCA change depends critically on the fluence, with a threshold around 1×10^{14} ion cm^{-2} for keV ions and saturation fluence above 1×10^{15} ions cm^{-2} (see Fig. 6) [24]. Polymers with a silicon-based backbone, such as polysiloxanes, generally undergo a more drastic wettability change under ion irradiation. For instance, 5-keV Ar^+ irradiation of polyhydroxymethylsiloxane (PHMS) produces a decrease of the WCA from about 90° to about 35°, suggesting that severe surface reorganization occurs under irradiation. There is a connection between surface reorganization, wettability change and the SFE components. The initial structure of PHMS (see Fig. 7) is characterized by the simultaneous presence of $Si–CH_3$ and Si–OH groups in the main chain, with an outward orientation of the methyl groups in air, leading to the surface hydrophobic character (WCA ∼90°). Irradiation with 5-keV Ar^+ completely modifies the PHMS surface structure as shown by ToF-SIMS analysis. The initial surface, containing both $Si–CH_3$ and Si–OH groups, is completely converted to a SiOH-containing surface at a fluence 5×10^{14} ions cm^{-2}. The wettability depends on the formation of a high SiOH concentration, as suggested by the linear relationship between the latter (deduced from the normalized yield of the ToF-SIMS signals) and the WCA decrease (Fig. 8) [56]. The drop in WCA, in turn, corresponds to the rise in total SFE γ^{tot}, essentially due to the marked increase of the acid–base component γ^{AB}, while the dispersive component γ^{LW} undergoes a small increase (see Fig. 9). In this particular case, the wettability increase is due to

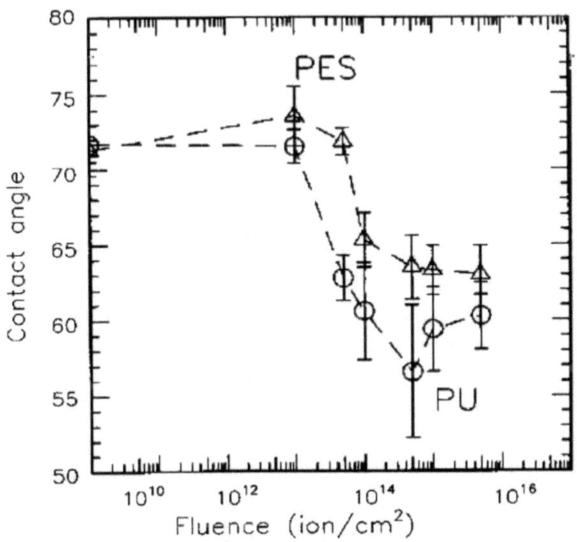

Fig. 6. Fluence dependence of PES (*open triangles*) and PU (*open dots*) static water contact angle after 35-keV Ar^+ irradiation (from [24])

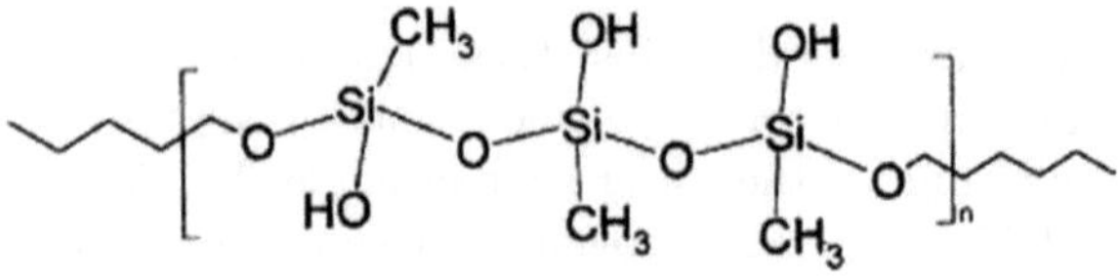

Fig. 7. Chemical structure of poly(hydroxymethylsiloxane) (PHMS)

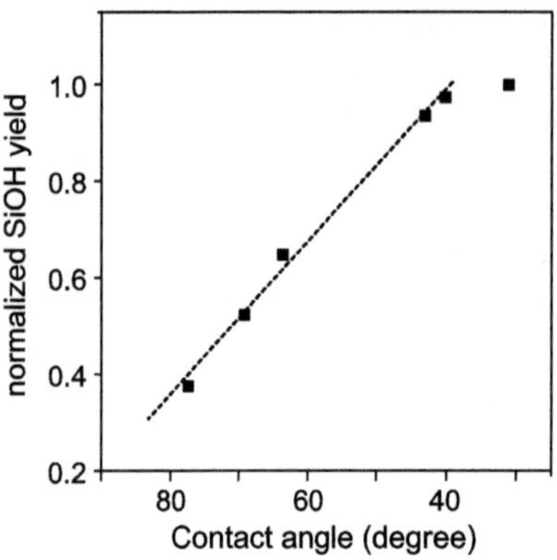

Fig. 8. Relationship between the Si–OH mass (signal from ToF-SIMS) and static WCA for PHMS irradiated with 5-keV Ar^+ at increasing fluence (from [56])

the increased acidity of the irradiated surfaces, corresponding to the increasing concentration of Si–OH groups [28]. Finally, some polymers, including polyesters such as PET and poly-ε-caprolactone (PCL), already have polar ester groups at surfaces and show the reverse behavior under ion irradiation, remaining unchanged or even increasing their WCA. This can be easily understood from Fig. 10, which shows the XPS spectra of O 1s and C 1s photoelectrons for PCL before and after 50-keV Ar^+ irradiation. Irradiation induces preferential elimination of the polar ester groups, as indicated by a decrease of the components at about 289.0 eV in the C 1s region (assigned to carbonyl groups) and at about 533.5 eV in the O 1s region (assigned to the ester hydroxyls) [55]. Accordingly, WCA undergoes a slight increase and the surface free energy exhibits a small decrease of the acid–base component γ^{AB} [55, 62].

Heterogeneous Nanometric Phases

Ion-track overlap in the fluence range $1\times10^{14}-1\times10^{15}$ ions cm^{-2} leads to the formation of three-dimensional amorphous carbon phases containing variable amounts of residual hydrogen: the H:C atomic ratio varies roughly from 0.2 to 0.8, depending on ion fluence, energy and predominant stopping mechanism [45]. These phases are carbon-rich domains embedded in polymer-like phases [57]. A signature of their formation is provided by a change in the samples' electrical properties: at a critical fluence the clusters interconnect and the irradiated layer becomes conductive. An example of an irradiation-induced insulating-to-conducting transition is shown in Fig. 11 [17] – samples irradiated to 1×10^{14} and 3×10^{14} ions cm^{-2} still have very high resistivi-

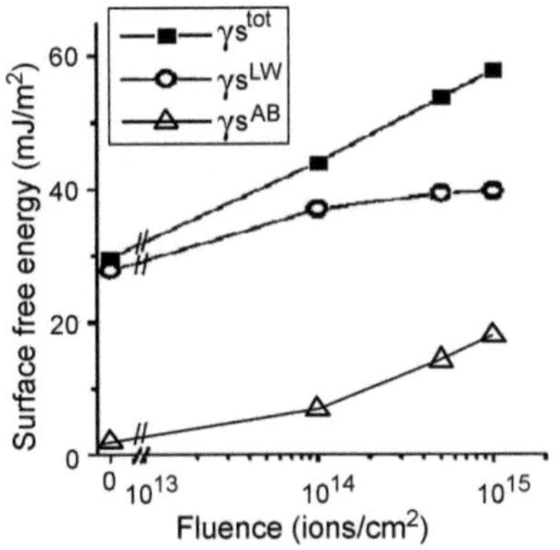

Fig. 9. Modification of surface free energy components for irradiated PHMS (5-keV Ar^+) at increasing fluences: γ^{tot} (*full squares*), γ^{AB} (*full triangles*) and γ^{LW} (*open dots*) (from [28])

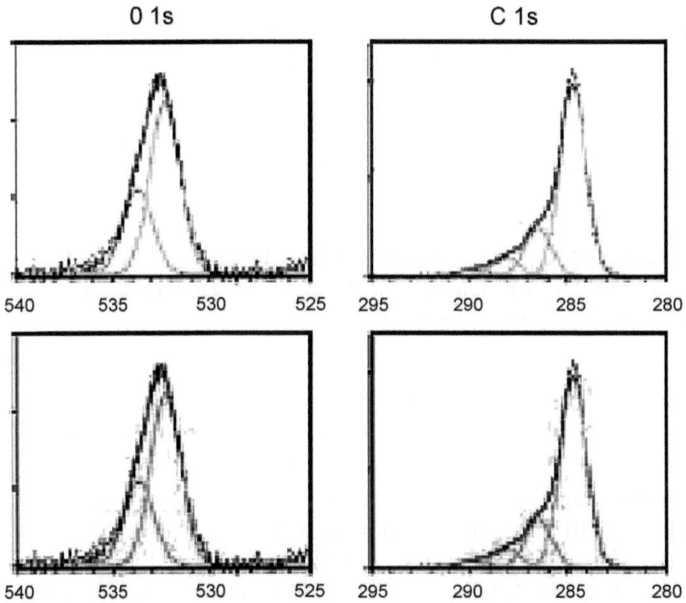

Fig. 10. C 1s and O 1s XPS regions of PCL surfaces before (*upper panel*) and after (*lower panel*) irradiation with 50-keV Ar^+ to 1×10^{15} ions cm^{-2} (from [55])

ties, resp. in the 10^{13}–$10^{10}\,\Omega\,\mathrm{cm}$ range, with linear I/V curves (Figs. 11a and b), whereas samples irradiated to 6×10^{14} ions cm^{-2} exhibit a characteristic Schottky diode nonlinear dependence on the applied voltage (Fig. 11c), and samples irradiated to 1×10^{15} ions cm^{-2} (Fig. 11d) exhibit ohmic conductivity with a relatively low resistivity, $\rho = 3.8 \times 10^2\,\Omega\,\mathrm{cm}$. Raman spectra confirm the existence of structural changes. Characteristic features of the samples' graphitic component are drastically enhanced at fluences where the samples are conducting. The Raman spectra also show that the graphitic regions are embedded in disordered regions at their edge [17]. Similar re-

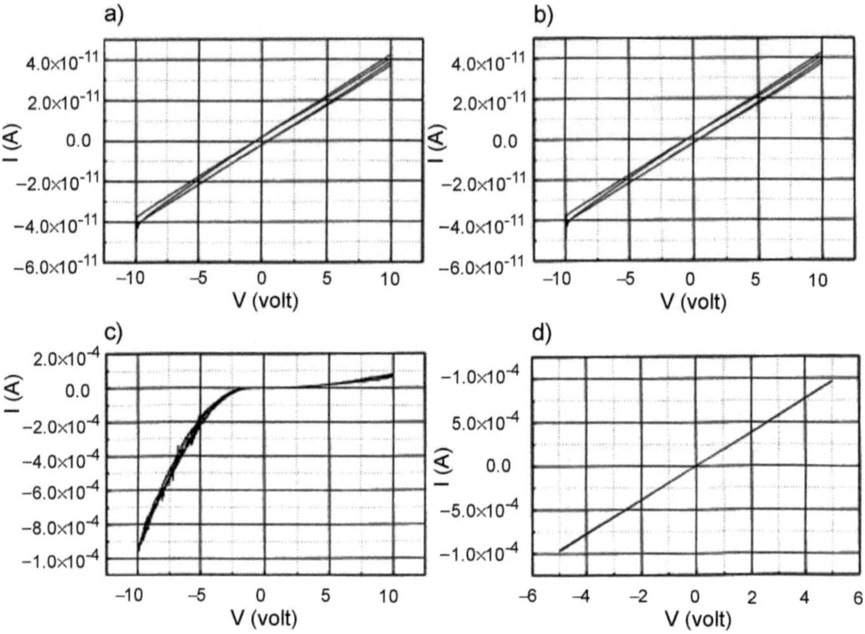

Fig. 11. Current/voltage (I/V) diagrams for polyisoquinoline (PIQ) irradiated with 600-keV Ar^+ to (**a**) 1×10^{14} ions cm^{-2}; (**b**) 3×10^{14} ions cm^{-2}; (**c**) 6×10^{14} ions cm^{-2}; (**d**) 1×10^{15} ions cm^{-2} (from [17])

sults were obtained for most irradiated biocompatible polymers, where the enhanced biocompatibility was related to the formation of amorphous carbon phases within the surface layer [50, 51, 58]. The formation of silicon- and carbon-rich clusters was also demonstrated for silicon-based polymers under ion irradiation [59–61]. Finally, irradiation with N_2^+, Ar^+ and He^+ at very low energies (0.5–2.0 keV) led to the formation of chemically heterogeneous nanostructures [18], which have been claimed to drastically affect the interaction of irradiated surfaces with biological systems [4, 62, 63].

Surface Grafting of Chemical Functionalities

A third way to efficiently functionalize polymer surfaces consists in using very slow polyatomic ions, including simple functional groups such as OH^+, NH_2^+, CF^+, etc; to attach them onto surfaces [64, 65]. Specifically, the polyatomic ion's kinetic energy overcomes the grafting reaction's activation energy [64]. The critical parameters of the process are the primary ion energy and the fluence. The primary ion energy must be high enough to activate the grafting reaction and sufficiently low to avoid cluster decomposition. The right energy to obtain the functionalization effect clearly depends on the ion mass: it is generally in the 3–20 eV range. Figure 12 shows the XPS spectra for CF^+

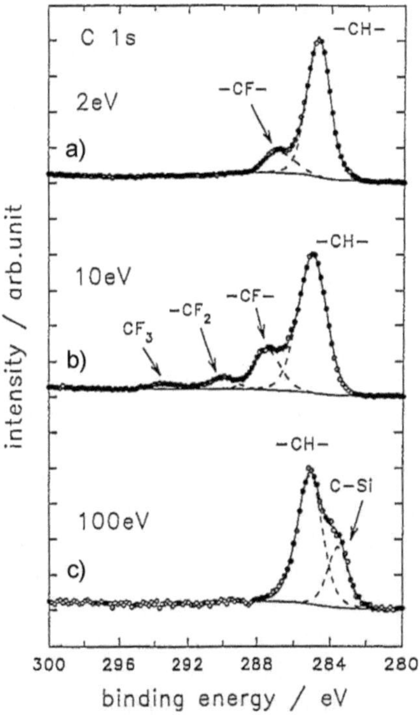

Fig. 12. XPS carbon 1s peaks from 2 nm hydrocarbon film onto Si substrate after exposure to 1×10^{16} CF^+ cm^{-2} at (**a**) 2 eV; (**b**) 10 eV; and (**c**) 100 eV (from [64])

ion impacts on a thin hydrocarbon layer (~2 nm) deposited on Si [64]. It can be seen that 2-eV CF ions are chemisorbed without any dissociation or disproportionation, while at 10 eV simple surface attachment starts to compete with dissociation processes, allowing the F radicals to react with attached CF groups, producing CF_2 and CF_3 species. The various CF_x groups on the surfaces are stabilized by absorbing hydrogen from the vacuum environment. Finally, at a primary energy of 100 eV, the CF^+ ions dissociate completely upon impact. The resulting C and F fragments then have enough energy to penetrate the subsurface region, and chemical potentials favor C–Si formation, (see Fig. 12c), and Si–F species that are lost in the gas phase. Similar reaction pathways have been observed for OH^+ beams bombarding polystyrene surfaces. In this case, below a primary energy of 10 eV only C–OH groups are formed, whereas above 10 eV the formation of C=O and COOH groups is observed [65]. Increasing the fluence also significantly affects surface chemistry. For example, for a 10-eV OH^+ beam on polystyrene surfaces, a fluence of about 1×10^{16} ions cm^{-2} leads only to the formation of COH groups, while at fluences 5×10^{16}–2×10^{17} ions cm^{-2}, more complex groups are formed, including C=O and COOH, as well as increasing amounts of C–O–C and C–OH groups. The process is statistical. It is noteworthy that very high fluence is needed to obtain significant concentrations of polar groups on surfaces. In summary, the "surface grafting" technique exploits the selectivity of reaction channels activated by tunable kinetic energies, but has the drawback of needing very high fluence to obtain significant concentration of polar groups. In fact, high fluence leads to the multiplication of species produced, due to the increasing number of consecutive reactions of bombarding ions with chemical groups previously attached to the surface.

4 Biological Response of Ion-Beam Modified Polymer Surfaces

Ion-engineered polymer surfaces generally display strong biological activity. Ion irradiation at fluences above 1×10^{14} ions cm^{-2} generally improves cell adhesion, spreading and proliferation on polymer surfaces. A typical result (Fig. 13) shows the adhesion of human fibroblasts BHK21 to poly(hydroxymethylsiloxane) (PHMS) surfaces, before and after irradiation with 5×10^{14} ions cm^{-2} of 5-keV Ar^+. Before irradiation, cells neither adhere nor spread at all. Thus, ion irradiation induces a strong increase of PHMS cytocompatibility, involving both a dramatic enhancement of short-term cell adhesion after 2 h incubation and complete cell spreading, with complete surface coverage after 48 h incubation [28]. The dramatic increase of cytocompatibility, in the case of PHMS, was related to a complex set of irradiation-induced modifications of the polymer surface composition and related wettability.

Figure 14 shows the correlation between the increase of adhered and spread cells (shown here in terms of the surface coverage parameter I.D.) after 48 h, versus the measured values of static contact angle for irradiated PHMS surfaces. XPS measurements showed that ion irradiation induced a progressive compositional modification of the polymer toward a SiO_x-rich phase, involving the loss of more than 50% of the original methyl groups and the transformation of the residual carbon-containing groups into nanometric hydrogenated amorphous carbon domains, so that the surface free energy underwent a drastic increase [28]. The latter was analyzed in terms of polar

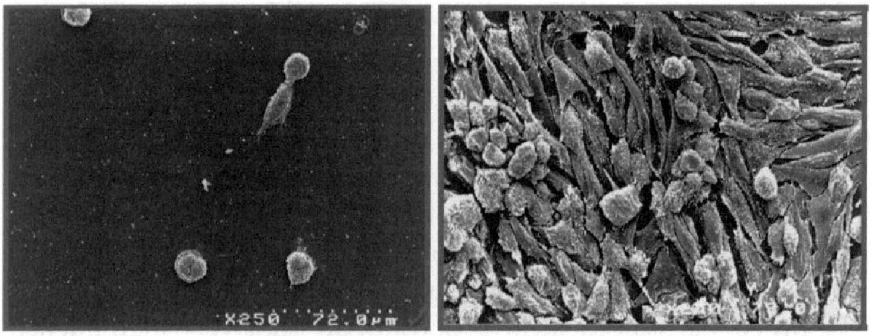

Fig. 13. Fibroblasts adhesion onto PHMS before (*left*) and after irradiation with 5×10^{14} ions cm^{-2} 5-keV Ar^+ (*right*) after 48 h of incubation (from [28])

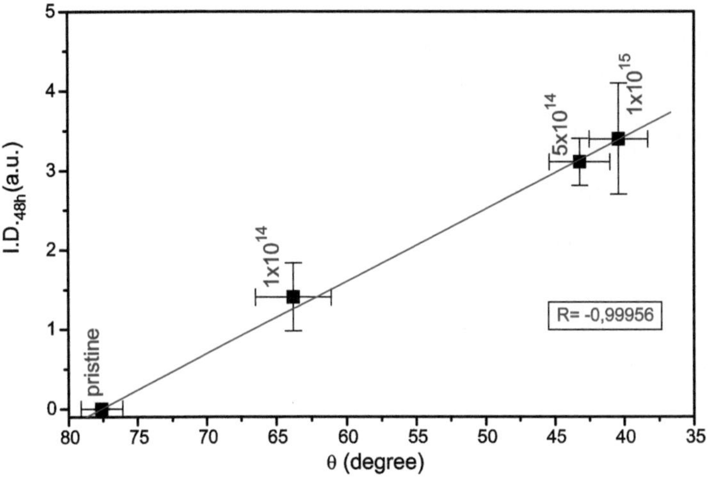

Fig. 14. Cell number values after 48 h incubation, reported in terms of the coverage parameter I.D., versus the static contact angle values for pristine PHMS surfaces and after irradiation with 1×10^{14}, 5×10^{14} and 1×10^{15} ions cm^{-2} (from [28])

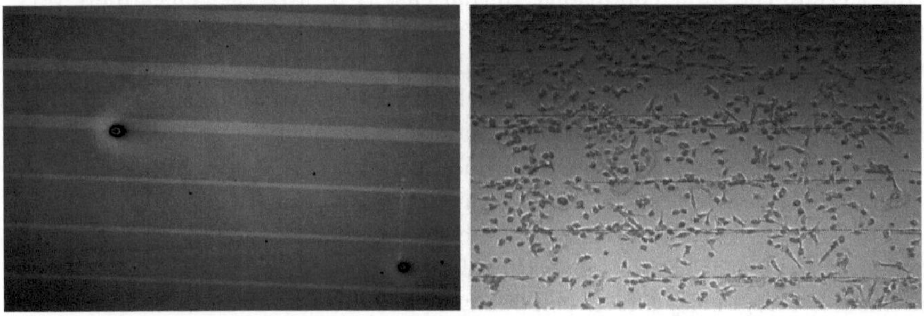

Fig. 15. PHMS surfaces after micropatterning with focused beams of 15-keV Ga^+ at 1×10^{15} ions cm^{-2} (*left*) and after incubation with VERO fibroblasts for 5 h (*right*) (from [31])

and dispersive contributions, showing that the strong enhancement of the hydrophilic character of the irradiated surfaces is mainly due to the enhanced polar acid–base force components (see Sect. 3, Fig. 9 above), the effect being due to the enrichment of the irradiated surfaces in silanol groups (Sect. 3, Fig. 8) and dipoles of the [SiO_4]-based network. This high selectivity in cell adhesion to irradiated surfaces was exploited to obtain cell patterns onto PHMS surfaces, by using focused ion beams (FIB) of 15-keV Ga^+ ions [31]. The FIB gun was used to draw stripes (width 10, 30 and 80 μm, resp.) on the surfaces. Figure 15 (left panel) shows the 30 and 10 μm stripes. It can be seen that cells already tend to align along the irradiated stripes after 5 h incubation. After 48 h the unirradiated regions are free of cells, suggesting

Fig. 16. PET surfaces after micropatterning with 15-keV Ga^+ focused beam at 1×10^{15} ions cm^{-2} (*left*) and after incubation with VERO fibroblasts for 5 h (*right*) (from [31])

that cells migrate on the surface towards the low contact angle (or high SFE) regions [31]. The driving force due to the SFE increase (or equivalently to the WCA decrease) is demonstrated by a counterexperiment performed by irradiating PET in the same conditions. In that case (Fig. 16), the cells adhere very effectively to the unirradiated surfaces and irradiation does not improve cell adhesion to the irradiated stripes. Analysis of the ion-induced surface modifications showed that, in spite of marked chemical modifications, the SFE values before and after irradiation remained very much the same [31]. These observations tie in with the general finding that relates a SFE threshold at about 40 mJ m^{-2} with surface cytocompatibility [35, 39].

But surface free energy is not the only parameter driving cell behavior. Recently, evidence showed that the aforementioned ion-induced nanoscale chemical heterogeneity affects cell behavior beneficially. Human-bone-derived osteoblasts were used to test the effects of ion-induced surface modification of PCL, a biocompatible polymer of outstanding importance for tissue engineering. Viability, morphology, and spreading of human osteoblasts and marrow stromal cells were studied, showing that ion irradiation of the PCL surfaces increased their adhesion, proliferation and spreading [55, 62, 63]. Furthermore, remarkable differences in the cytoskeleton organization within cells were observed: more stress fibers were produced in irradiated vs. untreated PCL surfaces and total cell adhesion was higher [55]. Surprisingly, the irradiated PCL surfaces do not show any significant change in roughness, wettability and surface free energy [55, 62, 63]. On the other hand, analysis of the advancing and receding contact-angle values for unirradiated and irradiated surfaces showed the occurrence of a remarkable WCA hysteresis. As the latter is related to surface chemical heterogeneity [33], this provides indirect evidence of polymer-surface nanostructuring under ion irradiation, confirming the characteristic carbon-cluster-like Raman spectral signatures mentioned in Sect. 3.2. Hence, since its SFE was not modified by irradiation, the cytocompatibility of irradiated PCL was assigned to nanostructuring, possibly via the formation of nanosized electrically active carbonaceous clusters [55, 62, 63]. The advantage of ion irradiation for chemical and electrical polymer-surface nanostructuring is obvious. It should contribute significantly to the ongoing discussion [13] on the general relevance of nanostructuring for biological applications, a difficult topic in view of the scarcity of direct structural and electrical nanoscale characterization techniques for very thin surface layers in polymers.

5 Conclusions

Ion beams induce complex surface modifications that are generally beneficial for cytocompatibility of polymer surfaces. Comparatively simple but diverse physico-chemical factors, such as the surface free energy and the presence of functional groups at surfaces, or the formation of still ill-defined nanoscale chemical and/or electrical domains, have been shown to produce similar changes of cell behavior. The unequivocal identification of those factors that determine cell behavior on irradiated surfaces is still a distant goal. Cell seeding on diverse material surfaces occurs in the presence of a very complex medium (often supplemented by foetal serum) containing proteins, salts and amino acids in different concentrations. The underlying problem in this field basically involves the very complex signaling pathways between cells and surfaces. It is thus quite surprising that simple surface modifications seem to determine cell behavior, in spite of the very complex mediating layers of biomolecules at the modified surfaces. The role of the irradiated surfaces probably consists in promoting a specific "ordering" of the medium components on the surfaces. Such "ordering" may involve either the selective adsorption of a given protein or biomolecule, or their aggregation state, or their conformational state. Accordingly, research in the area might focus (see for instance [20, 21]) on ways of organizing medium components at irradiated surfaces. This could be a very fruitful line of future research.

Acknowledgements

This is truly an area of which one might say "There is always a missing something that torments me" (Camille Claudel, sculptor, 1912). I thank Harry Bernas for challenging me to reflect further on these torments. Support by the EU Integrated Infrastructure Initiative "Ion Technology and Spectroscopy at Low Energy Ion Beam Facilities" (ITS LEIF), as well as by PRA 2007 (University of Catania) is gratefully acknowledged.

References

1. M. Tirrell, E. Kokkoli, M. Biesalski, Surf. Sci. **500**, 61–83 (2002) 345

2. H. Shin, S. Jo, A.G. Mikos, Biomaterials **24**, 4353–4364 (2003) 346
3. C. Satriano, G. Marletta, S. Guglielmino, S. Carnazza, in *Contact Angle, Wettability and Adhesion IV*, ed. by K.L. Mittal (2006), pp. 471–486 346
4. A.S. Curtis, B. Casey, J.O. Gallagher, D. Pasqui, M.A. Wood, C.D. Wilkinson, Biophys. Chem. **94**, 275–283 (2001) 346, 360
5. B. Finke, F. Luethen, K. Schroeder, P.D. Mueller, C. Bergemann, M. Frant, A. Ohl, B.J. Nebe, Biomaterials **28**, 4521–4534 (2007) 346
6. N. Facheux, R. Schweiss, K. Lutzow, C. Werner, T. Groth, Biomaterials **25**, 2721–2730 (2004) 346
7. J. Domke, S. Dannöhl, W.J. Parak, O. Müller, W. Aicher, M. Radmacher, Colloids Surf. B, Biointerfaces **19**, 367–379 (2000) 346
8. P. Fromherz, Chem. Phys. Chem. **3**, 276–284 (2002) 346
9. S. Britland, H. Morgan, B. Wojiak-Stodart, M. Riehle, A. Curtis, C. Wilkinson, Exp. Cell. Res. **228**, 313–325 (1996) 346
10. A.S. Curtis, B. Casey, J.O. Gallagher, D. Pasqui, M.A. Wood, C.D. Wilkinson, Biophys. Chem. **94**, 275–283 (2001) 346
11. I. Willner, B. Willner, Trends Biotechnol. **19**, 222–230 (2001) 346
12. N.J. Gleason, C.J. Nodes, E.M. Higham, N. Guckert, I.A. Aksay, J.E. Schwarzbauer, J.D. Carbeck, Langmuir **19**, 513–518 (2003) 346
13. A. Curtis, C. Wilkinson, Trends Biotechnol. **19**, 97–101 (2001) 346, 365
14. G. Marletta, Nucl. Instrum. Methods B **46**, 295–305 (1990) 346, 347, 352
15. G. Marletta, S.M. Catalano, S. Pignataro, Surf. Interface Sci. **16**, 407–411 (1990) 346, 347, 353, 356, 357
16. C. Satriano, S. Carnazza, S. Guglielmino, G. Marletta, Nucl. Instrum. Methods B **208**, 287–293 (2003) 346
17. A. De Bonis, A. Bearzotti, G. Marletta, Nucl. Instrum. Methods B **151**, 101–108 (1999) 347, 358, 359, 360
18. G. Marletta, I. Bertoti, A. Tóth, T. Minh Duc, F. Sommers, K. Ferenc, Nucl. Instrum. Methods B **141**, 684–692 (1998) 347, 360
19. C. Satriano, M. Manso, G.L. Gambino, F. Rossi, G. Marletta, Biomed. Mater. Eng. **15**, 87–99 (2005) 347, 353
20. C. Satriano, G. Marletta, S. Carnazza, S. Guglielmino, J. Mater. Sci., Mater. Med. **14**, 663–670 (2003) 347, 353, 365
21. C. Satriano, C. Scifo, G. Marletta, Nucl. Instrum. Methods B **166–167**, 782–787 (2000) 347, 353, 365
22. J.S. Lee, M. Kaibara, M. Iwaki, H. Sasabe, Y. Suzuki, M. Kusakabe, Biomaterials **14**, 958–960 (1993) 347, 353
23. L. Bačakova, V. Svorčik, V. Rybka, I. Micek, V. Hnatowicz, V. Lisa, F. Kocourek, Biomaterials **17**, 1121–1126 (1996) 347, 353
24. B. Pignataro, E. Conte, A. Scandurra, G. Marletta, Biomaterials **18**, 1461–1470 (1997) 347, 353, 356, 357
25. H. Sato, H. Tsuji, S. Ikeda, N. Ikemoto, J. Ishikawa, S. Nishimoto, J. Biomed. Mater. Res. **44**, 22–30 (1999) 347, 353

26. L. Bačakova, V. Mares, M.G. Bottone, C. Pellicciari, V. Lisa, V. Svorčik, Gen. Physiol. Biophys. **18**, 1–53 (1999) 347, 353
27. G. Xu, Y. Hibino, Y. Suzuki, K. Kurotobi, M. Osada, M. Iwaki, M. Kaibara, M. Tanihara, Y. Imanishi, Colloids Surf. B, Biointeraces **19**, 237–247 (2000) 347, 353
28. C. Satriano, E. Conte, G. Marletta, Langmuir **17**, 2243–2250 (2001) 347, 353, 356, 358, 359, 362, 363
29. G. Marletta, F. Iacona, in *Materials and Processes for Surface and Interface Engineering*, ed. by Y. Pauleau (Kluwer, Amsterdam, 1995), pp. 597–640 347, 353, 354, 357
30. J. Gierak, D. Mailly, G. Faini, J.L. Pelouard, P. Denk, F. Pardo, J.Y. Marzin, A. Septier, G. Schmid, J. Ferre, R. Hydman, C. Chappert, J. Flicstein, B. Gayral, J.M. Gerard, Microelectron. Eng. **57**, 865–875 (2001) 347, 352
31. C. Satriano, S. Carnazza, A. Licciardello, S. Guglielmino, G. Marletta, J. Vac. Sci. Technol. A **21**, 1145–1151 (2003) 347, 349, 363, 364
32. J.M. Goddard, J.H. Hotchkiss, Prog. Polym. Sci. **32**, 698–725 (2007) 347, 350
33. C.J. van Oss, M.K. Chaudhury, R.J. Good, Chem. Rev. **88**, 927–941 (1988) 348, 349, 364
34. W. Norde, in *Biopolymers at Interfaces*, ed. by M. Malmsten (Dekker, New York, 2003), pp. 21–44 349, 350
35. J.H. Clint, Curr. Opin. Colloids Interface Sci. **6**, 28–33 (2001) 350, 364
36. B. Kasemo, Surf. Sci. **500**, 656–677 (2002) 351
37. H. Wennerström, in *Physical Chemistry of Biological Interfaces*, ed. by A. Baszkin, W. Norde (Dekker, New York, 2000), pp. 85–114 351
38. E.H. Vogler, Adv. Colloids Interface Sci. **74**, 69–117 (1998) 351
39. R. Thull, Biomol. Eng. **19**, 43–50 (2002) 351, 364
40. G. Marletta, F. Iacona, Nucl. Instrum. Methods B **116**, 246 (1996) 352
41. F. Iacona, G. Marletta, Nucl. Instrum. Methods B **166/167**, 676–681 (2000) 352
42. M. Behar, D. Fink, in *Fundamentals of Ion-Irradiated Polymers*, ed. by D. Fink (Springer, Berlin, 2004), pp. 119–170. Chap. 4 352
43. L. Calcagno, G. Foti, A. Licciardello, O. Puglisi, Appl. Phys. Lett. **51**, 907 (1987) 352
44. R.M. Papaleo, M.A. de Araujo, R.P. Livi, Nucl. Instrum. Methods B **65**, 442–446 (1992) 352, 356
45. S. Pignataro, G. Marletta, in *Metallized Plastics 2 – Fundamentals and Applied Aspects*, ed. by K.L. Mittal (Plenum, New York, 1991), pp. 269–281 353, 354, 358
46. S. Carnazza, C. Satriano, S. Guglielmino, G. Marletta, J. Colloid Interface Sci. **289**, 386–393 (2005) 353
47. C. Satriano, N. Spinella, M. Manso, A. Licciardello, F. Rossi, G. Marletta, Mater. Sci. Eng. C **23**, 779–786 (2003) 353
48. J. Davenas, G. Boiteux, Adv. Mater. **29**, 521 (1990) 353

49. J. Davenas, P. Thevenard, G. Boiteux, M. Fallavier, X.L. Xu, Nucl. Instrum. Methods B **46**, 317 (1990) 353
50. A. Nakao, M. Kaibara, M. Iwaki, Y. Suzuki, M. Kusakabe, Surf. Interface Anal. **24**, 252 (1996) 354, 360
51. A. Nakao, M. Kaibara, M. Iwaki, Y. Suzuki, M. Kusakabe, Appl. Surf. Sci. **100/101**, 112–115 (1996) 354, 360
52. S.K. Koh, J.S. Cho, S.S. Yom, Y.W. Beah, Curr. Appl. Phys. **1**, 133–138 (2001) 355
53. E. Occhiello, M. Morra, P. Cinquina, F. Garbassi, Polymer **33**, 3007 (1992) 355
54. J.P. Biersack, A. Schmoldt, D. Fink, G. Schiwietz, Radiat. Eff. Defects Solids **140**, 63–74 (1996) 356
55. G. Marletta, G. Ciapetti, C. Satriano, S. Pagani, N. Baldini, Biomaterials **26**, 4793–4804 (2005) 357, 358, 359, 364, 365
56. A. Licciardello, C. Satriano, G. Marletta, in *SIMS XII Proceedings*, ed. by A. Benninghoven, P. Bertrand, H.-N. Migeon, H.W. Werner (Wiley, Chichester, 2000), pp. 897–902 357, 358
57. D. Fink, V. Hnatowicz, P.Yu. Apel, in *Fundamentals of Ion-Irradiated Polymers*, ed. by D. Fink (Springer, Berlin, 2004), pp. 309–340. Chap. 8 358
58. E.H. Lee, D.M. Hembree Jr, G.R. Rao, L.K. Mansur, Phys. Rev. B **48**, 15540–15551 (1993) 360
59. A. Tóth, I. Bertóti, G. Marletta, G. Ferenczy, M. Mohai, Nucl. Instrum. Methods B **116**, 299 (1996) 360
60. A. Tóth, V.S. Khotimsky, I. Bertóti, G. Marletta, J. Appl. Polym. Sci. **60**, 1883 (1996) 360
61. B. Pignataro, J.-C. Pivin, G. Marletta, Nucl. Instrum. Methods B **191**, 772–777 (2002) 360
62. G. Marletta, G. Ciapetti, C. Satriano, F. Perut, M. Salerno, N. Baldini, Biomaterials **28**, 1132–1140 (2007) 358, 360, 364, 365
63. I. Amato, G. Ciapetti, S. Pagani, G. Marletta, N. Baldini, D. Granchi, Biomaterials **28**, 3668–3678 (2007) 360, 364, 365
64. W.M. Lau, Nucl. Instrum. Methods B **131**, 341–349 (1997) 360, 361
65. W.M. Lau, R.W.M. Kwok, Int. J. Mass Spectrom. Ion Process. **174**, 245–252 (1998) 360, 361

Index

Index

H. Bernas (Ed.): Materials Science with Ion Beams,
Topics Appl. Physics **116**, 371–376 (2010)
DOI: 10.1007/978-3-540-88789-8,

MIX
Papier aus verantwortungsvollen Quellen
Paper from responsible sources
FSC® C105338

If you have any concerns about our products,
you can contact us on
ProductSafety@springernature.com

In case Publisher is established outside the EU,
the EU authorized representative is:
Springer Nature Customer Service Center GmbH
Europaplatz 3, 69115 Heidelberg, Germany

Printed by Libri Plureos GmbH
in Hamburg, Germany